Exploring Animal Behavior in Laboratory and Field

Exploring Animal Behavior in Laboratory and Field

Second Edition

Edited by

Heather Zimbler-DeLorenzo

Department of Life and Earth Sciences
Perimeter College at Georgia State University
Decatur, GA, United States

Susan W. Margulis

Department of Animal Behavior, Ecology, and Conservation
Department of Biology
Canisius College
Buffalo, NY, United States

ELSEVIER

ACADEMIC PRESS
An imprint of Elsevier

Academic Press is an imprint of Elsevier
125 London Wall, London EC2Y 5AS, United Kingdom
525 B Street, Suite 1650, San Diego, CA 92101, United States
50 Hampshire Street, 5th Floor, Cambridge, MA 02139, United States
The Boulevard, Langford Lane, Kidlington, Oxford OX5 1GB, United Kingdom

Notices

Knowledge and best practice in this field are constantly changing. As new research and experience broaden our understanding, changes in research methods, professional practices, or medical treatment may become necessary.

Practitioners and researchers must always rely on their own experience and knowledge in evaluating and using any information, methods, compounds, or experiments described herein. In using such information or methods they should be mindful of their own safety and the safety of others, including parties for whom they have a professional responsibility.

To the fullest extent of the law, neither the Publisher nor the authors, contributors, or editors, assume any liability for any injury and/or damage to persons or property as a matter of products liability, negligence or otherwise, or from any use or operation of any methods, products, instructions, or ideas contained in the material herein.

Library of Congress Cataloging-in-Publication Data
A catalog record for this book is available from the Library of Congress

British Library Cataloguing-in-Publication Data
A catalogue record for this book is available from the British Library

ISBN: 978-0-12-821410-7

For information on all Academic Press publications visit our website at https://www.elsevier.com/books-and-journals

Publisher: Charlotte Cockle
Acquisitions Editor: Anna Valutkevich
Editorial Project Manager: Cole Newman
Production Project Manager: Kiruthika Govindaraju
Cover Designer: Christian J. Bilbow

Typeset by TNQ Technologies

This volume is dedicated to the memories of Penny Bernstein and Susan Foster. Both were past chairs of the education committee of the Animal Behavior Society (ABS); both demonstrated a long-term commitment to the teaching of animal behavior, and to ABS; and they will be greatly missed.

Contents

PART 2 Theory of behavior

Appendices

Contributors

Eli A. Baskir
Department of Reproductive and Behavioral Sciences, Saint Louis Zoo, St. Louis, MO, United States

Carol M. Berman
Department of Anthropology, University at Buffalo, Buffalo, NY, United States

Penny L. Bernstein[†]
Kent State University, Stark Campus, Kent, OH, United States

Eduardo Bessa
Graduate Program in Ecology, Life Sciences Department, Campus of Planaltina, University of Brasília, Brasília, Brazil

Andrea M.-K. Bierema
Center for Integrative Studies in General Science and Department of Integrative Biology, Michigan State University, East Lansing, MI, United States

Alicia M. Bray
Biology Department, Central Connecticut State University, New Britain, Connecticut, United States

Rebecca Burton
Department of Biology, Alverno College, Milwaukee, WI, United States

Rachel Y. Chock
Recovery Ecology, San Diego Zoo Wildlife Alliance, Escondido, CA, United States

Afra Foroud
Department of Psychology, Department of Neuroscience, Institute of Child & Youth Studies, The University of Lethbridge, Lethbridge, AB, Canada

Robin L. Foster
University of Puget Sound, Tacoma, WA, United States; University of Washington, Seattle, WA, United States

Andrew Goldklank Fulmer
Department of Psychology, Lehman College, City University of New York, Bronx, NY, United States

James C. Ha
University of Washington, Seattle, WA, United States

[†]Deceased

Renee L. Ha
University of Washington, Seattle, WA, United States

Sylvia L. Halkin
Biology Department, Central Connecticut State University, New Britain, Connecticut, United States

Abigail L. Hines
Department of Animal Behavior, Ecology, and Conservation, Canisius College, Buffalo NY, United States

Chad D. Hoefler
Biology Department, Arcadia University, Glenside, PA, United States

Darren C. Incorvaia
Department of Integrative Biology, Michigan State University, East Lansing, MI, United States; Ecology, Evolution, and Behavior Program, Michigan State University, East Lansing, MI, United States

Elizabeth M. Jakob
Biology Department, University of Massachusetts Amherst, Amherst, MA, United States

Elizabeth Tinsley Johnson
Department of Integrative Biology, Michigan State University, East Lansing, MI, United States

Carolyn Loyer
University of Puget Sound, Tacoma, WA, United States; University of Washington, Seattle, WA, United States

Susan W. Margulis
Department of Animal Behavior, Ecology, and Conservation, Department of Biology, Canisius College, Buffalo, NY, United States

Jennifer Mather
Department of Psychology, University of Lethbridge, Lethbridge, AB, Canada

Robert W. Matthews
Department of Entomology, The University of Georgia, Athens, GA, United States

Janice R. Matthews
Department of Entomology, The University of Georgia, Athens, GA, United States

Kathleen Morgan
Department of Psychology, Wheaton College, Norton, MA, United States

Megan Murphy
Department of Biology, Indiana University Bloomington, Bloomington, IN, United States

Sergio M. Pellis
Department of Neuroscience, Institute of Child & Youth Studies, The University of Lethbridge, Lethbridge, AB, Canada

Elizabeth K. Peterson
Communities to Build Active STEM Engagement, Colorado State University-Pueblo, Pueblo, CO, United States; Department of Biology, Colorado State University-Pueblo, Pueblo, CO, United States

David M. Powell
Department of Reproductive and Behavioral Sciences, Saint Louis Zoo, St. Louis, MO, United States

J. Jordan Price
Department of Biology, St. Mary's College of Maryland, St. Mary's City, MD, United States

Eila K. Roberts
Department of Integrative Biology, Michigan State University, East Lansing, MI, United States

Domenic Romanello
Department of Anthropology, University of Texas at Austin, Austin, TX, United States

Brett Seymoure
Living Earth Collaborative, Washington University in St. Louis, St. Louis, MO, United States

Olivia S.B. Spagnuolo
Department of Integrative Biology, Michigan State University, East Lansing, MI, United States; Ecology, Evolution, and Behavior Program, Michigan State University, East Lansing, MI, United States

Philip K. Stoddard
Department of Biological Sciences, Florida International University, Miami, FL, United States

Malini Suchak
Department of Animal Behavior, Ecology, and Conservation, Canisius College, Buffalo NY, United States

Clara Voorhees
Department of Biology and Mathematics, D'Youville College, Buffalo, NY, United States

Douglas W. Wacker
Division of Biological Sciences, School of STEM, University of Washington, Bothell, WA, United States

Erin A. Weigel
Program in Evolution, Ecology and Behavior, University at Buffalo, Buffalo, NY, United States

Ken Yasukawa
Department of Biology, Beloit College, Beloit, WI, United States

Heather Zimbler-DeLorenzo
Department of Life and Earth Sciences, Perimeter College at Georgia State University, Decatur, GA, United States

Preface

In 2003, the first edition of *Exploring Animal Behavior in Laboratory and Field* was published. Editors Bonnie Ploger and Ken Yasukawa built upon educational workshops conducted at the annual conference of the Animal Behavior Society (ABS) and solicited further contributions from the Society members. The resulting volume has been an important pedagogical asset to the teaching of animal behavior for nearly 20 years.

Of course, times change, and methods and approaches change as well. In 2019, Academic Press approached the editors about creating a second edition of the book. The task of editing fell to the current and immediate past chairs of the Animal Behavior Society Education Committee. The Education Committee has been hosting workshops at the annual conference for many years. We felt we had a good handle on the needs of the animal behavior community and on the changes in the discipline and pedagogy that have developed in recent years. When considering what we hoped to see in this second edition, our aim was to encompass the full breadth of the discipline of animal behavior. With input and feedback from the Education Committee, we opted for a reorganization of the volume, with a greater focus on the application of animal behavior. We had the opportunity to pilot-test six of the activities as part of a virtual workshop during the 2020 virtual ABS conference.

Like the first edition, the book is divided into four sections, but in contrast to the first edition, which closely parallels the four questions developed by Niko Tinbergen (causation, development, adaptation, and evolution), we chose to focus on practice, theory, application, and communication. In addition, we opted to choose activities that (for the most part) would not require Institutional Animal Care and Use Committee (IACUC) review or extensive specialized equipment or skills. We avoided the use of vertebrates when possible to allow for these activities to be used in more classrooms; however, vertebrates have sometimes been included in observational activities or as optional additions.

Part 1 of this volume deals with describing behavior: what tools does a student need in order to conduct behavioral research? Activities include ethograms, methods, reliability, and formulating hypotheses.

Part 2 focuses on activities that investigate the theory of behavior and encompasses many of the topics that composed the bulk of the first edition. We have included revisions of a number of chapters that appeared in the first edition, as well as some new activities. Topics include game theory, foraging, communication, mate choice, agonism, and antipredator strategies.

Parts 3 and 4 are new to this edition. Part 3 focuses on the application of behavior and ways in which animal behavior may be used to solve problems. We reached out to other ABS committees, including Applied Animal Behavior and Conservation Behavior, to solicit relevant activities. Topics include animal welfare, positive reinforcement training, conservation behavior, and animal enrichment.

The final section, Part 4, focuses on communicating about animal behavior. It has become increasingly important to train students in how to share information with diverse audiences. Activities include communicating to the general public, reading and writing scientific papers, and submitting proposals.

In total, this new edition contains 25 activities. We have included appendices that review the basics of statistical analysis for animal behavior, data collection tools, and appropriate formats for citations.

We have structured all the chapters to include a student section and a separate instructor's section. The student section is divided into the following parts: Learning goals, Background, Purpose, Methods (including step-by-step instructions), Results/discussion, Conclusions, and Questions. The instructor's section includes Classroom management, Preclass preparation, Teaching the activity, Analytical approach, Areas of potential confusion, Recommendations for extensions, Sample results, and Answers to questions. Given that these activities may be used during standard 3-h laboratory periods, or as part of a lecture class, we have carefully described the time needed and how these activities may be used if you are teaching in a class that does not have a separate, scheduled laboratory period. Additionally, we have included in the supplementary material suggestions for how these activities could translate into an online format. Although this is not possible for all chapters, it is a viable option for many.

The supplementary material (e.g., data sheets and student handouts) is available online to facilitate ease of downloading and printing for classes. For some activities, videos demonstrating key skills or steps are provided with these supplementary materials.

Acknowledgments

We are indebted to Bonnie Ploger and Ken Yasukawa, editors of the first edition, for recognizing the importance of and need for this volume. We are grateful to the members of the Education Committee of the Animal Behavior Society, and ABS as a whole, for their insights and contributions. We note that much of the writing for this volume took place during an unprecedented pandemic, yet authors remained engaged and excited about working on this project. We thank Kiruthika Govindaraju for assistance and careful attention during the final editing stage.

PART

Describing behavior

1

CHAPTER

A question of behaviors: how to design, test, and use an ethogram

1

Olivia S.B. Spagnuolo[1,2], Darren C. Incorvaia[1,2], Elizabeth Tinsley Johnson[1], Eila K. Roberts[1]

[1]*Department of Integrative Biology, Michigan State University, East Lansing, MI, United States;* [2]*Ecology, Evolution, and Behavior Program, Michigan State University, East Lansing, MI, United States*

Chapter outline

Exploring Animal Behavior in Laboratory and Field. https://doi.org/10.1016/B978-0-12-821410-7.00023-6

Part I. Student instructions

Learning goals, objectives, and key concepts

- Formulate testable hypotheses and predictions about animal behavior.
- Construct, test, and revise an ethogram that allows you to test your predictions.
- Write empirical descriptions of behaviors.
- Differentiate behavioral states from events.
- Select an appropriate behavioral sampling method to test your predictions.

Background

An ethogram is one of the most fundamental building blocks of studying animal behavior. An **ethogram** is a species-specific catalog of the behaviors that form a species' basic behavioral repertoire—or a specific part of it—and the descriptions of these behaviors (Grier and Burk, 1992; Lehner, 1987; Martin et al., 1993; Tinbergen, 1963). Ethograms may be created using information from behavioral sampling (Altmann, 1974) or from the existing literature.

The discrete behaviors included in an ethogram should be well-defined, avoiding tautological descriptions. A **tautological description** is one that uses the name of the behavior category in its own description (Fig. 1.1).

tautological

adjective form of tautology

FIGURE 1.1

An example of a tautological description, where the name of the term is used to define/describe the term (it is not very helpful, which is why we avoid them in ethograms!).

Table 1.1 Examples of tautological, functional, and empirical (i.e., structural) descriptions for wolf (*Canis lupus*) behaviors. For this assignment, you must use empirical descriptions.

Behavior	✗ Incorrect Tautological	✗ Incorrect Functional	✓ Correct Empirical
Groom	Animal grooms its own skin or fur	Animal uses its tongue or teeth to clean itself	Animal repeatedly runs tongue across its own skin or fur or grasps fur in incisors, making quick, repetitive, nipping motions
Run away	Animal runs away from conspecific	Animal moves quickly to escape from a conspecific	Animal locomotes in opposite direction of conspecific. Front legs move together (in synchrony), and back legs move together. Once per stride, animal briefly has zero feet in contact with the ground.

The formal descriptions of behaviors in an ethogram should use empirical descriptions of focal-animal behavior. The **focal animal** is the subject that is being observed. An **empirical description** objectively describes the form and temporal pattern of the movements and postures associated with a given behavior and does not imply the focal animal's intent or the function (proximate or ultimate) of the behavior (Lehner, 1998). Someone who has absolutely no knowledge of animal behavior should be able to recognize this behavior, based on the description provided. By comparison, a **functional description** is more subjective and may describe the focal animal's intent or the effect(s) of the behavior on the focal animal, nonfocal animals, or the environment (Lehner, 1998). See Table 1.1 for a side-by-side comparison of tautological, functional, and empirical descriptions.

Behaviors can either be states or events. The key distinction between behavioral states and events is the duration over which they take place (Altmann, 1974). A **behavioral state** is a prolonged activity such as a body posture or proximity to something in the environment. It is best to think of states in terms of duration (how long the animal spends doing it). Examples of behavioral states of a wolf could include walking, running, grooming, resting, etc. A **behavioral event** is an instantaneous action such as a brief body movement or vocalization. It is best to think of events in terms of frequency (how many times the animal does it). Examples of behavioral events of a wolf could include barks, yelps, bites, etc. Both states and events can be included in an ethogram, but keep in mind that different **behavioral sampling methods** may be more appropriate for detecting either behavioral states or events (Altmann, 1974):

- *Focal-animal sampling* is appropriate for both behavioral states and behavioral events, but it is impractical to use on multiple animals simultaneously.
- *Scan sampling* (*instantaneous sampling* on groups) can be used to record multiple individuals' behavior simultaneously. This is appropriate for sampling

behavioral states and is often used for estimating the amount of time spent engaged in specific behaviors (i.e., time budgets; Altmann, 1974), but it is inefficient for detecting the quick occurrences of behavioral events.

- *All-occurrence sampling* (sometimes called *critical incident sampling*) is appropriate for recording behavioral events but not for measuring how much time is spent in various behavioral states.

Scientists design their methods around testing research questions, hypotheses, and predictions. It is important to formulate these before collecting data. A **research question** identifies the area of uncertainty you wish to address. A good research question is interesting, will yield useful information, is testable, and has a manageable scope. As an example, behavioral ecologists Holekamp and Sherman (1989) were studying Belding's ground squirrels (*Spermophilus beldingi*) when they asked: why do juvenile males disperse from their natal home range? A **hypothesis** is a possible answer to your research question and it must be testable and falsifiable (Lawson, 2004; McPherson, 2001; Popper, 1981). For a single research question a scientist may devise any number of hypotheses, and these may be mutually exclusive or non–mutually exclusive. A hypothesis is written in the present tense and is broadly applicable. For example, Holekamp and Sherman (1989) proposed 12 alternative hypotheses for the dispersal of juvenile male ground squirrels, one of which was that ground squirrels leave their natal home range to avoid food shortages. A **prediction** is a measurable outcome a scientist expects to see if a given hypothesis is true and, like a hypothesis, it must be testable and falsifiable (Lawson, 2004; McPherson, 2001; Popper, 1981). A single hypothesis may have one or multiple associated predictions. A prediction is phrased in the future tense, applies specifically to your expected results (e.g., what you expect your subjects to do under certain conditions), and is measurable. When formulating a prediction, it may prove helpful to draw a graph of what you expect your results to look like. You may sometimes see hypotheses and predictions written in the format, "if [hypothesis], then [prediction(s)]." For example,

> *If natal dispersal occurs because of food shortages, then juveniles whose natal burrow is surrounded by abundant food will be more philopatric than those from food-poor areas; immigration to food-rich areas will exceed emigration from them; dispersing individuals will be in poorer condition (perhaps weigh less) than males of the same age residing at home; and, based on the strong sexual dimorphism in natal dispersal, food requirements of young males and females should differ.*
>
> **(Holekamp and Sherman, 1989).**

In this case the hypothesis is "natal dispersal occurs because of food shortages." This hypothesis would be supported if evidence was found that any of the following predictions were true:

1. Juveniles whose natal burrow is surrounded by abundant food will be more philopatric than those from food-poor areas.
2. Immigration to food-rich areas will exceed emigration from them.

3. Dispersing individuals will be in poorer condition (perhaps weigh less) than males of the same age residing at home.
4. Food requirements of young males and females should differ.

A single study never conclusively proves a hypothesis to be correct. Rather, if your results do not match your predictions, you may "reject" your hypothesis. If your results do match your predictions, you have failed to reject your hypothesis, or, in other words, you have supported your hypothesis. There is absolutely nothing wrong with results that are inconsistent with what you predicted. In fact, sometimes these results are the most interesting!

When designing an ethogram to answer a specific research question, it is important to keep your research question in mind when delineating your behavioral categories. For example, if you are interested in constructing a general activity budget for wolves, you may include the most common behaviors and group all social behaviors into a single category. Alternatively, if you are interested in how different age/sex classes varied in their rates of aggression, submission, and play, you may divide social behavior into multiple detailed categories.

Purpose

Your assignment is to put yourselves in the shoes of an animal behaviorist who is starting a research project. First, you will select a study species and devise a study question, hypothesis, and prediction(s) about its behavior based on background research and *ad-lib* behavioral sampling. Second, you will create your own ethogram, in which you describe the behavioral repertoire of your species. Third, you will swap ethograms with your project partner, collect pilot data on their study species, and provide them with feedback on their ethogram. Likewise, your project partner will provide you with feedback, which you will use to improve your ethogram. Finally, your instructor(s) may have you use your new and improved ethogram to collect behavioral data to test your predictions.

Methods

Species and subject selection

Individually: When selecting your study species and subjects, you should consider your interests and accessibility of the subjects. Your specific study subjects should be easily accessible for observation by you and your project partner over the entire course of the project. You will need to obtain demographic information on your study subjects, including sex, age, and, if possible, relatedness. For example,

- The subjects may be visible via a live cam. In this case, the entire enclosure should be in view of the camera. You will need to distribute the URL to your instructor(s) and project partner.

- The subjects may be captive and accessible in person, such as on exhibit at a zoo. If it is not practical for your project partner to visit your subjects in person, you will need to record a 30-minute video to send to them. Check with the zoo to ensure that these individuals will be on public exhibit for the duration of your project.
- The subjects may be wild and accessible in person. This method is only an option for students working or interning at a site where they can easily and consistently locate subjects and can tell their approximate age (juvenile or adult) and sex visually. Unless your project partner is on-site with you, you will need to record a 30-minute video to send to them.

You should choose a study species that is interesting to you—one that you hope to work with in the future and/or one that is a good model species to test a study question that interests you. However, keep accessibility in mind when you make this selection. Consider the ease of accessibility, the visibility, the activity level of your species (e.g., in a given time frame, you will observe more diverse behavior from a chicken than from an alligator), whether you are willing to record a video to send to your partner, etc. Your instructor(s) will need to approve of your study species before you go any further!

Materials needed, including variations based on species selection

The materials required include supplies for note-taking (clipboard, notebook, writing utensil, etc.), a timer (can be your cell phone), binoculars (if needed), video camera or camera phone (if needed), and internet access.

Step-by-step instructions

1. Do some background research on the study species and behavioral sampling.
   - **i.** Behavioral sampling: If you have not learned about these methods, read Chapter 4, "A matter of time: comparing observation methods," of this textbook for descriptions of behavioral sampling methods. For actual data collection, you will choose either *focal-animal sampling* or *instantaneous (scan) sampling*, so keep the available methods in mind later as you decide on a study question, hypothesis, and prediction(s). You may conduct all-occurrence sampling in addition to focal-animal or scan sampling, if necessary.
   - **ii.** Study species: Use peer-reviewed literature to determine the geographic range, conservation status (see iucnredlist.org), lifestyle (e.g., terrestrial, arboreal), habitat (e.g., savannas, rain forests), activity pattern (e.g., nocturnal, diurnal), diet, social structure, and any other relevant ecological or behavioral traits of your study species. Type up a summary with references, which will constitute **Part A of Assignment 1.** If you happen to find an ethogram that was used for this species by another research group, then save it; you may refer to this when developing your own ethogram.

iii. Study subjects: Determine the group size, ages, sexes, and relatedness of your subjects. Type up this information as **Part B of Assignment 1.** If captive, also describe the exhibit design, relevant husbandry practices (e.g., diet, feeding times), and any other relevant information. You may visit the website of the zoo or institution, talk to zookeepers, etc. to obtain this information. If you are observing wild animals, you will describe the geographic location and habitat. Each time you collect observations, you will record the specific location, approximate size of the group, and the age (juvenile vs. adult or more specific) and sex (at least for adults) directly observed.

2. Conduct ***ad-lib* sampling** on your study species for half an hour. You may observe the entire group (recommended) or one individual. Keep detailed notes; for *ad-lib* sampling, these notes may be formatted however you wish. If individuals are identifiable (preferred), then be sure to note identifying features (e.g., color, stripe or spot pattern, collar) of individuals. Do not rely on never glancing away from your subjects or on always having them side-by-side to compare; if you look down to write something and when you glance up one animal has gone out of sight and the other has moved, you should easily be able to determine which one you are looking at. It is acceptable to collect data on groups of animals that are not individually identifiable, but you will be limited to scan sampling and so should be sure that your hypothesis is testable using this method.
   Before starting, record
   - date, start and end times, and location (e.g., zoo name and specific exhibit);
   - individual ID(s), identifying characteristics, sex(es), age(s), and relatedness between individuals (if applicable);
   - weather: temperature, sunlight, wind, and precipitation;
   - exhibit: few/many visitors, anything loud nearby, any food/enrichment in exhibit, etc.;
   - anything else that could affect behavior.
3. Develop a study question, hypothesis, and prediction(s). Your instructor(s) will provide you with additional guidelines regarding these components. Label each accordingly (e.g., Study Question, Hypothesis, Prediction 1, Prediction 2). This will constitute **Part C of Assignment 1.**
4. Create an ethogram with 5–10 behaviors. Make sure to include any behavior(s) you will need to observe in order to test your prediction(s). Use the following format:
   Code: Assign each behavior an abbreviation or one-word name for convenience.
   Description: Write a detailed, objective, **empirical** description of each behavior.
   State/event: Explicitly state whether each behavior is a state or an event. Note that if one of the behaviors you need to observe to test your prediction(s) is a behavioral event, you will likely need to use all-occurrence sampling (simultaneously with focal-animal or scan sampling).

It is advisable to include an "Other" behavior to catch behaviors not included in your original categories, as well as an "Out of View (OOV)" behavior for when an animal moves out of your sight. These two categories do not count towards your minimum of five behaviors.

Consider whether you want multiple behaviors to be able to co-occur. If not, make sure that descriptions of similar behaviors are mutually exclusive. For example, should an individual be able to "eat" and "stand" at the same time or should these be separate behaviors? Write the descriptions thoughtfully.

Type your ethogram up as a table with the following format.

Code	Description	State or event
[Behavior 1: e.g., rest]	[Description of behavior 1]	[Example: state]
[Behavior 2: e.g., stand]	[Description of behavior 2]	[Example: state]

5. Decide on how you will test your predictions.
   - Which sampling method will you use? First choose focal-animal or instantaneous sampling. You may choose to additionally and simultaneously conduct all-occurrence sampling (see Altmann, 1974).
   - How many sessions will you need to conduct and how long will each be? Your instructor(s) may have specific guidelines. If you are doing instantaneous sampling (1) are you observing an individual or the whole group (i.e., scan sampling) each time and (2) how long will the intervals between "scans" be?
   - What exactly will you compare?
   - What do you expect your results to look like? Draw a graph of what your hypothetical results will look like if they support your prediction(s).

   Part D of Assignment 1 will consist of your ethogram, the sampling method you plan to use, and the number and duration of observation sessions you plan to conduct for each condition (i.e., treatment).
6. Exchange the following materials with your project partner. Do not talk about them or explain them; your partner should rely entirely on what you have written. This will allow your partner to determine whether you have explained it clearly enough or not.
   - Study question, hypothesis, and prediction(s);
   - Ethogram;
   - The sampling method(s) you plan to use (if instantaneous then specify group/individual and length of intervals).

   Note: If you are using an online live cam, then provide your project partner with the URL for the live cam. Otherwise, your partner may visit your subjects in person (if feasible) or you will need to record a 30-minute video and email it to them. In this case, it is your responsibility to ensure they receive a quality video in a format they can easily view and analyze. This video should capture a different 30 minutes than that from which you collected behavioral data.

7. Review the materials you have received from your partner. Using your partner's ethogram, conduct 30 minutes of behavioral sampling (using the sampling method they specified) on their study subjects (using the live cam or video footage they provided you with). Record your data.
8. Compile the following, which will constitute **Assignment 2.**
   - Behavioral data collected (original and complete notes, not a summary of results).
   - Written: Explain what worked well about your partner's ethogram and what did not. For example, were any of the behaviors' descriptions unclear? Did you observe any behaviors that were not included in the ethogram? Did you see any behaviors that would fall into more than one category? Did a large proportion of observed behaviors fall into the category "Other"? Did you notice any other problems in the ethogram? Will this ethogram and sampling method(s) allow your partner to address their prediction(s)? Do their study question, hypothesis, and prediction(s) need revisions (e.g., are their hypotheses really hypotheses and their predictions really predictions)? Do you have any other suggestions?
9. Incorporate your partner's feedback and revise your ethogram accordingly. You may also need to revise your study question, hypothesis, prediction(s), and/or behavioral sampling methods.
10. Your instructor(s) may assign **Assignment 3.** In this case, collect your data using the finalized ethogram using the live cam or in person (do not reuse videos used earlier in the assignment). Your instructor(s) will give you guidelines on the minimum time requirements.

Results/discussion

Follow the guidelines provided by your instructor(s) for format (full paragraphs vs. bullet points, font and text size, word or page count, etc.), citation style (APA, MLA, other), and other requirements.

Assignment 1

Part A: Provide the common name and scientific (Latin) name of your study species. Describe your species' geographic range, conservation status (see iucnredlist.org), lifestyle (e.g., terrestrial, arboreal), habitat (e.g., savannas, rain forests), activity pattern (e.g., nocturnal, diurnal), diet, social structure, and any other relevant ecological or behavioral traits. Include in-text citations for each piece of information and a reference section, in the citation format specified by your instructor(s). Put everything into your own words and avoid direct quotes, even when including in-text citations. It is insufficient to keep the same sentence structure and swap out a few words. It may help you use your own words to write your text without looking at the original source and then check the source afterward for accuracy.

Part B: *For captive animals* describe the age, sex, and relatedness/relationship (e.g., unrelated, mother-offspring, mated pair) of your subjects in as much detail as possible. State how many individuals are in their social group (e.g., how many animals in this zoo exhibit). Describe the exhibit, relevant husbandry practices (e.g., diet, feeding times), any health problems or ongoing medical treatments, and any other relevant information.

For wild animals describe the age (juvenile vs. adult or more specific) and sex (at least for adults) of your subjects for every occasion on which you collect behavioral data. If you have reason to believe they are related, then state their relationship and explain your reasoning (e.g., if you see a young animal nursing from an adult female in a species for which allonursing is not common, you may deduce that this is a mother-offspring pair). State how many conspecifics are seen with them. Describe the geographic location (distance and direction from point of interest and/or geographic coordinates) and habitat type.

Part C: Provide your study question, hypothesis, and prediction(s), each labeled as such. For example, "My study question was … My hypothesis was … I predicted that …" Attach a copy or photo of your *ad-lib* sampling data.

Part D: Provide your ethogram, typed as a table. Your table should follow this format:

Code	Description	State or event
[Behavior 1: e.g., rest]	[Description of behavior 1]	[Example: state]
[Behavior 2: e.g., stand]	[Description of behavior 2]	[Example: state]

Include 5–10 behaviors. You may include "Other" and "Out of sight" but these do not count toward the five-behavior minimum. Most of your subjects' behaviors should fall into your original categories; "Other" should be used rarely.

Codes are shorthand names of behaviors, which you can write quickly (one word or an abbreviation). Descriptions must be empirical (not functional); should be clear, detailed, and objective; and should avoid tautologies. Finally, specify whether each behavior is a state or an event.

Briefly, state the behavioral sampling method(s) you plan to use and the number and duration of observation sessions you will conduct for each condition (i.e., treatment).

Assignment 2

Using your partner's ethogram and selected behavioral sampling methods, conduct 30 minutes of behavioral sampling on their study subjects (using the live cam or video footage they provided you with). Record your data.

Submit your behavioral data and a write-up explaining what worked well and what did not as well as specific suggestions for improvements. These could address their ethogram, study question, hypothesis, prediction(s), and/or methods. See #8 in the section Step-by-step instructions.

Assignment 3 (if assigned)

Revise your ethogram, study question, hypothesis, prediction(s), and behavioral sampling methods according to your partner's feedback. Collect your data using the live cam or in person (do not reuse videos used for previous parts of this assignment). Follow your instructors' guidelines for how many hours of data you should collect.

After your sampling period, you should have copious notes describing the behavior of your subjects. You should organize these notes into a summary table of behaviors observed (i.e., total duration for focal-animal sampling, total percentage for instantaneous or scan sampling, or total count for all-occurrence sampling) for each subject per treatment. For an example, see the supplementary material.

Paper instructions

i. Introduction: Include background information on a certain species or behavior and gradually narrow the focus, leading to your particular research question and hypothesis. Include in-text citations. (one page maximum)

ii. Methods: Describe your study subjects, their environment or exhibit, and the conditions (weather, etc.) when you observed them. Include your revised ethogram as a table. Describe your behavioral sampling methods and, if you used instantaneous/scan sampling, the time interval used. (one page maximum, excluding ethogram)

iii. Results: Describe your results objectively and without interpretation (save that for the discussion!). For example, summarize data between conditions (e.g., time of day, before or after eating) or subjects that you compared and, if you were instructed to, specify if any differences were statistically significant. Include your graph(s) with descriptive captions. (one page maximum, excluding graphs)

iv. Discussion: Interpret your results. Specify whether these results aligned with your predictions or not. Provide a possible hypothesis for any other notable similarities or differences between your compared conditions/subjects, based on the experimental conditions or the biology of your species. Relate these to your hypothesis and study question and bring this story out to a broader context; how could this information be useful in the real world? (one page maximum)

v. Reflection: Describe any limitations of your study and how you would improve your project if time (and money) were no object. Describe what you learned along the way—if you had to change something midway through the project, any unexpected issues (or positive things!) that cropped up, etc. (one page maximum)

vi. References: All sources cited in text appear correctly formatted in the references section.

Conclusions

This activity has given you experience in one of the most important parts of animal behavior research: correctly classifying and describing different behaviors. It is vital for researchers to use consistent, objective descriptions of behaviors when conducting observations in order to test their hypotheses. Following are some points you should use to reflect on your experience while completing this exercise:

1. Did you need to modify your study question, hypothesis, or prediction(s) based on your partner's feedback? Why or why not?
2. Did you need to modify your ethogram based on your partner's feedback? What modifications were needed and why?
3. Why is it important to describe behaviors empirically rather than functionally?
4. What is the difference between a behavioral state and a behavioral event? When collecting behavioral data, how should states and events be treated differently?
5. What sampling method did you choose and why? In what scenario would your chosen sampling method have been inappropriate?

References

Altmann, J. (1974). Observational study of behavior: Sampling methods. *Behaviour, 49*(3–4), 227–266.

Grier, J. W., & Burk, T. (1992). *Biology of animal behavior.* Mosby-Year Book.

Holekamp, K. E., & Sherman, P. W. (1989). Why male ground squirrels disperse: A multilevel analysis explains why only males leave home. *American Scientist, 77*(3), 232–239.

Lawson, A. E. (2004). The nature and development of scientific reasoning: A synthetic view. *International Journal of Science and Mathematics Education, 2*(3), 307.

Lehner, P. N. (1987). Design and execution of animal behavior research: An overview. *Journal of Animal Science, 65*(5), 1213–1219.

Lehner, P. N. (1998). *Handbook of ethological methods.* Cambridge University Press.

Martin, P., Bateson, P. P. G., & Bateson, P. (1993). *Measuring behaviour: An introductory guide.* Cambridge University Press.

McPherson, G. R. (2001). Teaching & learning the scientific method. *The American Biology Teacher, 63*(4), 242–245.

Popper, K. (1981). Science, pseudo-science, and falsifiability. In R. D. Tweney, M. E. Doherty, & C. R. Mynatt (Eds.), *Scientific thinking* (pp. 92–99). New York: Columbia University Press (Original work published in 1962).

Tinbergen, N. (1963). On aims and methods of ethology. *Zeitschrift für tierpsychologie, 20*(4), 410–433.

Part II. Instructor notes

Classroom management/blocks of analysis

This activity can be used for in-person courses as well as online courses. Instructors teaching an in-person class may wish to have students visit subjects in person rather than relying on live cams, depending on the proximity of the class to a zoo or wildlife area with readily accessible animals. This activity can be adjusted to various time constraints (though it cannot be done in a single lab period) and Assignment 3 is optional. Portions of this chapter (Assignments 1–2) could be used as an exercise to prepare students for a different behavioral sampling project or this entire activity (Assignments 1–3) may serve as a final project, which students can work on throughout the semester.

Teaching the activity

Preclass preparation

Before assigning this activity, students should be familiar with behavioral sampling techniques, including *ad-lib*, focal-animal, instantaneous/scan, and all-occurrence sampling (assigned reading: Chapter 4 of this textbook). Ensure that students understand how to conduct *unbiased* instantaneous sampling (i.e., recording behavior that occurs precisely at the predetermined time interval, even if something more interesting happens right before or after that time point).

Three assignments are described in this chapter. **Assignment 1** consists of four parts: (1) Study Species Background; (2) Study Subjects' Demographic Information; (3) Study Question, Hypothesis, and Prediction(s); and (4) Ethogram and Sampling Methods. **Assignment 2** consists of pilot data collected by each student using their project partner's subjects and ethogram and then providing feedback to the ethogram author. **Assignment 3** (optional) requires students to collect data to test their predictions and present their findings in the form of a short paper. Whether to assign Assignment 3 or not (optional) can be decided by the instructor(s); that is, Assignments 1–2 may be assigned to focus specifically on ethograms or as a precursor to a more in-depth behavioral sampling project. Alternatively, Assignments 1–3 can be assigned as a short-term project (using instructions provided) or the requirements for Assignment 3 may be adjusted to constitute a semester-long or final project. For example, to expand the scope of this project, the instructor(s) may increase the required hours of data collection, require more advanced statistical analyses of results, require more details in the paper, or change the presentation of the results of the paper to another format. *For each assignment, instructors should specify a word minimum and maximum as well as desired format, including citation style (where applicable).*

The Final Assignment described here is a paper, but it could be adapted to a presentation, video, podcast, blog post, or news article. We highly recommend making this plan visible for students at the start of the semester, preferably on the syllabus. Grading rubrics should be included.

If taught online, it is preferable for students to use subjects that are accessible via a live cam throughout the entire duration of the project. However, one benefit of an online Animal Behavior course is that it may allow students to focus their study on animals they are working with remotely, such as through an internship at a zoo or field site. Therefore, the instructor(s) may choose to allow students to study subjects in person, either in situ or ex situ. In both cases, students should ensure that they will be able to locate their study subjects easily and consistently. For example, might the zoo animals be off exhibit in the near future? Do the animals they study in situ take a long time to locate?

If students choose to observe animals in person, they may be required to perform an additional step; that is, if it is not feasible for their project partner to visit their subjects in person, they must record a 30-minute video to email to their project partner. This will allow their project partner to test their ethogram by observing animals in the video. That being said, it is easier to use a live cam or subjects that are locally accessible to both project partners.

The instructor(s) may provide specific guidelines for acceptable study species (e.g., domestic vs. nondomestic, specific taxa). We recommend reviewing each student's selection of a study species and subjects for approval as early as possible so as to redirect them as necessary before they commit a great deal of time toward studying a given species.

The instructor(s) may allow students to choose project partners or they may choose to assign them. Assigning partners with similar study species or study questions will allow each student to provide higher quality feedback to their project partner. On the other hand, assigning partners with dissimilar interests may broaden each student's experience.

The instructor(s) should specify any behavioral sampling requirements. Specifically, they should provide examples of acceptable treatments (e.g., compare two sexes, compare two age groups, compare the same individuals before and after feeding), a minimum number of sampling sessions per treatment (could be as low as one each), and a minimum duration per sampling session. The instructor(s) may also wish to specify acceptable intervals for scan sampling.

The instructor(s) may choose to have project partners peer-review each other's Assignment 3 and provide feedback for credit.

In-class preparation

For analysis of their data, it would be ideal for students to use statistics. Instructors may want to incorporate statistics practice into other activities in preparation for this aspect of the project. If expected to use statistics, students must collect multiple replicates of behavioral data per treatment. For example, if comparing age or sex classes, a student could (1) observe each of two individuals on multiple sampling occasions (sampling occasions are equal in duration) or (2) observe multiple animals from each of two age/sex classes for equal durations. If comparing between two different times of day, a student could perform multiple sampling occasions (equal in duration) per each time of day. Note that the spotted hyena example (supplemental material) provided merely represents results as percentages per category.

Students may struggle to correctly identify whether literature is peer-reviewed or not and may need help finding and accessing relevant scientific literature. Students may also struggle with choosing between the different sampling methods, as well as devising a good hypothesis with accompanying predictions. Finally, students should be given examples of different experimental treatments. Instructors should offer to provide feedback to students when they need it in order to ensure they are on the right track.

The instructor(s) may base the required hours of behavioral sampling on the objectives of this activity. When testing each other's ethograms, we recommend that each student collect 30 minutes of data. If Assignment 3 is assigned with the goal of giving students some preliminary experience in designing and executing a study, 1–2 hours of data collection is sufficient. If, however, the goal of the instructor(s) is to give the students substantial experience in collecting behavioral data, this could be expanded into a semester-long project. In this case, we recommend assigning 20–30 hours of behavioral sampling.

See the supplemental material for a sample project. The instructor(s) may choose to make this sample project available to students. The instructor should also provide students with copies of rubrics for each project assigned to them.

Part III. Supplemental material

Supplementary data related to this chapter can be found online at https://doi.org/10.1016/B978-0-12-821410-7.00023-6.

CHAPTER

2 Consistency in data collection: creating operational definitions*

Heather Zimbler-DeLorenzo

Department of Life and Earth Sciences, Perimeter College at Georgia State University, Decatur, GA, United States

Chapter outline

* Adapted from Terry Glover's chapter "Developing Operational Definitions" in the first edition.

Exploring Animal Behavior in Laboratory and Field. https://doi.org/10.1016/B978-0-12-821410-7.00017-0

Part I. Student instructions

Learning goals, objectives, and key concepts

1. Students will be able to design operational definitions for ethograms.
2. Students can explain why interobserver reliability is important.
3. Students will be able to modify their definitions to improve reliability.

Background

People have been fascinated by animal behavior for centuries. Many people base their knowledge of animals on informal observations. Although these informal observations are a good beginning, they cannot provide information that is reliable and accurate enough to advance the science of animal behavior. However, reporting of data reliability is lacking based on a review of leading animal behavior journals by Burghardt et al. (2012). Our current knowledge of interesting phenomena such as dominance hierarchies, mating behavior, and habitat preferences has been based on careful, systematic observations (Martin & Bateson, 1993; Siiter, 1999). Systematic observations depend on clear definitions of the behavior observed and an assurance that these definitions are used consistently for all the data collected.

Operational definitions enable researchers to translate concepts into more concrete terms so that they can be measured reliably by anyone using the definition (Ray, 1997). For example, an operational definition of aggression might state that each time one animal bites another, it will count as one aggressive encounter. Controversies can result if definitions are not used in identical ways by all researchers (Carroll & Maestripieri, 1998). Once a concept is operationally defined, it must be consistently used by all observers. Often the study of animal behavior, whether in the field or in the laboratory, requires more than one observer, particularly if the observations are made over an extended period. Consistency between two observers can be tested by having both observers simultaneously observe the same animal. They then compare results using some statistical procedure that will give them a measure of interobserver reliability. Reliability is affected by practice, experience, training, and clarity of the operational definitions (Kaufman & Rosenthal, 2009).

There are several methods to measure interobserver reliability, such as comparing total counts, percentage agreement, Pearson's correlation, and kappa coefficient.

- The simplest method, although least exact, for calculating interobserver reliability is to divide the smaller total number of counts by the larger total number of counts and multiply by 100. However, you should be cautioned because there is no guarantee that the observers are recording the same instances of the behavior.
- A more exact way is to use exact count per interval values. This is the percentage of intervals in which observers record the same count divided by the number of intervals multiplied by 100.
- You can use Pearson's correlation coefficient (r) to compare between partners' data. There is an assumption that the scale being recorded is continuous (Kendell's τ or Spearman's ρ for ordinal data). If the correlation is close to +1.00, it indicates a strong agreement between observers. Martin and Bateson (1993) state that a correlation of +0.70 is the lowest acceptable level of agreement. If the correlation value is lower, it may be necessary to reconsider the operational definitions used.
- Kappa coefficient corrects for how often ratings might agree by chance. You can use Cohen's kappa, for two observers, or Fleiss' kappa, for a fixed number of observers, to improve upon percentage agreement. Kappa measures tend to fall on a scale of 0–1, with 0 indicating agreement no better than chance and 1 indicating perfect agreement. There can be negative kappa statistics if the agreement is worse than chance, but this occurs rarely (Cohen, 1960). As a rule of thumb, values of kappa from 0.40 to 0.59 are considered moderate; 0.60 to 0.79, substantial; and 0.80, outstanding (Landis & Koch, 1977). Most statisticians prefer for kappa values to be at least 0.6 and most often higher than 0.7 before claiming a good level of agreement. McHugh (2012) recommends $\kappa > 0.80$ agreement as the minimum acceptable interrater agreement. Therefore there is only a 20% or less erroneous data.

Purpose

This activity will introduce the use of operational definitions and a technique for measuring whether there is consistency between two observers who have observed the same animal, i.e., interobserver reliability. You and a partner will use Pearson's correlation coefficient r to measure interobserver reliability (Spatz, 2000).

Methods

This activity uses the common house cricket, *Acheta domesticus*, which is used extensively in North America as fishing bait and as food for reptiles. This species can survive and breed in large groups pretty easily, as they are generalists and eat a variety of foods (Walker & Masaki, 1989). Male crickets produce songs by rubbing their hind wings together. The species-specific song is used to attract females. Adult females have an ovipositor, a long, thin tube that protrudes from the tip of the abdomen. Males lack an ovipositor. When another cricket approaches a male, he uses chemoreceptors in his antennae to identify whether it is a male or female. During mating, males transfer a jellylike spermatophore to the female. The female uses her ovipositor to lay eggs in damp soil (Loher & Dambach, 1989). Both male and female crickets have been used to study how much locomotion is affected by habitat disruption (With et al., 1999). In general, crickets avoid bright light and seek enclosures or corners.

Each pair of students will receive two adult crickets, a clear plastic container, a blank sheet of paper, a timer, and a small piece of food.

Step-by-step instructions

Step 1: preliminary observations

You and your partner will begin by doing a 4-minute observation of ONE cricket; sex does not matter. During this time, you will record all the behaviors you observe. List each behavior you see in the order in which it occurs (For example, if the animal moves, grooms itself, and moves again, write "move, groom, move"). Do not worry about using technical names for the behaviors you observe. The animal should not be aware of your presence. During the observations adhere to the following guidelines:

do not touch or move the container,
do not stand or lean over the container,
minimize your movements,
do not talk to your partner.

Begin recording when instructed and stop when indicated. You should wait a few minutes after your cricket is introduced to the environment before observing.

Questions for discussion

1. Did you and your partner find it easy to record the sequence of every behavior?
2. Would it have been easier to focus on a few behaviors instead of recording every behavior?
3. Did your list of behaviors agree with that of your partner's?
4. Can you quantify how much the two of you agreed?
5. Did you see any patterns of behavior or develop any hypotheses on the basis of your observation?

Step 2: creating an operational definition for locomotion

You may have found it difficult to record all behaviors. Although recording all behaviors gives a good overview of the animal's behavior repertoire, it can provide more information than is necessary. Research usually focuses on one of more target behaviors and frequently uses time sampling instead of recording a continuous sequence. Time sampling is where you record a subject's behavior at different time intervals (Bakeman, 1997). It is important for observers as well as readers of research articles to know exactly what was being recorded. Therefore it will be necessary for you to develop clear and quantifiable operational definitions for behaviors, such as locomotion. You will then be able to record the cricket's behavior and measure interobserver reliability.

How would you define locomotion so that the locomotion level of the cricket can be compared with that of the crickets other students are observing? A classic way is to use an enclosed space that has been divided into a grid and to count one movement each time the animal crosses over a grid line. This open-field technique (Calhoun, 1968) has been used with a variety of species such as rats (Renner & Seltzer, 1991), guppies (Budaev, 1997), and angelfish (Gómez-Laplasa & Morgan, 1991).

First, you will need to divide the area, into sections, that the cricket has to move around. How many is up to you. Each section should be neither so small that a cricket is larger than a single section nor so large that the cricket can be active and never cross a line. Once there is a decision on the number of sections draw grid lines on a piece of paper that can go under your container. Slide it under without disrupting the cricket.

Then you must define how much of the cricket's body must cross a line for it to be counted as one movement. The operational definition could require as much as the entire body crossing into another section or as little as the head crossing a grid line.

You and your partner will each independently record your cricket's behavior in the data sheet using this operational definition. Each time a target behavior (locomotion) is observed make a mark on the data sheet. You will record all occurrences of only the target behavior. You MUST NOT talk to your partner or let your partner see your data sheet. The instructor will announce the beginning of each time interval (1-minute blocks) until the recording session is over after 6 min. Make sure to record your tally marks in the appropriate time interval row in the data sheet. At the end, total up the number of tally marks for each time interval and fill in your partner's data into your data sheet.

Step 3: creating an operational definition for contact

You will receive a second cricket into your container. You are to record the number of times the original cricket is contacted by the newly introduced cricket, in addition to recording the locomotion of only your original cricket. Again, the class must agree on what constitutes one contact. For example, is touching with antennae sufficient or must some other part of the body be used? You only need to watch your original cricket. If you need to be able to identify it, you can add a marking with a paint pen or white-out.

You and your partner will independently record your cricket's behavior in the data sheet. Each time a target behavior (locomotion and contact) is observed make a mark in the appropriate section on the data sheet. You will record all occurrences of only the target behavior. You MUST NOT talk to your partner or let your partner see your data sheet. The instructor will announce the beginning of each time interval (1-minute blocks) until the recording session is over after 6 min. Make sure to record your tally marks in the appropriate time interval row in the data sheet. At the end, total up the number of tally marks for each time interval and fill in your partner's data into your data sheet.

Step 4: creating an operational definition for feeding

A piece of food will be placed in the center of the container and you will record the number of times the original cricket eats, in addition to recording the locomotion of your original cricket and contact with the new cricket. Again, the class must agree on what constitutes feeding. For example, is touching the food with only the antennae enough or must the head be over the food item? You still only need to observe your original cricket.

You and your partner will each independently record your cricket's behavior in the data sheet. Each time a target behavior (locomotion, contact, and feeding) is observed make a mark in the appropriate section on the data sheet. You will record all occurrences of only the target behavior. You MUST NOT talk to your partner or let your partner see your data sheet. The instructor will announce the beginning of each time interval (1-minute blocks) until the recording session is over after 6 min. Make sure to record your tally marks in the appropriate time interval row in the data sheet. At the end total up the number of tally marks for each time interval and fill in your partner's data into your data sheet.

Results/data analysis

At the end of the session convert your tally marks into minute-by-minute totals so that you have 18 locomotion numbers, 12 cricket contact numbers, and 6 feeding numbers. Make sure you have your partner's data as well.

The data can be presented in graphical form where the frequency of each behavior is plotted as a function of successive minutes. A sample is shown in Fig. 2.1. It is possible for two behaviors to have the same frequency, so different symbols or lines should be used for each behavior. When drawing your graph, whether by hand or by computer, keep in mind that three parts of this activity were not continuous but successive; therefore, there should be a break between the trials. Label your figure appropriately with axes titles and provide a figure caption.

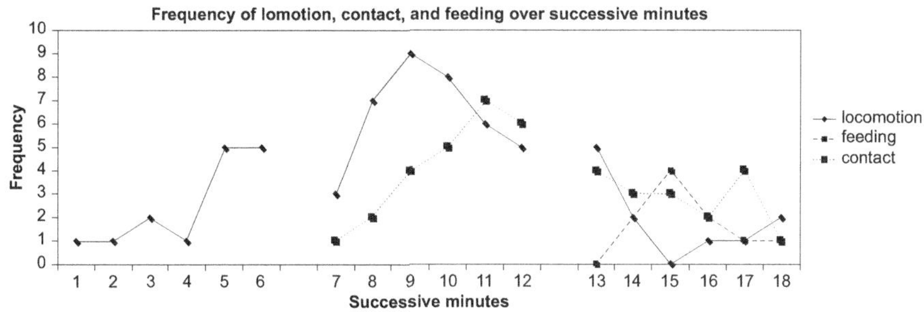

FIGURE 2.1

Frequency of locomotion, contact, and feeding over successive minutes.

Step 5: Create a scatter plot using your data and your partner's data for locomotion. For each minute put one point on the graph. If you saw two movements during the first observation and your partner saw three, then plot the point where 2 on the x-axis intersects with 3 on the y-axis. Do this for each minute of the observation. If you and your partner record identical data, this would indicate perfect interobserver reliability and the points will form a line going diagonal.

Interobserver reliability

There are three commonly used interobserver reliability indices. You can calculate your reliability using more than one method for comparison.

- Using percentage agreement
  The simplest calculation of interobserver reliability just requires the number of times the raters agree, divided by the total number of observations and multiplied by 100. However, percentage agreement does not account for agreement by chance (Cohen, 1960).
- Using Pearson's correlation
  The Pearson's correlation helps determine how consistent data recorders are in their rank orderings of data observations. Perfect interobserver reliability will result in a correlation coefficient of +1.00. If there is no relationship between the two sets of observations, the points will be scattered, and the correlation will be close to 0.00. If one partner reports high numbers and the other partner reports low numbers, you will have a negative correlation, ranging between 0.00 and −1.00 (Fig. 2.2).
  Pearson's correlation coefficient (r) can be done by hand using the formula for Pearson's r by spreadsheet or using a statistical package. If your correlation value is below +0.70, you would want to modify your operational definitions and test your interobserver reliability with your partner again.
- Using kappa coefficient (or Cohen's kappa)
  The kappa coefficient is a modification of percentage agreement that includes chance. It is expressed as a proportion, rather than as a percentage. Kappa

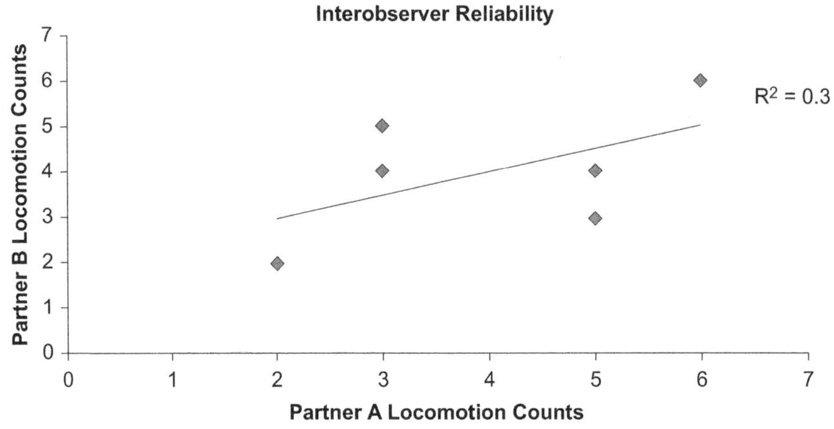

FIGURE 2.2

Interobserver reliability.

coefficient values ≤0 indicate no agreement between raters. Values >0.61 are substantial agreement and >0.81 are almost perfect. A negative kappa represents disagreement, or agreement worse than expected (McHugh, 2012).

Step 6: Calculate interobserver reliability values for locomotion (this is the behavior for which you have the most data) between your data and your partner's data.

Discussion questions

1. An important question for observational studies is whether your data matched your partner's. Were your operational definitions clear enough so that each of you measured the same behaviors?
2. What value did you obtain for your interobserver reliability?
   A. If using Pearson's correlation, did you get an r of 1.00? If your definitions were clear and both of you recorded identical data, you should obtain a Pearson's correlation coefficient r of 1.00. If you did not, why?
3. In terms of locomotion, did your cricket increase or decrease locomotion after the introduction of another cricket and the introduction of food? Do you think there would have been more or less activity if the crickets had been put in the individual containers right earlier before observations began?
4. Did your cricket contact the other cricket?
5. Did it eat the food?
6. Do you think the feeding measure should have been either the duration of time spent eating or how long it took the cricket to begin eating (latency)?
7. In general, how do you feel about frequency as a measure?
8. If you were able to do this activity over, what would you do differently?

References

Bakeman, R. (1997). *Observing interaction: An introduction to sequential analysis* (pp. 50–51). Cambridge: Cambridge University Press.

Budaev, S. V. (1997). "Personality" in the guppy (*Poecilia reticulata*): A correlational study of exploratory behavior and social tendency. *Journal of Comparative Psychology, 11*, 399–411.

Burghardt, G., Bartmess-LeVasseur, J., Browning, S. A., & Morrison, K. E. (2012). Perspectives - minimizing observer bias in behavioral studies: A review and recommendations. *Ethology, 118*(6), 511–517.

Calhoun, W. H. (1968). The observation and comparison of behavior. In A. W. Stokes (Ed.), *Animal behavior in laboratory and field* (pp. 7–10). San Francisco: Freeman.

Carroll, K. A., & Maestripieri, D. (1998). Infant abuse and neglect in monkeys – a discussion of definitions, epidemiology, etiology, and implications for child maltreatment: Reply to Cicchetti (1998) and Mason (1998). *Psychological Bulletin, 123*, 234–237.

Cohen, J. (1960). A coefficient of agreement for nominal scales. *Educational and Psychological Measurement, 20*, 37–46.

Gómez-Laplasa, L., & Morgan, E. (1991). Effects of short-term isolation on locomotor activity of the angelfish (*Pterophyllym scalare*). *Journal of Comparative Psychology, 105*, 366–375.

Kaufman, A. B., & Rosenthal, R. (2009). Can you believe my eyes? The importance of interobserver reliability statistics in observations of animal behaviour. *Animal Behaviour, 78*(6), 1487–1491.

Landis, J. R., & Koch, G. G. (1977). The measurement of observer agreement for categorical data. *Biometrics, 33*, 159–174.

Loher, W., & Dambach, M. (1989). Reproductive behavior. In F. Hauber, T. E. Moore, & W. Loher (Eds.), *Cricket behavior and neurobiology* (pp. 1–42). Ithaca, NY: Cornell University Press.

Martin, P., & Bateson, P. (1993). *Measuring behavior: An introductory guide* (2nd ed.). Cambridge: Cambridge University Press.

McHugh, M. L. (2012). Interrater reliability: The kappa statistic. *Biochemical Medicine, 22*(3), 276–282.

Ray, W. J. (1997). *Methods towards a science of behavior and experience* (5th ed.). Pacific Grove, CA: Brooks/Cole.

Renner, M. J., & Seltzer, C. P. (1991). Molar characteristics of exploratory and investigatory behavior in the rat (*Rattus norvegicus*). *Journal of Comparative Psychology, 105*, 326–339.

Siiter, R. (1999). *Introduction to animal behavior.* Pacific Grove, CA: Brooks/Cole.

Spatz, C. (2000). *Basic statistics: Tales of distribution* (7th ed.). Pacific Grove: Brooks/Cole.

Walker, T., & Masaki, S. (1989). Natural history. In F. Hauber, T. E. Moore, & W. Loher (Eds.), *Cricket behavior and neurobiology* (pp. 1–42). Ithaca, NY: Cornell University Press.

With, K. A., Cadaret, S. J., & Davis, C. (1999). Movement responses to patch structure in experimental fractal landscapes. *Ecology, 80*, 1340–1354.

Part II. Faculty instructions

Classroom management

This activity can be performed in a variety of contexts, in class or laboratory. If executed as a complete activity, it takes about an hour and 15 minutes. It is designed in several parts, each one building off each other. To be able to actually test consistency and interobserver reliability, all parts of the activity need to be completed.

Teaching the activity

Preclass preparation

This activity is designed to be done with crickets; therefore, the Institutional Animal Care and Use Committee (IACUC) approval is not necessary. You can order these online from scientific companies such as Carolina Biological or Wards or obtain them from a pet store or bait shop. You can also use other organisms that are easy to handle and do not require IACUC approval, such as mealworms or cockroaches. If this activity is modified to be used with vertebrates, IACUC approval will be required.

Before preliminary observations with students begin make sure to remove food from the cricket-holding container so that the crickets have not eaten for 24 h; however, water should be available. This will encourage the crickets to show interest in the items being presented during this activity. If possible mark each cricket with a paint pen or white-out for easy identification during this activity. You only need to mark enough crickets for each pair to have one marked cricket.

Each pair of students will need two adult crickets, a clear plastic container (can be as large as a shoebox or smaller like a petri dish), a blank sheet of paper, a timer, and a small piece of food. If possible, a divider of some variety that can be placed between students will help them not look at their partner's data. There is a data sheet available in the instructor materials.

Modifications to this activity

Although this activity is designed to perform "live" observations in the classroom, it can be adapted to utilize video. You will need to find a clip of an organism that lasts 10 min. However, the conversation about creating sections will need to be modified.

Areas of potential confusion or difficulty for students

For any kind of observation, students often confound observations from inferences. This activity is meant to be simple and straightforward; however, students can get bored (if their organism is not doing anything) or try to encourage activity. Crickets are perceptive organisms and will not behavior "normally" if students are hovering above them. Also, the crickets will need an adjustment period when moved from their holding environment to the student containers. Students will be recording three behaviors at the same time, at the end of this activity.

In-class preparation

Step 1: preliminary observations

Each group of students should receive ONE cricket, a container, and the *Cricket Observation Data Sheet* (available to download). The students will observe the cricket for 4 min, recording every behavior exhibited and the order in which they are exhibited. The purpose of this part of the activity is to demonstrate to students that it is difficult to observe organisms with consistency among individuals if there is no preparation beforehand. At the beginning of each time interval (1-min blocks) announce to the class the next time period, such as stating begin now, next time block, etc. The discussion of the questions posed will lead into step 2.

Step 2: creating an operational definition for locomotion

This activity has three types of animal behavior that will be observed and is set up in a stepwise fashion. We begin with just watching for locomotion. Students are asked to create an operational definition for locomotion. This can be done as a class or within pairs of students. All members of the class/pair must use the same operational definition. Students will watch their cricket for 6 min, recording tally marks in the data sheet. After the time is up, students will record their partner's data on their data sheet.

Step 3: creating an operational definition for contact

The next part of this activity will require a second cricket given to each pair of students. They will now do the same procedure in step 2, creating an operational definition for contact. The students will only need to watch their original cricket (possibly marked). They will record how many times during each minute of observation that the two crickets come into contact, as well as locomotion for their original cricket.

Step 4: creating an operational definition for feeding

Each student pair will need to receive a small piece of food (such as apple, orange, potato). This item should be placed in the middle of the container. It may take some time for the cricket to find the food. The students will, again, only need to watch their original cricket and record its behavior.

Analytical approach

Students should calculate interobserver reliability values, and it is up to the instructor to choose a method to use. It is suggested that students at least use percentage agreement, which is easily done by hand, and then Pearson's correlation. Students can utilize Excel or a statistical software to calculate r. There is an *Interobserver Reliability Calculations and Graph Assignment* handout available to download.

In addition to calculating interrater values, students should create a visual representation of their agreement in partner data collection. An example is provided in the student section (Fig. 2.2). This can be done manually or by using a computer.

Recommendations for extensions or continuations for more advanced classes

As this activity is utilizing only focal continuous animal sampling, you can extend or modify it by incorporating different sampling methods, such as scan sampling. For example, you can record the behavior that the cricket is performing every 10 s and then compare results among methods.

Answer key

Preliminary questions

1. You may have found it difficult to record all behaviors.
2. Although recording all behaviors gives a good overview of the animal's behavior repertoire, it can provide more information than is necessary. Research usually focuses on one of more target behaviors and frequently uses time sampling instead of recording a continuous sequence.
3. You may have found that you and your partner did not always record the same behaviors. You may have recorded contact at the same moment your partner recorded movement.

End-of-activity questions

These questions are to get the students to think about their experiences during the activity. There are no correct answers. They are fully dependent on the student's experience. Some students' answers may be different than those of other students depending on how their pairs interacted.

Part III. Supplemental material

Supplementary data related to this chapter can be found online at https://doi.org/10.1016/B978-0-12-821410-7.00017-0.

CHAPTER

Observation and inference in observing human and nonhuman behavior

3

Susan W. Margulis[1], **Penny L. Bernstein**[2,†]

[1]*Department of Animal Behavior, Ecology, and Conservation, Department of Biology, Canisius College, Buffalo, NY, United States;* [2]*Kent State University, Stark Campus, Kent, OH, United States*

Chapter outline

[†] Deceased

Exploring Animal Behavior in Laboratory and Field. https://doi.org/10.1016/B978-0-12-821410-7.00030-3

Part I. Student instructions

Learning goals, objectives, and key concepts

- To distinguish observation from inference in the study of animal behavior.
- To explore the use of facial expression in human communication.
- To apply observation techniques to videos of nonhuman species.

Background

When people think about human communication, they usually think about talking. Vocalizations are so pervasive in our everyday life that we tend to overlook the ongoing silent communication signals that surround us. Yet these nonverbal signals are crucial to our daily lives; conscious of them or not, we are quite good at giving them and responding to them appropriately. These signals often support and reinforce our talking, but they can also act as independent sources of communication. Smiles comprise a set of nonverbal graded signals, running through a continuous set of changes from a simple closed-mouth smile all the way to a full, open-mouthed grin (for overviews see Ekman, 1992; Ekman & Friesen, 1982; Ekman et al., 1988; Fridlund, 1994; Russell, 1994; Smith, 1977; for discussion of the evolution of smiles see Chevalier-Skolnikoff, 1982; van Hooff 1972; Preuschoft, 1992; Preuschoft & Preuschoft, 1994; Preuschoft & van Hooff, 1997). Smiles communicate information about motivation and intent and shape our responses (Martin et al., 2017; Senft et al., 2016). People use this set of signals abundantly every day, in many different situations, and recipients respond in what seem to be organized, meaningful ways. Keltner and Haidt (1999) and Martin et al. (2017) described three categories of smiles, based on the presumed intent or desired outcome of the smile: *reward smiles* reinforce a particular behavior in the recipient, *affiliation smiles* function to establish or strengthen social bonds, and *dominance smiles* play a role in navigating social hierarchies. The physical appearance of these three smile categories has been analyzed and studied, and thus one can categorize smiles based on their appearance and infer the function (Fig. 3.1). Similarly, nonhuman animals use a range of signals to communicate. Although we often *think* we understand the meaning of such signals, our interpretations are often laden with inferences—using our preconceived ideas about what a behavior means to interpret it. Anthropomorphism can also bias our interpretations of our observations. Such inferences can often color our interpretations of behavior and lead to inaccurate

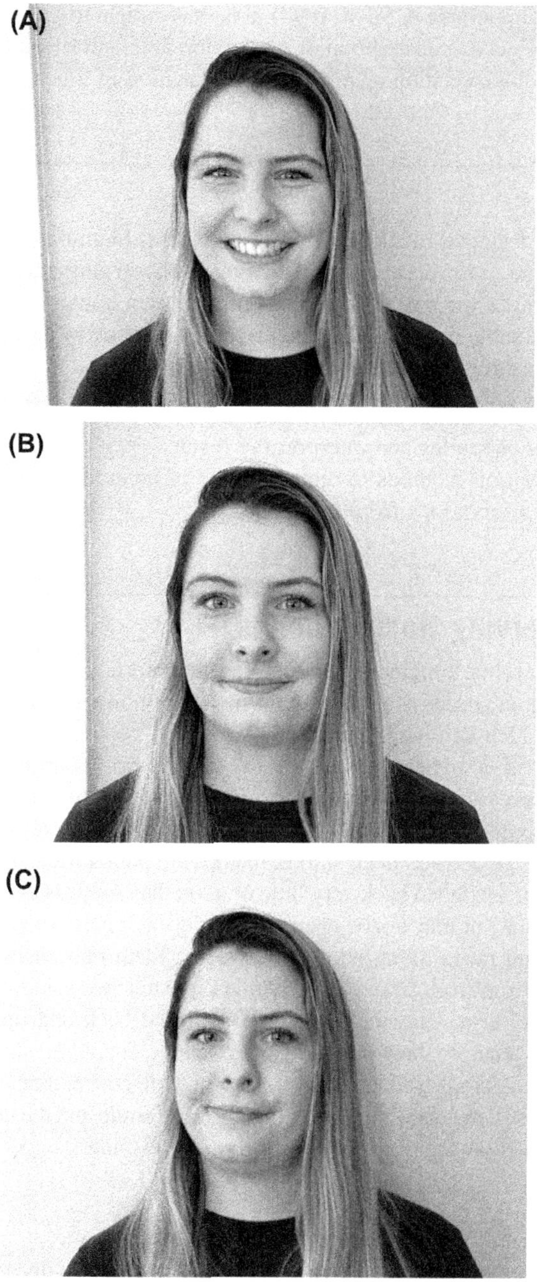

FIGURE 3.1

Three types of smiles as described by Martin et al. (2017). (A) The upper smile, the "reward smile," is characterized by open mouth and teeth showing. (B) In the middle smile, designated the "affiliation smile," teeth are generally not visible. (C) Finally, the lower smile is called the "dominance smile" and is more asymmetric.

interpretations (Timberlake & Silva, 1994). It is thus critical to distinguish between an *observation* (an objective description of a behavior, i.e., its form and appearance) and an *inference* (an interpretation or functional explanation of a behavior).

Purpose

This exercise is designed to address two key concepts in animal behavior. First, we will explore human nonverbal communication by observing smiles in a variety of contexts and analyze the role these signals may play in human interaction. Second, we will apply the same basic observation principles to observe digital videos of other species, distinguishing between observation and inference.

- Collect data on smiling in humans in a range of environments.
- Analyze data on smiles and interpret the results.
- Apply observation methods to digital videos of nonhuman species.
- Distinguish observations from inferences.

Part 1: Observing Human smiles

You will be observing humans as they interact in natural settings. Specifically, you will be focusing on smiles as an important means of nonverbal communication and the context in which smiling occurs.

You will need a paper and pencil to write down your observations. Your instructor may provide data sheets or you may be free to take your own notes. You will be relying on your ability to focus and observe and to take detailed notes.

Smiles form a graded set. In all smiles, the lips are pulled back and somewhat compressed. They may be pulled back very little or a lot; they may also be raised toward the ears. The mouth may or may not be opened, and if open, it may be open slightly or open widely (with a full range of variation in between). Teeth may or may not be visible. Generally, the set runs from closed-mouth smile (lips stretched somewhat but not parted, no teeth showing) to a wide-open grin (lips stretched back and up, lips open, teeth showing), with variations between the two extremes, including the closed grin and the half-smile. To simplify data collection, you will concentrate on observing the most obvious smiles, the open-mouthed grin or reward smile and the closed-mouth affiliation smile, as well as the degree of symmetry of the smile.

Procedure

Data for this exercise are gathered outside the classroom, either during class time or on your own time. You will work on your own or with a partner, depending on the instructor's directions. You can look for smiles in any situation, although some situations are better than others (because of the way the signals work and the information they provide), and hints about good situations might be given or discussed in class before you begin observing (e.g., the library, the dining hall, or a sporting event).

Step 1: The hypothesis. As a class decide on specific hypotheses that you aim to test. Remember, you will be collecting data on smiling, the types of smiles, and the context in which they occur. Your hypotheses should be testable, given these criteria. Your instructor will help shape the hypothesis.

Step 2: Data collection on your own. You will be observing 3, 5, or 10 smiles, each from a different individual, as directed by your instructor. These data will be pooled and discussed in class, and this should provide enough data for patterns to begin to emerge, which will allow you to make some reasonable inferences about the function of smiles. Your instructor may direct you to have all your cases be of the same kind of smile (such as all closed-mouth smiles or all wide-open grins) or a mixed set. But the cases observed by the class as a whole should be mixed so that the class can determine whether and how patterns vary between the two ends of the continuum. To avoid sampling problems, you should observe a different individual for each smile.

In this type of observational procedure, it is critical that you be an unobtrusive, uninvolved observer. You will need to find a place to observe where you will not be playing any role in the interactions as they unfold and will not be readily noticed by the people you are observing.

Observations involve describing behaviors and the contexts in which they occur. Be sure to write down observations, not interpretations, that is, describe rather than infer (e.g., if you think someone was happy, then describe the behavior and contextual aspects that you observed that led you to this interpretation). Furthermore, it is critical to focus on the behaviors of the communicator before, during, and after the signal is given, the responses that occur, and the context. Here is an example: "Two people are talking in low voices in a cafeteria. One then takes the other person's hand, gazes at the person, kisses the person, gives a wide-open grin (lips stretched back and up, mouth open, teeth showing) while still holding the person's hand, and initiates a long conversation while continuing to hold hands and look directly at the other person; the other person listens, talks, and smiles as well." This description illustrates the sort of detail necessary for analyzing a signal. Some sample descriptions are provided in the sample data sheet given in Table 3.1. This sort of description might sound complicated, but you will find it is easier to do it than to read about it. The observing process is described in greater detail below.

Use the data sheet provided by your instructor or developed by your class. When you observe a smile, it is critical that you write down the following pieces of data. (1) *Form*. Record what the signal looked like: lips closed or open, pulled back or not and generally how far, teeth showing or not, symmetry, position of eyebrows, etc. If you define your terms ahead of time, you might be able to write down specific terms instead of a general description (such as closed or open). (2) *Communicator behavior*. Record what the communicator was doing before, during, and after the signal. Be as specific and descriptive as possible of the smile, of when it occurred, and of the variety of other behaviors with it, such as body, head, and eye orientation.

Table 3.1 Sample data sheet completed for six smiles.

Form	Type	Before	During	After	Context	Duration
Closed	Affiliation	Looks at girlfriend. She is talking.	Nods up and down, closed smile as he looks at girlfriend (lips back, not open, small)	Looks down, looks at hands as he fiddles with napkin. She stops talking, they sit in silence.	Boyfriend and girlfriend. She is talking loudly complaining about another friend.	1 min
Closed	Affiliation	Hurrying toward class, looks at approaching person	Gives smile (lips back, not open), nods head and continues to walk at same rapid pace	Continues down hall and into room	In hall just before next class period	10 s
Closed	Dominance	Kid looks at mother. Asks what is for dinner.	Gives smile (small, closed) as mom tells him they will have liver for dinner	Thanks mother and turns and leaves room	My brother, dinner time, is hungry, does not like liver	30 s
Open	Reward	Looks at girlfriend, she is talking	Nods up and down (lips back, open, teeth show)	Talks animatedly as girlfriend listens. Continues for over 5 min and observer moves on.	Discussing plans for dinner date	>6 min
Open	Affiliation	Approaching friend in hallway	Stops and begins talking to friend, friend stops, engage in conversation as smile continues (lips back far, open, teeth show)	Says goodbye and moves on down the hall	Two friends meet in hallway not at a class time	4 min
Open	Reward	Two friends sitting together in conversation	Smile while talking, and then as listen to response, back and forth (lips back far, grin, open, teeth)	Continues until observer leaves	Gestures are slow, conversation not rapid. Do not seem to be in a hurry.	>6 min

It is especially important to keep track of when the smile actually occurred and of the behavior after the smile ends. Try to keep in mind the following kinds of questions: Was there an interaction before, during, or after the smile? Did the communicator initiate interaction as he or she smiled or soon after? Did someone else initiate interaction? Did the interaction end? Was he or she looking at a person or away from the person before and during the smile? Was the body facing the person or away? Did he or she continue to interact after the smile or leave the interaction? (3) *Context*. Record what else was going on besides the signal and communicator behavior. What were the responses? Where did the interaction take place? What else was happening? Who was involved? Who else was there? Did the people know each other or not? The relationships? The ages? The sexes? See the sample description quoted in step 2 for the kind of detail you should be trying to include. (4) *Duration*. You may also wish to time the interactions that occur, if possible. See Table 3.1 for examples.

Results/discussion

For analyzing data the first step involves compiling data from the whole class. You will need to add new cases to your data sheets as they are discussed; that is, add information about form, behavior before, during, and after; and context (see Table 3.1 for sample descriptions). Once this is completed, you can begin to test the data.

Analysis of the whole-class data requires us to examine two alternative statistical hypotheses in each case (note that these are different from the biological hypotheses for the function of smiling). The first is traditionally called the statistical null hypothesis (H0). For example, if your hypothesis is that smiling is associated with interaction, then the null hypothesis would be that smiling should be equally likely when there is no interaction and when an interaction is occurring. The alternative hypothesis (the alternative to the null hypothesis, H1) is that smiling is associated with interaction. If this alternative is true, then smiling should be more common during interactions than when individuals are not interacting.

To test these hypotheses, raw data (see Table 3.1 for sample cases) are examined for patterns of communicator behavior in context. For example, when an individual gave a closed smile, was she already in an interaction or not, did she initiate an interaction when she smiled, did she continue or maintain an interaction? Data could be grouped by smile type (independent variable) and examined for four main dependent variables: interaction (was there an interaction context, and did interaction occur in conjunction with the smile or not?), type of interaction (friendly or unfriendly), initiation (did the signaler initiate further interaction during or after the smile or not, or did someone else initiate further interaction or not?), and continuation of interaction (did the signaler continue the interaction or not, or did another individual continue the interaction or not?). If the class timed the durations of interaction, these results should also be recorded. Duration provides a more objective measure of continuation of interaction and thus offers another way to test the last hypothesis.

Analytical approach

Means generated for smiles during interaction may be tested with a binomial or sign test (Siegel & Castellan, 1988). However, you should be aware that your sample may be biased toward interactive situations. It is difficult to find smiles in noninteractive situations; by looking for smile situations, you may find yourself noticing only interactive situations because that is where smiles are most likely to occur. Means generated by interaction type (e.g., friendly vs. unfriendly) may be tested with a chi-square test for goodness of fit (Dunn, 2001). A chi-square test for independence may be used to compare initiation of further interaction by smile type (Dunn, 2001). If duration data were collected, the Mann–Whitney U test or Kruskal–Wallis test may be employed to test differences in duration of social interaction following different types of smiles.

Questions

1. Based on your data, what information do you think smiling provides?
2. Why might it be important for a communicator to provide interaction information with smiles?
3. Why might it be important for a signaler to be able to use a graded set of signals to provide more finely tuned information about interaction?
4. Was there any evidence to suggest that there might be a gender difference in frequency of smiling?
5. Did you find any evidence to suggest that smiles might not necessarily always be associated with friendly interaction?
6. Do smiles function primarily for the communicator, for the recipient, or for both? Must both benefit in the same way or to the same extent? At what point or under what conditions would you expect recipients to cease responding to this display?
7. Can you identify and distinguish observations in your data from functional inferences? Are these inferences reliable?

Part 2. Observation and inference when observing nonhuman animals

Now that you have carefully observed, collected data, and analyzed data on a nonverbal signal from a very familiar species, it is time to apply this approach to observe unfamiliar species. Before beginning any research study in animal behavior, it is critical to become familiar with the species and be able to objectively define and describe the behaviors that you will observe (see chapter 1). Such preliminary, or "reconnaissance," observations help us develop objective definitions of behaviors and avoid making inferences about function. It is important to note that the more familiar the species, the easier it is for us to jump to functional inferences. In many cases, when species are familiar and well-studied, there is enough support to indicate that our functional inferences are accurate. With less familiar species, however, it may take considerable time to make accurate inferences. Therefore it is important to describe and define behaviors objectively and *operationally*; that is, what does the behavior look like? Here, we will use readily available digital videos of familiar and unfamiliar species and generate definitions that avoid inference.

Procedure

Watch one or two videos together in class. It is advisable to turn the sound off for two reasons: (1) the behaviors you are observing are nonverbal and (2) some videos have narration, which oftentimes is loaded with inferences and presumed functional descriptions. Describe what you are seeing. Write it down. Watch each video three times. The first time just watch. The second time take notes.

Results/discussion

Now take 5–10 min to reorganize your notes into a table of behaviors and definitions. Compare definitions with a lab partner or fellow student. Next, work as a class to compile a set of clear, objective definitions that reflect your observations, not your inferences.

Questions

1. Were the behaviors of some species easier to describe objectively than others? Why?
2. To what extent does your familiarity with a species make it easier, or more difficult, to be objective in your descriptions?
3. Why is it important to refrain from inferring about the functions or causes of behavior?
4. In what situations are functional definitions or descriptions appropriate?

Acknowledgments

This activity incorporates a portion of the original lab activity written for the first edition of this book by the late Dr. Penny Bernstein. We are indebted to all that Dr. Bernstein contributed to the field of animal behavior, particularly with respect to education. Thanks to Megan Miller (photographer) and Caeley Robinson (smile model).

References

Chevalier-Skolnikoff, S. (1982). A cognitive analysis of facial behavior in old world monkeys, apes, and human beings. In C. T. Snowdon, H. Brown, & M. R. Peterson (Eds.), *Primate communication* (pp. 303–368). Cambridge: Cambridge University Press.

Dunn, D. (2001). *Statistics and data analysis for the behavioral sciences*. New York: McGraw-Hill Higher Education.

Ekman, P. (1992). Facial expressions of emotion: An old controversy and new findings. *Philosophical Transactions of the Royal Society of London - B, 335*, 63–70.

Ekman, P., & Friesen, W. V. (1982). Felt, false and miserable smiles. *Journal of Nonverbal Behavior, 6*, 238–252.

Ekman, P., Friesen, W. V., & O'Sullivan, M. (1988). Smiles while lying. *Journal of Personality and Social Psychology, 54*, 414–420.

Fridlund, A. (1994). *Human facial expression: An evolutionary perspective*. New York: Academic press.

Hooff van, J. A. R. A. M. (1972). A comparative approach to the phylogeny of laughter and smiling. In R. A. Hinde (Ed.), *Nonverbal communication* (pp. 209–241). Cambridge: Cambridge University Press.

Keltner, D., & Haidt, J. (1999). Social functions of emotions at four levels of analysis. *Cognition and Emotion, 13*, 505–521.

Martin, J., Rychlowska, M., Wood, A., & Niedenthal, P. (2017). Smiles as multipurpose social signals. *Trends in Cognitive Sciences, 21*, 864–877.

Preuschoft, S. (1992). "Laughter" and "smile" in Barbary macaques (*Macaca sylvanus*). *Ethology, 91*, 220–236.

Preuschoft, S., & Preuschoft, H. (1994). Primate nonverbal communication: Our communicative heritage. In W. North (Ed.), *Origins of semiosis: Sign evolution in nature and culture* (pp. 61–102). Berlin: Moiuton de Gruyter.

Preuschoft, S., & van Hooff, J. A. R. A. M. (1997). The social function of "smile" and "laughter": Variations across primate species and societies. In U. Segerstrale, & P. Molnar (Eds.), *Nonverbal communication: Where nature meets culture* (pp. 171–190). Mahwah, NJ: Lawrence Erlbaum.

Russell, J. A. (1994). Is there universal recognition of emotion from facial expression? A review of cross-cultural studies. *Psychological Bulletin, 115*, 102–141.

Senft, N., Chentsova-Dutton, Y., & Patten, G. A. (2016). All smiles perceived equally: Facial expressions trump target characteristics in impression information. *Motivation and Emotion, 40*, 577–587.

Siegel, S., & Castellan, N. J., Jr. (1988). *Nonparametric statistics for the behavioral sciences* (2nd ed.). New York: McGraw-Hill.

Smith, W. J. (1977). The behavior of communicating, after twenty years. In D. H. Owings, M. D. Beecher, & N. S. Thompson (Eds.), *Perspectives in ethology* (Vol. 12, pp. 7–53). New York: Plenum Press.

Timberlake, W., & Silva, F. J. (1994). Observation of behavior, inference of function, and the study of learning. *Psychonomic Bulletin & Review, 1*, 73–88.

Part II. Instructor notes

Classroom management/blocks of analysis

Time needed: This activity can be performed in a variety of contexts, depending on the emphasis. In a standard 3-h lab period, we recommend starting with the setup to observations on smiling (15–30 min). Next, have students pair up and go out around campus to collect their data on smiling (45–60 min). Have students return to the classroom and enter their observations into a spreadsheet that you have previously set up (20 min). Discuss the analytical approach that you will use, then have students work in groups to conduct the statistical analyses (20 min). Finally introduce the video portion of the activity, and watch one to two videos as a class (20 min). Have the students complete at least two additional videos as an assignment to be submitted the following week. In a "standard" class structure, this activity can be completed over the course of a week (either two or three class periods, depending on the schedule). During class 1 allocate approximately 20–30 min to explain the activity and provide details on the assignment. Students then collect data on smiles as a homework assignment and enter their observations into a spreadsheet that is available on the learning management system or shared drive. If enough detail is provided, you may have students complete the analyses on their own and submit this as an assignment; alternatively, you may have students complete the analyses during the next class. This should take 20–30 min. Use the remainder of the time to introduce the video observation portion of the lab, completing at least two videos as a class and then assigning at least two additional videos for homework. This activity works well in preparation for any sort of inquiry-based observations and data collection.

Teaching the activity

Preclass preparation

Because the observations on nonhuman subjects are done using video, no Institutional Animal Care and Use Committee (IACUC) review is required. Observing humans for a class assignment, and in which no individual identification information is recorded, should not require Institutional Review Board (IRB) review.

Table 3.2 lists a number of links to YouTube videos that work well. There is no guarantee that the links will remain functional in the long term, and as a result, it is advisable to preview these links and search for others if needed. Because such links provided may not be permanent, they provide a good idea of the sorts of videos to

Table 3.2 Examples of suitable YouTube videos.

Tigers https://youtu.be/TbMtFSdC8GQ
Orb weaver courtship https://youtu.be/eTeMqcKuTw4
Pilot whale https://youtu.be/jwqOruNZxj8
Lizard dewlap display https://youtu.be/BeFKTpgRpzQ
Crane courtship dance https://youtu.be/renwuZnKeOg
Giraffe necking https://youtu.be/UgfP-jup6Dw
Dogs https://youtu.be/W5xob2NL-1E

search for. The key components on which to focus when identifying videos are (1) diverse species that vary in the degree of familiarity that students are likely to have, (2) lengths of 1–6 min, and (3) no narration. In any case, it is advisable to watch videos without sound.

Modifications to the activity

Although it is quite feasible to complete this exercise using "live" observations of any species that are readily available (e.g., at the zoo or aquarium, dog park, or in the lab, if organisms are available for other activities), using video is actually preferable in this case. If individual instructors have video from their own research studies, this can be readily used instead of videos available via the internet or social media.

In-class preparation

Analytical approach

For the smiling activity, several statistical approaches are suggested. Using a binomial or sign test is quite simple, and it can be done using a variety of online analysis tools without specific software or detailed statistical knowledge. For example, GraphPad offers a simple online calculator for sign and binomial tests (https://www.graphpad.com/quickcalcs/binomial1.cfm). Chi square tests can be easily calculated using Excel. Other software is required for the Mann-Whitney or Kruskal–Wallis test, so this should only be conducted if such software (e.g., SPSS) is both *available* and *familiar* to students.

Areas of potential confusion or difficulty for students

For all observations (human and nonhuman), students often confound observations with inferences. For human observations, providing examples of the descriptions and information that should be recorded will be helpful. For the observations of nonhuman species via video, students initially may confound observations and inferences. It is critical to review the difference and go over examples to clarify. A common error, for example, is to describe two animals as "playing." In most cases, this is an inference. It is clearer to describe what is observed: one individual approaches another, hits the other with a paw, and the second individual follows the first. They roll around on the ground for 20 s and then the first individual runs after the second. Play is an inference. Reviewing a number of examples, and seeking consensus on descriptions, can help alleviate this confusion.

Recommendations for extensions or continuations for more advanced classes

This exercise can serve as a good introduction to an inquiry-based, semester-long data collection project. In conjunction with other chapters, it emphasizes the importance of reconnaissance observations and being objective in describing behavior. Observations of humans can serve as an excellent starting point to studies on human behavior. The observations on smiling provide a solid foundation for more detailed observations on nonverbal human behavior.

Answer key (smiling)

1. Based on your data, what information do you think smiling provides?

   Answers may vary based on the nature of the data, but in general, students should realize that smiles convey information about intent and affiliation and represent an "honest" signal in that "false smiles" are often lacking some of the diagnostic features associated with smile types.
2. Why might it be important for a communicator to provide interaction information with smiles?

   Smiles serve as signals and convey information about the nature of the interactions. Although this is an inference, given the extensive body of literature on smiling, we have high confidence in our inferences.
3. Why might it be important for a signaler to be able to use a graded set of signals to provide more finely tuned information about interaction?

   Humans rely on verbal and visual communication. Visual signals such as smiling are "understandable" independent of language and therefore provide an important reliable indicator of intent. As noted, smiling is a continuum, and as such, more specific information can be conveyed via subtle variations in the nature of the signal.
4. Was there any evidence to suggest that there might be a gender difference in the frequency of smiling?

   These results will vary based on the data collected.
5. Did you find any evidence to suggest that smiles might not necessarily be always associated with friendly interaction?

These results will vary based on the data collected. Dominance smiles are not necessarily associated with friendly interaction.

6. Do smiles function primarily for the communicator, for the recipient, or for both? Must both benefit in the same way or to the same extent? At what point or under what conditions would you expect recipients to cease responding to this display?
   By definition, the sender of a signal must benefit; the recipient may or may not benefit. In the context of smiling in humans, the recipient is likely to benefit via the increased ability to interpret intention and predict subsequent behaviors.
7. Can you identify and distinguish observations in your data, from functional inferences? Are these inferences reliable?
   These results will vary based on the data collected. Observations should focus on the form of the smile and physical characteristics. Context provides some degree of inference. Categorizing smiles (reward, affiliation, dominance) represents an inference but one based on extensive validated data analysis.

Answer key (videos)

1. Were the behaviors of some species easier to describe objectively than others? Why?
   These results will vary based on the videos selected. Generally, species that are more familiar to students (e.g., dogs and cats) or that demonstrate behavior more similar to humans (e.g., nonhuman primates) tend to be easier to describe.
2. To what extent does your familiarity with a species make it easier, or more difficult, to be objective in your descriptions?
   Greater familiarity with species often leads to more inferences.
3. Why is it important to refrain from inferring about the functions or causes of behavior?
   Inferences represent our interpretations about behavior and are influenced by our biases.
4. In what situations are functional definitions or descriptions appropriate?
   The less familiar we are with a species, and the less data that are available on the species, the more tenuous are our interpretations. Functional observations are more conservative and should be used.

Part III. Supplemental material

Supplementary data related to this chapter can be found online at https://doi.org/10.1016/B978-0-12-821410-7.00030-3.

CHAPTER

A matter of time: comparing observation methods

4

David M. Powell, Eli A. Baskir

Department of Reproductive and Behavioral Sciences, Saint Louis Zoo, St. Louis, MO, United States

Chapter outline

Exploring Animal Behavior in Laboratory and Field. https://doi.org/10.1016/B978-0-12-821410-7.00029-7

Part I. Student instructions

Learning goals and objectives

- Students will be able to select the appropriate data collection method based on the scientific question.
- Students will learn how to collect scientifically rigorous and objective behavioral data using different sampling strategies.
- Students will be able to discuss the advantages, disadvantages, and resource needs associated with each sampling strategy by comparing data collected from the same animal(s) using different methods.

Background

When parents or friends ask "What did you do today?", there are many ways you can answer the question, with differing levels of detail. They may want to know your activities throughout the day (e.g., swimming, studying, eating) and how exactly you spent your time (e.g., you were at the pool for an hour, then studied for 3 h, and then went to a restaurant for 45 min); if you are in trouble, they might want an accurate accounting of what you were doing every minute, including who you were with and where you spent your time. In much the same way, the question being asked provides different types of insights and answers in studies of animal behavior. Scientists can ask a multitude of questions about animals' behaviors and may need different levels of detail to find appropriate answers. The logistics of how observations are performed can impact what questions can be asked and what methods we use to answer them as well. For example, the number, activity level, speed, and habitat of the animal(s) in question can have a significant influence on what method you should use. First, let's look at some of the most commonly used techniques for gathering behavioral observations. These were first formalized by Jeanne Altmann (1974) and further expanded on by Martin and Bateson (2007).

Continuous sampling involves observing a single animal or, in rare cases, multiple animals for a predetermined period. You will record the start and stop time of every behavior the subject completes, for every instance of the behavior. Depending on your question, you might also add modifiers to the behaviors you record; for example, to whom the behavior was directed or where it happened. It is critical that you be able to individually identify the animals you are trying to observe using this method so you know who is doing what and to or with whom. This method allows you to gather data on behavioral states (behaviors that last for a measurable length of time and for which we aim to calculate durations) and events (behaviors with very short durations for which we tabulate frequencies instead of durations). This method gives you not only the most complete, accurate, and detailed record of the animal's behavior but also information about the sequences of behaviors. Depending on your research question and ethogram, this method can be very labor intensive and may require the use of behavior scoring software, handheld computers, or video review to collect the desired data. Animals that are very active and fast-moving, live in large social groups, or change behaviors frequently can be challenging to study with continuous sampling. It is possible to use continuous sampling even for simple research questions such as "How long do lions sleep in a typical 24 h period?", but there are other, less intensive methods that can provide good data in this case.

Instantaneous scan sampling can also be used on individuals or groups of animals. This method involves recording behavior *only* at predetermined intervals (e.g., every minute) to provide an *estimate* of how long animals perform various behaviors. In most cases, the scientist uses a timer to note intervals and signal by alarm when it is time to record behaviors. Think of this method as taking photos every minute for 10 min and then describing what the animal did based only on those photos with nothing else observed. You can also have behavioral modifiers with this method. Depending on your research question, animals don't have to be individually identifiable to use this method. This method works well for behavioral states, but it is not suited for recording events because they are short in duration and unlikely to occur exactly when the timer sounds. Your sampling interval indicates how often you record behavior, and it depends on your research question and characteristics of the animal. If the animal's behavioral states typically last a long time, and they don't change behavior very often, longer sampling intervals will give you just as accurate an estimate of how long the behavior truly lasts and will be less labor intensive to record than short intervals. If animals are likely to perform many different behaviors in short periods, however, shorter intervals are needed to more accurately create estimates of how much time they spend performing various behaviors.

All-occurrence sampling involves keeping a tally of the number of times an animal performs a behavior, which allows you to generate a behavioral frequency (e.g., how many times per minute); this method will not, however, give you an estimate of the behaviors' durations. This method can be used to study individual animals or groups. Depending on the research question, you may need to be able to individually identify the animals or to record the exact times that behaviors occur. You can use behavior modifiers, if needed. This method is ideal for studying behavioral events, but it is not recommended for studies in which you want to estimate how much time is spent performing each behavior. If, however, you were only interested in how many naps a lion takes each day but did not care about the duration of those naps, you would use all-occurrence sampling instead of continuous or instantaneous scan sampling to collect those data. Continuous sampling can provide similar data to this method, but continuous sampling could be unnecessarily labor intensive compared with all-occurrence sampling. Very often, scientists will use all-occurrence and instantaneous scan sampling simultaneously to study behavior when continuous sampling is not possible or necessary.

One-zero sampling can be used with individuals or groups, individually identified or not, and it can be used to study behavioral states or events. It is similar to instantaneous scan sampling in that it relies on timed sampling intervals, but it differs in several important ways. In one-zero sampling, at the end of the sampling interval, you score a "1" for any behavior (including states AND events) that occurred *during* the interval, regardless of how many times it actually occurred or how long the behavior lasted. You score a "0" for any behavior that did not occur *during* the interval. Instantaneous scan sampling gives you an estimate of how much time the animal spent in various behavioral states, whereas one-zero sampling gives you an estimate of the number of intervals in which behavioral states or events occurred. It is a less accurate record of the animals' behaviors than the other methods discussed. Despite this limitation, one-zero sampling may be the method of choice when there are many animals to observe, when the ethogram contains a large number of behaviors, and when behaviors change rapidly and frequently. It is also useful for research questions that don't require much detail or for converting other kinds of data (e.g., log books, health records, daily diaries) into quantifiable data. For example, if we wanted to know how long female lions demonstrate reproductive behavior during their estrous period and we were relying on zookeepers to collect the data, one-zero sampling would likely be the best method to use. Zookeepers see their animals throughout the day multiple times, but they don't have hours to observe animals using a formal behavioral data collection method. Instead, they write down their observations in a daily log. We can review those logs and score "*1*'s" for days in which they observed reproductive behavior and "*0*'s" on days when these behaviors were not seen. This information allows us to estimate how many days (intervals) the reproductive behavior lasts during the estrous cycle.

Purpose

In this exercise, we will cover the major methods scientists use to collect objective data on animal behavior, talk about the advantages and disadvantages of each method in terms of the questions for which they are useful, and discuss how to select an observation method based on considerations other than to the scientific question. For example, the number, activity level, speed, and habitat of the animal(s) in question can have a significant influence on which method you should use. You will observe several video clips and score each using multiple data collection methods. Afterward, you will compare your findings to help decide which method provides the best data for the research question.

Behavioral "rules"

As the examples will demonstrate, your selected method of scoring can provide different results, even when using the same ethogram, subjects, and video sample.

For this reason, your ethogram and your data collection methodology must be considered together.

When calculating activity budgets with either method, it is important to be mindful which behaviors are mutually exclusive and which are not. If your ethogram contains "laying down" to indicate when an animal is prone while resting and a "grooming" behavior, both should be defined such that overlap is not possible when scoring. You can accomplish these distinctions in multiple ways, with varying effectiveness and acceptability. Writing careful descriptions in the ethogram can make it clear that an animal laying down is not engaged in any other behavior, although this decision could lead to overly complex and wordy definitions. You should be wary, however, of including definitions on the ethogram that are just combinations of two or more other behaviors, for example, using "laying down," "grooming," and "grooming while laying down." Instead, you may decide when developing your ethogram and scoring behaviors that the more active behavior will be the one scored, but be careful of behavior combinations where relative activity is difficult to judge—i.e., an animal eating while moving or interacting with enrichment while performing social behaviors. Having to determine a hierarchy of what to score when overlap occurs usually reveals a weakness in your design.

Finally, depending on your study question, you may decide to record multiple behaviors that are grouped and analyzed differently. For example, if you wanted not only to examine how often an animal is walking and how stationary it is while alert, resting, or performing other locomotor activities but also to separately track investigative behaviors (e.g., sniffing, digging, etc.), you could record both an investigative and a locomotor activity at each scan. Take care that this method does not lead to recording too much information for each member of a large group, or it will affect your accuracy. Note that in some cases, especially if data are collected continuously on a single subject, it may be easy to record start and stop times of multiple states; however, this method will almost certainly require the use of video or an additional observer if the ethogram is large.

The amount of detail and attention required for your subject means observing multiple animals at the same time is difficult when using continuous sampling. If collecting data from recorded video, it is possible to follow and continuously score animals separately by watching the footage multiple times, paying attention only to an individual subject each iteration. If data collection is performed live, we make use of "focal" animals: selected subjects that will occupy our attention for either part of or the entirety of the observation period. Once a focal animal is selected, only behaviors performed by that subject are recorded; however, social behaviors—that is, ones with an initiating subject and a receiving subject—are also usually recorded if the subject is part of the behavior, even if only the recipient. Focal animals should be selected randomly for observation, and if multiple focal animals will be observed in a session, the order of observations should be randomized. Departure from random selection of focal animals is possible in cases where the study design requires it (e.g., a study that focuses only on female lions).

To use continuous sampling with a group of subjects, you can rotate which animal is focal, either between or within observation periods. For example, if you are observing a group of six waterfowl for an hour each day, you may decide that only a single member of the group of six is observed for the hour-long session. Alternatively, you may choose to observe each animal for only 10 min, rotating throughout the group.

Consistency is the key to collect good data, both within and across observation periods. Establish a pattern for how scans will proceed before any data are collected: When conducting instantaneous scan sampling, each scan must be done in the same way every time (e.g., always scan the area from left to right or scan sample individuals in the group in the same order each time).

Similarly, before data collection starts, you should decide how you will account for time when a subject is not visible. These periods are important to note because observers cannot collect data or determine the location of and activities performed by out-of-sight animals. In continuous sampling, it is easy to mark the start and end of periods when a subject is no longer visible. These out-of-sight periods can be totaled and subtracted from the total observation time to get a modified time that better reflects when observers were actually able to collect data from the subject. This modified time serves as the denominator for any activity budget calculations. For time interval sampling, observers can record if an animal is not visible for individual scans, but this omission can lead to overestimation of how long an animal was out of sight. Ultimately, it is up to the observers to decide how to handle these out-of-sight instances before any data are collected, as well as how much of an animal can be visible before observers decide it is out of sight.

Other general guidelines for data collection that will help observers from becoming overwhelmed include the use of a "5-second rule": when an animal is performing a state behavior, if it pauses but later returns to that behavior within 5 seconds of the start of the pause, then observers need not record that the behavior stopped and then resumed. This shorthand is helpful when an animal is performing a sporadic behavior over a long period in which it takes frequent but short breaks.

Methods

Open the video clip for the crane exercise and review the instructions, ethogram, and data sheet provided. After reviewing these materials, watch the video clip at least once before collecting your data to familiarize yourself with the behaviors. In your groups, one student will score the video using continuous sampling. The other students will simultaneously score the same video using instantaneous scan sampling, but one student will use a 5-second interval and the other will use a 10-second interval. The digital clock on the screen should be used to identify when the intervals have passed. After all three students have scored the clip, each one should calculate an activity budget that indicates what percentage of time the crane spent performing each behavior during the video clip. For continuous sampling, the student will sum the total seconds spent performing a particular behavior and divide by 95 seconds (the total duration of the clip). The student will repeat this calculation for each behavior in the ethogram. The student who scores the clip using instantaneous scan sampling at 5-second intervals will sum the number of scans performing each behavior and divide those sums by 19 (the number of 5-second intervals in the clip). The student scoring the clip using instantaneous scan sampling at 10-second intervals will sum the number of scans performing each behavior and divide those sums by 9 (the number of *complete* 10-second intervals in the clip). Be prepared to share the activity budget you generated using your method with your groupmates and the class.

Now, open the video clip for the tiger exercise and review the instructions, ethogram, and data sheet provided. After reviewing these materials, watch the video clip at least once before collecting your data to familiarize yourself with the behaviors. This time, we will compare one-zero sampling to all-occurrence sampling using licking behavior in tigers. Working again in groups of three (or two, if necessary), one student will use one-zero sampling to score licking behavior in the tiger while the other students use all-occurrence sampling. The one-zero sampling student will be recording which 10-second intervals contain licking behavior; the all-occurrence sampling students will attempt to count the number of times they can clearly observe the tiger licking the snow or other substrate. If the students are in groups of three, one student can be the all-occurrence "caller" (i.e., announces licks every time they see one) and the other will be the all-occurrence "recorder" who keeps the tally for how many times their partner indicates licking. At the end of the exercise, the one-zero sampling student will tally the number of intervals in which licking was observed and compare that to the all-occurrence data provided by the other students.

Results and discussion

The video review exercises you completed should highlight the suitability of different scoring methods based on the types of behaviors you recorded and the study questions you asked. In your groups, go through the following questions and discuss your experiences with the different methods.

For the crane video, answer the following questions:

1. Compare the activity budgets you generated using continuous sampling and 5- and 10-second instantaneous scan sampling. What differences do you note? Do all three methods perform comparably in this scenario?
2. Which methods were the easiest or most difficult?
3. What do you think would happen if the crane being observed was moving faster or more active? Would the job of observing that crane become more or less difficult for each of the three observation methods?

For the tiger video, answer the following questions:

1. Which of the two methods was easiest in this scenario?
2. Are the results of these two methods directly comparable? If not, how could you make them directly comparable?

3. Using the tiger's licking behavior in this video as an example, construct a scenario in which one-zero sampling would be adequate for assessing this behavior and a scenario in which all-occurrence sampling would be recommended.

General questions include the following:

1. Think of a research question but tailor it to each of the sampling methods discussed earlier. Are there any sampling methods that could not be used with your question?
2. Suppose you had a set of data collected using continuous sampling. Could this information be converted into an all-occurrence sampling format? What about one-zero? How does the output change with each change to sampling method?

Hopefully, this chapter has illustrated that along with species and ethogram, determining your sampling method is an important, nontrivial part of your study design. The selection of an observation method depends on the level of detail needed to best answer the question being asked. Some methods may over- or underestimate certain behaviors, if not miss them entirely. While scientific rigor should guide these decisions, resources available for doing the study and observer ability are important considerations. Remember that the amount of time needed to physically record behaviors can be important when choosing a sampling method—sometimes more detail can quickly become too much detail. Observers may not be able to adequately track animals' behaviors if there are too many animals, too many behaviors to score, or if scoring the behaviors diverts attention from the animals, leading to inaccurate data. Technology (video cameras, remote sensors, audio recordings, observation software) can minimize some of these challenges (although using these tools can sometimes lead to other challenges), but it is more important that the observer has a clearly defined research question and attempts to only collect the data they will reasonably need to answer the question. A final consideration is what kind of data each method actually produces and how or whether it can be used in various statistical tests during data analysis. All ethograms and observation methods should be tested during pilot observations and reviewed by peers in advance to help refine them.

References

Altmann, J. (1974). Observational study of behavior: Sampling methods. *Behaviour, 49*(3/4), 227–267.

Martin, P., & Bateson, P. (2007). *Measuring behavior: An introductory guide* (3rd ed.). Cambridge U.K: Cambridge University Press.

Part II. Instructor notes

Classroom management

The introductory text of this exercise is easily made into a mini-lecture that briefly covers each of the major observation methods, how they work, and some basic pros and cons of each method. The Behavioral "rules" section is not critical to conduct the exercise itself and could be used as supplemental reading, used as material for discussion, or incorporated into the main lecture if time permits. The mini-lecture could take anywhere from 15 to 30 min depending on how much detail the instructor wants to provide or whether all of the exercise will be completed in one class period or not.

The exercises work most efficiently if the class can be set up in groups of three so that multiple observation methods can be used at the same time by different students. Larger groups can also be accommodated: students can either take turns playing a role in the exercise or more than one student can do each role. The crane video itself is 95 second long and is played three times. The tiger video is similar in duration. Students probably need 5–8 min to read the instructions for the activities, ask questions, and become familiar with data sheets. A good 10–15 min is probably needed after the video scoring is done for the students to perform calculations and answer the discussion questions. In total, the crane and tiger activities together probably require ~45 min. The instructor can determine which of the exercise-specific or general discussion questions the students should explore. One or two groups of students could do the crane exercise while another group or two does the tiger exercise simultaneously if class time is limited. Students could also conduct the assignment individually and convene in groups to discuss results. Ideally, all students complete both exercises to get exposure to more observation methods and experience with their unique features. Depending on the time available, this activity in its entirety could be done during a lab period or over the course of two classes, with the mini-lecture and sorting into groups in one class period and the data collection and discussion in the next class period. If more than one class period is available to cover the material, the tiger and crane exercises can be split into two class periods while other information is covered, and there are other video clips, ethograms, and data sheets in the DVD or digital tutorial that can be used for the students to get more practice with these and other methods (e.g., scan sampling a group of animals). It is also possible to adapt this exercise for observing animals directly (see later discussion).

Teaching the activity (preclass preparation)

The full video tutorial is available for purchase as a DVD or as digitally downloaded files (https://www.aza.org/methods-for-animal-behavior-research-dvd). Go to the link indicated in the exercise for ordering instructions and payment options. It is very economical at $8 per copy. Alternatively, to only view the videos associated with this exercise, visit the following Elsevier link, https://doi.org/10.1016/B978-0-12-821410-7.00029-7. Students will need to be able to clearly see the screen/monitor(s) where the video is projected to make observations. Audio is also necessary so they can hear the cues to make behavioral observations. Exercise instructions, data sheets, and a key are provided in the online supplementary materials. Calculators or access to calculator apps on phones is useful, but no math beyond basic arithmetic is required. Results for the crane activity (see later discussion) could also be graphed in Excel for a visual presentation that compares the methods, if desired.

This activity doesn't require any Institutional Animal Care and Use Committee (IACUC) approval and could also be done by observing local animals, but live observations may require adjusting the ethogram and data sheets and may become challenging if animals leave the observation area. To get around this issue, observing animals at a local zoo or aquarium would be a good option as would observing dogs in a dog park. If live animals are going to be manipulated in any way to facilitate observations (e.g., by setting up a feeding station or broadcasting food for birds or squirrels) an IACUC should be consulted. Humans could also be observed as subjects using these exercises, but an institutional human subject review board should be consulted first.

Teaching the activity (in-class preparation)

Each video is shown once before the students actually start collecting data in order for them to get familiar with the clip. During this first showing, students are likely to be comparing video to the ethogram and data sheets multiple times so they can be ready when the video is played a second time. If students seem overwhelmed after watching the video the first time, then replay it until they are ready to actually collect real data during the second showing of the video. It is fine if they want more "practice" before they do the "real" data collection.

Crane video activity

The crane video activity highlights the differences between continuous sampling and instantaneous scan sampling at two sampling intervals. The data sheet provided in the tutorial for this activity is for continuous sampling; the data sheet provided in the online supplementary materials for this book combines continuous and scan sampling.

Answers to crane activity questions

1. The activity budget for the crane obtained via continuous sampling should be 80% Feed/Forage, 12.6% Alert/Head Up, and 7.4% Scratch/Preen. The activity budget based on 5-second scan sampling is 78.9% Feed/Forage, 15.8% Alert/Head Up, and 5.3% Scratch/Preen. With 10-s scan sampling, the activity

budget is 66.7% Feed/Forage, 33.3% Alert/Head Up, and 0% for scratching. These figures may vary slightly if the students were off by a few seconds, but minor variation is acceptable unless students are significantly off in their recordings of start/stop times. Slight variations from these values may also occur because at least one of the scans happens at a time near a behavioral transition, and the observer(s) could score one behavior or the other. This discrepancy is not so important for the comparison of the activity budget but can highlight a very real situation in which a scan happens at the time an animal is transitioning its behavior. Take this opportunity to discuss what the observer should do in that circumstance. Acceptable possible answers are to score the behavior being transitioned from, to score the behavior being transitioned to, or to use an "unclear/unknown" category. If the observer decides to actually select one of the behaviors (transitioned to or from), they should be consistent about this "rule" across all their observations. Make it clear to students that they should never score both behaviors if a scan occurs during a transition because additional complications will be introduced into the calculations of the activity budget.

The continuous sampling activity budget for the crane should be considered the gold standard, because it is the most accurate and complete record. The 5-second scan activity budget compares quite favorably. It slightly overestimates Alert behavior and slightly underestimates Scratch behavior. The 10-second scan activity budget deviates dramatically from the other two and misses the Scratch behavior altogether.

2. Students will likely report that the continuous sampling method was the most difficult here, followed by the 5-second and then 10-second scan methods. Students may also relay that they were concerned about "what they were missing" when using either of the scan sampling methods. This concern is valid and allows for the opportunity to discuss selecting an appropriate sampling interval. A set of pilot observations will help refine an adequate sampling interval. At the extremes, a very short sampling interval is almost the same amount of work as continuous sampling and could be less advantageous (especially if the scientist wants to include event behaviors, which scan sampling is highly unlikely to capture). A very long sampling interval will miss so many behaviors that it underestimates rare or even somewhat common behaviors and greatly overestimates very common behaviors.
3. More active animals are not necessarily more difficult to score using scan sampling because observers are still only recording behaviors at preset time intervals; however, if an animal is changing behaviors rapidly, the chance that an interval occurs during a behavioral transition increases, as discussed earlier. Observers using continuous sampling may note more difficulty with more active animals because they will have to take more notes on behavior (requiring taking eyes off the animal to record behavior or to enter key strokes on a computer or tablet). With very active animals, if a continuous sampling method must be used, it may be useful to conduct observations using video, which allows for multiple reviews to capture all the behavioral dynamics.

Tiger video activity

The tiger video activity highlights the differences between all-occurrence and one-zero sampling. The data sheet provided in the tutorial for this activity is for one-zero sampling; the data sheet provided in the online supplementary materials for this book combines one-zero and all-occurrence sampling.

Answers to tiger activity questions

1. In this video, students will find that one-zero sampling is definitely easier than all-occurrence sampling. If a behavior isn't very frequent, however, all-occurrence sampling can be relatively simple as well. If an ethogram contains many behaviors and the animal is active, one-zero sampling can become challenging. Both methods can also be applied to groups of animals, although additional subjects will understandably make observations more challenging.
2. Initially, the results of these two methods are not directly comparable. One-zero provides a percentage of time intervals in which a behavior is observed (in this case 6/8 intervals or 75% of intervals), whereas all-occurrence sampling gives both a sum and a rate, which can be calculated by dividing the sum by the duration of the observation (in this case the sum is ~32 or 0.4 licks per second). An hourly rate of licking from both data sets can be calculated by the following: One-zero sampling: For each interval in which a lick is indicated, you assume at least one lick occurred. As licking is indicated in 6/8 intervals, the rate of licking is at least 6 per 80 second (the length of this observation). There are 45 80-second intervals in an hour, so multiply 6 by 45 to get an hourly rate of 270 licks per hour using this method. Please note that this calculation only provides a *minimum* rate of licking. One-zero sampling produces small counts by design and sometimes does not adequately capture how frequently a behavior occurs. All-occurrence sampling: You have 32 licks per 80 second (the length of this observation). Again multiply by 45 and you get a rate of 1440 licks per hour. Obviously, a very big difference! For behaviors that are frequent, one-zero sampling is a poor choice for trying to back-calculate a rate of behavior.
3. The answers to the third question about scenarios in which licking behavior should be assessed with one-zero as opposed to all-occurrence sampling may vary. It really depends on the question and the resources that can be put to answer the question. If the tiger is recovering from oral surgery and the keeper only has time to do casual intermittent observations throughout the day to just check on the tiger, one-zero sampling is the method of choice. This would allow the keeper to determine if the irritation resulting from the surgery is going away. Alternatively, if the keeper needs to know if an antibiotic injection has helped a wound on the tiger's paw to heal, the keeper can use one-zero sampling to get a general sense of how irritating the wound continues to be for the tiger, inducing it to lick its paw. Conversely, if an intern or additional staff member can spend focused time just watching the tiger, all-occurrence sampling will give a finer-scale answer that is probably more reliable. If the question is which flavored

enrichment items are more desirable to the tiger, all-occurrence sampling is probably a better choice because it provides greater resolution. In general, for questions when a more accurate accounting of rate is needed, one-zero sampling may not be sufficient. For more general questions—e.g., do any courtship behaviors occur between a breeding pair—one-zero sampling is adequate; more complicated questions that have a greater degree of granularity require all-occurrence sampling—e.g., does the frequency with which a male performs courtship displays directed toward a female predict breeding success between pairs?

Answers to general questions for students

1. The first discussion question should reinforce the student's understanding of strengths of various observation methods and to what scenarios they are best suited. This question is self-guided, but the goal is to help the student to integrate question and method in science. When research questions are very general or vague (e.g., how do cranes spend their time? or do tigers prefer the flavor of blood or meat?), many different observation methods can be used, but they may not provide the kind of data or resolution needed to resolve a specific question or test a specific hypothesis. This discrepancy is why all aspects of good study design are linked.
2. The second question helps students understand how their choice of observation method affects the "back end" of the study as well as which kinds of analyses are actually possible with each kind of data. Choice of analysis should be made before any study begins as part of the selection of observation method.
   Some "back calculation" is possible among some of the methods. For example, feeding behavior, typically a state behavior with measurable duration, scored using continuous sampling can be converted to an all-occurrence data point (a rate) by just counting *bouts* of feeding behavior in the continuous behavior record. Continuous sampling data can also be converted to instantaneous scan sampling by noting the behavior that is occurring in the continuous record at the desired scan sampling interval. As demonstrated, one-zero sampled behavior can at least be converted to minimum rates of behavior. Scan samples can also be considered as occurrences of behavior and converted to minimum rates.

Part III. Supplementary materials

Supplementary data related to this chapter can be found online at https://doi.org/10.1016/B978-0-12-821410-7.00029-7.

CHAPTER

Who is taking whom for a walk? An observational study of dog–human interactions

5

Jennifer Mather

Department of Psychology, University of Lethbridge, Lethbridge, AB, Canada

Chapter outline

Part I. Student instructions

When you complete this laboratory exercise, you will be able to

a) explain how the ethological method of behavior analysis works,
b) describe how two individuals can move not only in space but also with relation to one another,
c) predict what motivation might underlie the actions you saw,
d) have a basic understanding of theory of mind (what an animal might know about another's motivation).

Exploring Animal Behavior in Laboratory and Field. https://doi.org/10.1016/B978-0-12-821410-7.00005-4

Background

Because dogs are so common around people, we know a lot about them, and there are several books specifically on their cognition and behavior (Horowitz, 2014; Miklosi, 2015). Dog ownership is very common in many countries, from around 15% to 40% of households (Bauman, Christian et al., 2011). Dogs and people are bonded, in the sense that each likes and pays attention to and tries to please the other (Higgins et al., 2013), and dogs pay close attention to what their owner wants and knows (Reid, 2009). Such responsiveness is the result of the gradual domestication of dogs from their wolf ancestors, who are not similarly attentive. This is likely due to a predisposition to attend to humans, fast learning and early contact of pups with people, and lifelong social training (Reid, 2009; Udell et al., 2010). Legally, this is a one-way relationship, as people are owners of dogs, but most of them think of their dogs not as possessions but as companions or even family members (Cavanaugh et al., 2008).

Although dogs are responsive to their owner, not all people and dogs are the same, and both have distinct personalities (Cavanaugh et al., 2008), so the balance of this relationship is also not the same. One of the responsibilities of owners is the pressure to take the dog for a walk, presumably on a leash. Not all dog owners walk their dog regularly: 40%–80% do so (Bauman et al., 2011), depending on the health and motivation of both members of the pair. Who decides to go for a walk? Higgins et al. (2013) report that a dog's behavior is often the stimulus to go out, and Lim and Rhodes (2016) found that although many owners felt an obligation to take the dog walking, characteristics of the dog, such as size and activity level, predicted whether they would actually go out. Researchers also focused on the benefits to people of doing so (Bauman et al., 2011; Curl et al., 2017; Higgins et al., 2013), including physical health and activity, bonding, and social contact (no one asked the dogs).

Research is missing on what the dog and person are doing when walking together. To get information about this, you have to think of the person and the dog as a unit, two objects at each end of a line. They move together, but they move because one or the other (now the dominant) starts off. The dominant one could pull on it and the subordinate would be pulled. A dominant one would be out ahead and the subordinate would follow, or stop and the other would wait. Also, the dominant would "decide" where to go and the other would follow (see the sample Table in Supplementary materials). You could score a 1 every time you see each of these behaviors.

However, we have a problem with simple observation of movement. The owner might tell the dog what to do, should we have an entry for Command? The dog might disobey, should we have an entry for Ignore? The dog and owner may have a customary route and follow it. How do they know, and how could we find out? Dogs and owners are socially bonded. How do they "read" one another as they move along together? No one pushes or pulls because they both know where they expect to go. Do they have a "theory of mind" (Reid, 2009; Schaafsma et al., 2014) about what the other wants and knows? Does the person know from the dog's tail wagging, upright posture, and sniffing that the dog has been cooped up all day and really wants enrichment to explore every little bit of the neighbors' front yards? Can the dog tell from the person's posture and pace that it has been a bad day at the office and they are tired but going for a walk because it is expected. Will that make a difference to who is in charge?

Purpose

The objective of this observational study is to look at what humans and their dogs are doing when they are out for a walk. They are connected by a leash, but who leads and who follows and who controls route taking and stopping–starting? What might they know about their partner's knowledge and intentions? You will gain experience creating a limited behavioral repertoire, which means noticing who is doing what when you watch the dog and the person (see the sample Table). This gives you a chance to try the approach of ethological analysis (Lehner, 1998) and to start by looking and seeing what is happening. You will have to decide what is going on to cause these behaviors and what the interactions that you see might "mean." You should notice how dogs and their owners know a lot about one another, and you might wonder about the dog's (and the person's) "theory of mind" (Horowitz, 2011; Reid, 2009).

Methods

Students can observe dogs on videotape or watch in their natural environment, especially in neighborhood parks.

Materials: If students are watching dog–people pairs in an outdoor location such as a city park, they will need appropriate clothing for the weather and stable "sensible" shoes. Whether outdoors or in the lab, they will need note-taking equipment, either a spiral notebook and a pen or a laptop computer or tablet, sheltered from possible rain. If they are watching video footage in the lab, they will need stop, slowed, or frame-by-frame motion through the video so that they can make sure of what they are seeing.

Step-by-step instructions

Start with the hypothesis: The person is dominant and is "taking" the dog for a walk.

1. Think about what behaviors you will look for. If you are watching a video, you can run through it a couple of times to see what happens, then slow it down and watch for specific behaviors. You want to decide on *dominance* (i.e., who appears to be in charge), so begin by deciding what behaviors indicate this. Make up a catalog or "ethogram" of these behaviors in each individual and decide which behaviors are dominant (positive) and which are subordinate (negative). Consider the "dominant" behaviors to indicate the leader and the "subordinate" behaviors to indicate the follower.

2. Can you tell by the leash: does the dog pull at it or does the owner pull the dog? Who goes first as they walk? Who waits for the other when one pauses? Watch and record the behaviors indicating dominance and control (positive value). See the Table for an example.
3. Give either the dog or the person a score of 1 each time you see a "dominant" behavior and enter a mark in the appropriate box (notice we are assuming that a dominant by one results in a submissive for the other, so we do not need to record that). Dominant dog behaviors might be pulling on the leash or being first along the path. Dominant persons would be the dog following "at heel", with the owner calling and dog coming. After you have entered a check mark every time you see these behaviors, you total them up in the boxes and then across all boxes. Look at the totals for dog and person and subtract the smaller from the larger to get a difference score. You may find a few difficulties with these simple scores; for example, how would you score if a person called and the dog did not come?
4. Total up a difference score for each dog–person pair and compare this among several different pairs. An example of possible results is found in Table 1 in Supplementary materials. According to the table, the dog is dominant in the first entry and the person in the second one. The third entry shows a dog "not under control" not doing what the person asked—a puppy? The fourth entry shows a dog and person walking together "in harmony." This is a chance to think about "theory of mind;, how do they know what each other wants?

Results/discussion

By adding up the scores, at least for your observations and possibly for all the class, you can end up with average dog–person scores. Because no one has done a formal research study on this, there is no clear-cut answer. Careful to record not only the scores but also in which column. You might think that as people "take their dogs for a walk", it should be obvious that the person is in control and that your data will *support your hypothesis*. But maybe not; remember that Higgins et al. (2013) found that dogs stimulated owners to take them out. If you do this as a class, then record the dog breed, size, and age to see which makes a difference and compare observations. Quite likely a German shepherd will have more control than a Pekinese, a puppy than a grey-haired senior—remember the Lim and Rhodes (2016) study. How do you tell how closely they are bonded? See the Reid (2009) and Udell et al. (2010) studies about social sensitivity, whether the dog has had obedience training as a puppy and if the person reminds him or her of this. Most of all practice what the ethologist does (Lehner, 1998), which is to look and measure. Find out as you do this that you are *noticing* how much there is to see in this everyday activity.

Questions

1. Legally, the persons are described as owners and, as long as they are not abusive, they can do what they like with the dog. But does "owning" describe the interactions that you see?
2. When we watch many different interactions, we take an average of all of them to describe what is "typically" seen. But dogs and people are different from one another; think about their personalities. Should we see them as individual pairs and not as averages?
3. Theory of mind is a controversial area in animal behavior right now, mostly because it is so difficult to prove. How can we know what an animal, even another human animal, is thinking? A lot of researchers say that we cannot but that we can only see behaviors and these can be interpreted in several different ways.

References

Bauman, A., Christian, H. E., Thorpe, R. J., & Macniven, R. (2011). International perspective on the epidemiology of dog walking. In R. A. Johnson, A. M. Beck, & S. McCune (Eds.), *The health benefits of dog walking for people and pets* (pp. 25–42). Purdue University Press.

Cavanaugh, L. A., Leonard, H. A., & Scammon, D. L. (2008). A tail of two personalities: How canine companions shape relationships and well-being. *Journal of Business Research, 61*, 469–479.

Curl, A. L., Bibbo, J., & Johnson, R. A. (2017). Dog walking, the human-animal bond and older adults' physical health. *The Gerontologist, 57*, 930–939.

Higgins, J. W., Temple, V., Murray, H., Kumm, E., & Rhodes, R. (2013). Walking sole mates: Dogs motivating, enabling and supporting guardians' physical activity. *Anthrozoos, 26*, 237–252.

Horowitz, A. (2011). Theory of mind in dogs? Examining method and concept. *Learning & Behavior, 39*, 314–317.

Horowitz, A. (Ed.). (2014). *Domestic dog cognition and behavior.* Springer Publishers.

Lehner, P. N. (1998). *Handbook of ethological methods.* Cambridge University Press.

Lim, C., & Rhodes, R. E. (2016). Sizing up physical activity: The relationship between dog characteristics, dog owners' motivation and dog walking. *Psychology of Sport and Exercise, 24*, 65–71.

Miklosi, A. (2015). *Dog behaviour, evolution, and cognition* (2nd ed.). Oxford University Press.

Reid, P. J. (2009). Adapting to the human world: Dogs' responsiveness to our social cues. *Behavioural Processes, 80*, 325–333.

Schaafsma, S. M., Pfaff, D. W., Spunt, R. P., & Adolphs, R. (2014). Deconstructing and reconstructing theory of mind. *Trends in Cognitive Sciences, 19*, 65–72.

Udell, M. A. R., Dorey, N. R., & Wynne, D. D. L. (2010). What did domestication do to dogs? A new account of dogs' sensitivity to human actions. *Biological Reviews, 85*, 327–345.

Part II. Instructor notes

Classroom management

Time needed for activity: this is fairly flexible. For a standard 3-h lab exercise, students can watch a video on the web or videos taken by the instructor. Although part of ethology is learning what to see, in a single 3-h lab, the instructor can provide a data sheet so that students already know what behaviors to look for. For two or more lab sessions, the instructor might spend the first class talking about the ethological methods, getting students to describe their own dog-walking experiences, and working out what to look for. Then they could go to locations where dogs and owners walk. This will have to be scouted first, and it might be independent of the lab meeting time, as dog-walking tends to be both a weekend and fair-weather activity. In the second (or third) lab meeting, students could compile their results, compare with others' results in the class, and work out a consensus of "who walks whom." In this class, they can be introduced to the idea that one can read motivation behind the actions, and if there is time, they can think about whether and how the owner and dog anticipate each other's needs (Theory of Mind) and plan for ways that they can fulfill them.

Preclass preparation: if the students watch videos, no Institutional Animal Care and Use Committee (IACUC) or Canadian Council on Animal Care (CCAC) research permission is necessary. If students are watching people in a public space, most institutional review boards and animal care committees agree that no permission is needed, particularly if information on individuals is not being used specifically. Even so, it would be a good idea to confirm this at your own institutions and to have students prepare a single-page information sheet, which is to be handed out if someone asks for it. Dog-walking is everywhere, but city parks, weekends, and during good weather (never a guarantee) are the best. If the instructor chooses to videotape walks for presentation in class, permission would be necessary, but then it could be reused each year.

In-class preparation: although the ethological method is simple, just watch, it takes some practice. A simple way to display the interactions would be to have a couple of rocks joined together with a piece of string. They can be moved around in the air to demonstrate leading and following, stopping and pulling. Students can start with a simple activity of watching each other (e.g., watching a pair having a conversation and deciding who dominates) to get used to the practice of observing and making conclusions. If there is time, they can be introduced to making a data sheet, converting ideas into numbers, and then drawing conclusions on the basis of these numbers. There may be areas of difficulty when students assume motivation from a quick observation without gathering a quantitative background, relying too much on their own past experience. A tightly bonded dog–person pair might quietly walk side by side, generating minimal apparent interaction and making it difficult for students to observe "control." In this case or in a more advanced class, students might be given ideas about and might discuss the "Theory of Mind" before doing the observations and they might be looking for subtle cues that suggest each of the pair is well tuned to the other. This could lead to a class discussion of domestication, contrasting the activity of walking one's dog with the more challenging task of walking a cat. Similarly, the instructor could link to classtime on the idea of "ownership" of other animal species and the ethical issues of the care of dogs and owner responsibilities.

Because no research study has asked this specific question, there are no sample results. If the lab exercise is done over several years, an instructor can build up a sample and tell students "that's different from last year's", for instance. But this is an excellent example of how new ideas are being generated all the time, how research can be initiated by a student lab. For this reason, a class could generate a hypothesis as to who is walking whom at the beginning of the exercise and then see whether the assumption was supported (this author's guess is that over a large-enough sample, it would turn out equal).

Question answers

1. Legally, the persons are described as owners and, as long as they are not abusive, they can do what they like with the dog. But does "owning" describe the interactions that you see?

   Note that when you have a positive score for the dog, you are not supporting the suggestion that the owner is "in charge."
2. When we watch many different interactions, we take an average of all of them to describe what is "typically" seen. But dogs and people are different from one another; think about their personalities. Should we see them as individual pairs and not as averages?

   A big variation in scores would lead us to suggest that the mean is not a very useful way to predict what any particular pair will do.
3. Theory of Mind is a controversial area in animal behavior right now, mostly because it is so difficult to prove. How can we know what an animal, even another human animal, is thinking?

   A lot of researchers say that we cannot but that we can only see behaviors and these can be interpreted in several different ways.

Part III. Supplementary material

Supplementary data related to this chapter can be found online at https://doi.org/10.1016/B978-0-12-821410-7.00005-4.

CHAPTER

Movement analysis: expanding the resolution of analysis in animal behavior

6

Afra Foroud[1], **Sergio M. Pellis**[2]

[1]*Department of Psychology, Department of Neuroscience, Institute of Child & Youth Studies, The University of Lethbridge, Lethbridge, AB, Canada;* [2]*Department of Neuroscience, Institute of Child & Youth Studies, The University of Lethbridge, Lethbridge, AB, Canada*

Chapter outline

Exploring Animal Behavior in Laboratory and Field. https://doi.org/10.1016/B978-0-12-821410-7.00008-X

Part I. Student instructions

Learning goals, objectives, and key concepts

- To identify how one's ideas, thoughts, intentions, preconceptions, and knowledge can influence what one sees when looking at a behavior
- To make the distinction between what the animal is doing rather than what one sees or expects
- To describe the sequence and timing of behavior using techniques from movement analysis
- To discuss how movement analysis is used for understanding how behavior is organized

Background information

One of the challenges of studying behavior, whether in the field or in the laboratory centers on the clarity of our initial observations (Fentress, 2008). How does one identify the measures that capture what the animal is doing rather than what one expects to see? *Standardized measures* may be relied upon to reduce *observer bias* and may provide information on the frequency of a behavioral event. For example, during rough and tumble play (RTP), rats compete to nuzzle the nape of their partner (Pellis & Pellis, 1987), and during the ensuing wrestling, they often adopt the "pin" configuration, with one playmate lying on its back against the ground and the other rat standing on top (Panksepp, 1981). Pinning is a readily identifiable behavior pattern that has been used extensively to measure the frequency of RTP in a variety of experimental conditions (Himmler et al., 2013).

The problem is that standardized measures often fail to identify patterns that are key to the behavior. By limiting behavioral analysis to document pinning actions, both the rules of engagement and how these rules vary in their broader contexts are likely to be missed. What are the actions that lead up to the play bout, and how does it begin? What do the animals do that enable the play bout to continue? In what ways does the play bout change over time, before, during, and after the pinning? Moreover, as pinning can arise from multiple actions by both play partners, are all cases of pinning the same? A unitary, snapshot measure, such as pinning, conveys the impression that when there is a numerical difference between two experimental conditions, the difference is along one dimension. However, if pinning can arise in multiple ways, such a causal interpretation may be incorrect (Himmler et al., 2013, 2016).

To investigate these types of questions, one needs a framework from which the sequence and interactions of the movements that produce the behavior over time can be skillfully observed and described (Fentress, 2009). *Movement analysis* provides such a framework and enables scientists to gain information on what the animal *experiences*, what kinds of *predictions* the animal makes, and which of the arising outcomes, or *consequences,* are important for the animal (Pellis & Bell, 2020). This approach helps researchers to focus on the behavior of the animal, rather than on the measure(s) applied, thus providing insight on the underlying processes.

Continuing with our example of rodent play, movement analysis has shown that rats alter both their "attack" and "defense strategies" (Foroud & Pellis, 2002, 2003; Pellis & Pellis, 1987, 1998) during RTP in ways that increase their own vulnerability. For example, juvenile rats tend to vary the placement of their hind limbs when pinning their partner. Instead of keeping their hind limbs anchored on solid ground, they may place them onto the torso of their play partner. This leads to greater instability in the attacker and provides the defender with greater opportunity for successful counterattacks. Such self-handicapping requires both rats to continuously gauge their actions, and those of their partners, as they shift back and forth, blurring the lines between attack and defense roles (Foroud & Pellis, 2003). It is likely these early perceptual motor experiences lead to the development of effective decision-making necessary to navigate the social complexities of adult life (Foroud & Pellis, 2003; Pellis & Pellis, 2017; Pellis et al., 2005).

For the rat, pinning appears to be the goal of the behavior, yet the experiences gained from the subtle shifts in strategies, along with their consequences, provide the animal with information about itself, its play partner, and when and how much to challenge itself and its partner, all of which are crucial for navigating complex social situations including access to safety, food, and sex. As a researcher, by seeking to understand what the animal experiences, expects, and seeks, one begins to gather information on how the behavior is organized. This approach can help address questions such as whether or not the rats are engaged in a predetermined set of actions or are making strategic decisions within a dynamic activity.

Movement analysis has been used to solve a variety of behavioral problems in many species, some examples include grooming in mice (Golani & Fentress, 1985), combat in sage grouse (Pellis et al., 2013), fighting in wolves (Moran et al., 1981), developmental trajectories in infants (Foroud & Whishaw, 2012), changes in skilled movement after stroke (Foroud & Whishaw, 2006), and animal models of obsessive-compulsive disorder (Szechtman et al., 2017), as well as in the field of robotics (e.g., Lourens et al., 2010).

There are two forms of handwritten movement analysis used for the study of behavior, namely, the *Laban Movement Analysis* (LMA) (Laban, 1960) and the *Eshkol–Wachman Movement Notation* (EWMN) (Eshkol & Wachmann, 1958). Each offers a unique lens with emphasis on different characteristics of movement. Both systems can provide either a shorthand form of notation that captures essential aspects of the performed movement or a highly detailed notation with which a reader can fully reenact the sequence without ever having seen it performed.

Purpose

You will be introduced to a few introductory applications of EWMN. Working individually, with a partner, or in small groups, you will examine two videos of cricket combat. Each video shows the same interaction in a pair of crickets recorded either from above (aerial view) or from the side (lateral view). Using the applications of EWMN, you will describe the sequence of interactions between the two crickets over time and, time permitting, the timing of the movements that produce the behavior.

Methods

Species selection

The videos in this chapter are of two male, banded crickets (*Gryllodes sigillatus*). These crickets are commonly sold in pet stores as prey for pet reptiles. Studies of cricket combat have been made in a variety of species including *Acheta domesticus*, *Gryllus bimaculatus*, and *Teleogryllus oceanicus*.

Materials needed, including variations based on species selection

This activity involves looking at the same video, or set of videos, repeatedly including *frame by frame*. You will need controls that enable rewinding and reviewing shorter sections of the video(s) repeatedly. It is recommended to view the video(s) on a computer monitor, or at least a tablet, rather than on a phone.

Other materials required include the following:

- Pencil (not pen) and graph paper are required to create the EWMN sheets (refer to Figs. 6.1, 6.5, and 6.8 for examples).
- Clear transparent sheets and permanent markers will be used to measure distances between animals in the video.

FIGURE 6.1 The Eshkol–Wachman Movement Notation (EWMN) sheet.

An example of a blank notation sheet is shown. In EWMN, body parts are represented in horizontal rows, with columns marking units of time (e.g., video frames). Double bar lines depict the beginning and ending of the notation and single bar lines depict each time unit. Starting positions are notated between the opening double bar lines. Movements are notated along the columns. Notations are read from left to right and from bottom to top.

- Thirty-two sticky notes, one pen or marker, and tape.
- Seven plain pipe cleaners, four of a different color and three of the same color.

Step-by-step instructions

1. *Initial observations of videos.* The following two videos will be used for the activities in this chapter: Video A and Video B. Both videos are recordings of the same event within the same time frame, each from a different perspective. Your first step is to look at the videos and make your initial observations. First, watch Video A and write what you see step by step in bullet points. Repeat this step using Video B.
2. Next, you will learn to use "EWMN" developed by choreographer Noa Eshkol (1924–2007) and architect Abraham Wachmann (1931–2010). EWMN uses a spherical coordinate system (Fig. 6.2) to describe the *spatiotemporal*

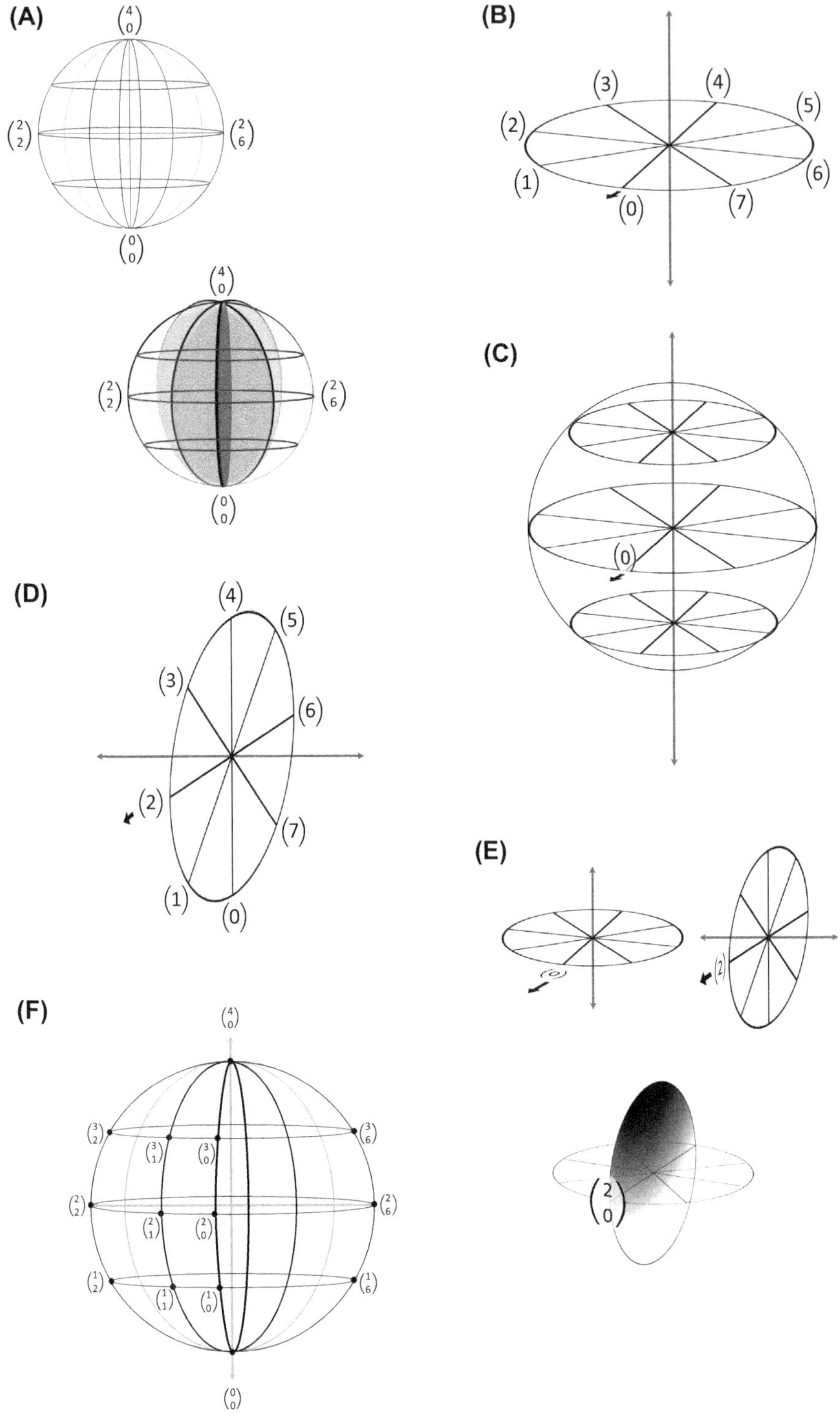

FIGURE 6.2 The Eshkol–Wachman Movement Notation (EWMN) system of reference: the sphere.

(A) A representation of the EWMN sphere is presented in two drawings to help conceptualize the sphere. The sphere is made up of a horizontal *axis* and a vertical *axis components of movement* made by the body, including changes in *spatial relations* between bodies (e.g., two or more animals) and parts of the body (e.g., head, torso, and limbs). The sphere provides a *system of reference* (Fig. 6.2, Video C) and is used to extract spatial coordinates that describe the relations and changes in relations between parts of the body. The system envisions the entire body inside a sphere (Fig. 6.3A and B) and each limb segment and joint inside their own smaller spheres (Fig. 6.3B and C). A larger sphere can be used for reference of *fixed (immovable) space* around the animal(s) (Fig. 6.3A).

Learning exercise 1: the Eshkol–Wachman Movement Notation sphere

Begin each of the following exercises by standing in a *neutral position* (face forward with your legs slightly apart for stability, and feet pointing forward; leave your arms and hands resting downward at your sides) (Fig. 6.2A). As you move through the exercise keep your body facing forward in the neutral position. Try to limit movement to your arm.

1. Primary *horizontal plane* (Fig. 6.2B and C)
   Keep your arm straight throughout this exercise.
   a. Lift up your right arm until it is parallel to the ground and extended in front of you.

and intersecting horizontal and vertical *planes*. Each point at which the planes intersect has a *coordinate* value. Four examples of the EWMN coordinates are shown. Each coordinate is read from the bottom number first. 0,0 and 0,4 are the opposing ends of the vertical axis and 2,2 and 6,2 are the opposing ends of the horizontal axis. Imagine you are standing in the middle of the sphere facing out of the page toward the reader, 2,2 is on your right, 6,2 on your left, 0,4 directly above your head, and 0,0 directly beneath your feet. (B) The *horizontal plane* with numerical values. The long gray arrow represents the vertical axis, and the small black arrow indicates the front of the plane. Imagine the body in place of the gray axis facing out toward the front (0). In EWMN, the unit of measurement is often defined by the 45-degree angle (the angle between each of (0), (1) (2), (3) (4), (5) (6), (7) is 45 degrees). (C) The horizontal plane exists at multiple levels, the numerical value for each level is represented by the vertical plane. (D) The *vertical plane* with numerical values, with the long gray arrow representing the horizontal axis. The small black arrow indicates the front of the plane. Imagine the body standing in the center of the plane, perpendicular to the axis facing out toward the front of the plane (2). (E) The horizontal and vertical planes intersect at multiple levels. Each intersection forms a point on the sphere. This figure shows horizontal (0) intersecting with vertical (2). The coordinate for this point is 0,2. The horizontal value is written on the bottom inside the ( ) brackets and read first. For example, 0,2 is a different point than 2,0. (F) Additional coordinate values of the EWMN sphere. Each point at which the planes intersect has a *coordinate* value named from its horizontal and vertical planes.

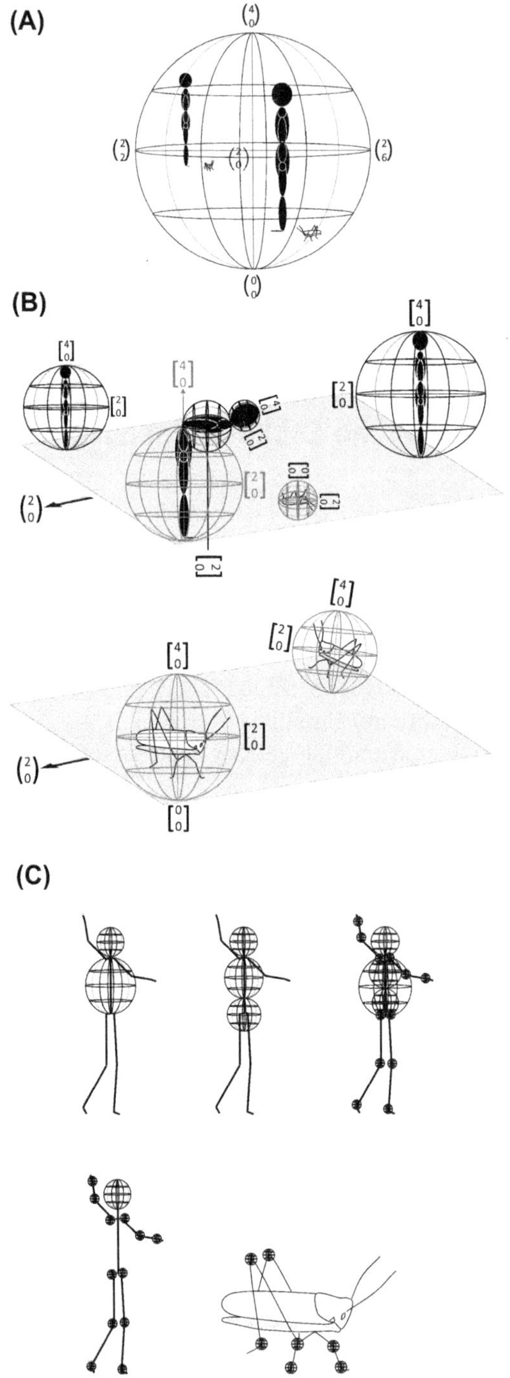

FIGURE 6.3 The Eshkol–Wachman Movement Notation (EWMN) system of reference: absolute space and bodywise.

(A) The EWMN sphere fixed in space. Points on the sphere are used to notate movement relative to the sphere fixed in space, and this is referred to as *absolute* space. Notated scores use sphere coordinates in round brackets ( ) to reference absolute space. Imagine

b. Imagine a line is drawn from your torso to the end of the fingertips on your right arm. The end point of this line is referred to as (0).
c. The round brackets ( ) around the number indicate a frame of reference that is fixed in the space around you. This means that for this exercise, you are using the larger sphere that is fixed in the space around you.
d. Keep the same arm straight and parallel to the ground as you move it to the right. Stop at the 45-degree mark and imagine a new line drawn from your torso to the end of fingertips. The end point of this line is referred to as (1).
e. Move the same arm for another 45 degree unit. The end of this imagined line, drawn from your torso to the end of your fingertips, is referred to as (2).
f. Continue the exercise by moving the same arm in the same direction. Stop at each 45 degree unit to identify its numerical reference.
g. Continue until you reach (0) again. At some point, you will have to switch arms to complete the exercise.
h. Once complete fill in the numeral references in the worksheet 1 *Horizontal Plane in the EWMN Sphere*.

2. Primary *vertical plane* (Fig. 6.2C and D)
   Keep your arm straight throughout this exercise.
   a. Begin in the neutral position.
   b. Imagine a line is drawn from your torso down to the end of your fingertips. The end point of this line is referred to as (0).
   c. Keep your arm straight while lifting it up directly in front of you. At the 45-degree mark imagine a new line drawn from your torso to the end of the fingertips on the same hand. The end point of this line is referred to as (1).
   d. Move the same arm upward for another 45 degree unit. Now your arm should be parallel to the ground. The end point of this imaginary line, drawn from your torso to the end of your fingertips, is referred to as (2).
   e. Notice that in the horizontal plane, this point is referred to as (0).

you are inside a large sphere, you may move on the spot or travel anywhere. Similar to the walls in the room, the sphere does not move, it remains stationary while you and your classmates move inside the sphere. (B) Smaller EWMN spheres attached to each body and limb segments. For this type of reference, called *bodywise*, notated scores use sphere coordinates in square brackets [ ]. The bodywise sphere travels with the body and can be used to notate movements of the body relative to the body itself rather than the larger sphere that is fixed in the space. Imagine you are wearing the sphere and the sphere moves with you regardless of which direction you travel. Note that for bodywise, 0,2 is always in front of the body regardless of where the body may be in absolute space. If you are inside the sphere, bodywise 0,2 will always be in front of you irrespective of where you may be facing in absolute space. (C) Additional representations of the various ways the EWMN sphere may move with the body (bodywise). These examples help illustrate that a sphere sits on each limb segment and at each joint.

f. Continue the exercise, moving your arm upward. Stop at each 45 degree unit to identify its numerical reference until you reach (4).
g. Notice that at (4), your arm is pointing straight up, parallel to the length of your torso or to the wall next to you.
h. To continue along the vertical plane, you will have to move your arm down behind you for another 45 degree unit. Try to keep facing forward, keeping your arm straight, and prevent it from moving out to the side (you will need to rotate at the shoulder). The end point of this imaginary line, drawn from your torso to the end of your fingertips, is referred to as (5).
i. Continue the exercise until you reach (0) again.
j. Once complete fill in the numeral references in the worksheet 2 *Vertical Plane in the EWMN Sphere.*

3. Intersecting planes (Fig. 6.2E and F)
   a. Take a look at Fig. 6.2E and F. There are several points where the horizontal and vertical planes intersect in the EWMN sphere. These points are identified by their coordinates using horizontal and vertical values. For example, in learning exercise 1.1a, when your arm was held straight in front of you and parallel to the ground, you were at (0) in the horizontal plane and (2) in the vertical plane (Figs 6.2E,F and 6.5).
   b. There was one more point at which the horizontal and vertical planes intersected in learning exercise 1.2, what is the name of that coordinate?
   c. When naming an EWMN coordinate, the horizontal plane is always referred to before the vertical plane. If someone says "move your arm to zero two," you will understand that to mean directly in front of you and parallel to the ground.
   d. If you are asked to "move your arm to two zero," where should you place your arm?
   e. If you are asked to "move your arm to two three," where should you place your arm?
4. Use Videos C and D to review what you have learned from learning exercise 1. Copy the performer as you watch the video. Learning to use EWMN involves a tangible knowledge of the system of reference that can only be gained by performing the movements oneself. Copy the performer in Video C to study the difference between absolute and bodywise (Fig. 6.3) and in Video D to help embody the sphere and the horizontal and vertical planes and their intersections.

In EWMN, the body is defined as a system of *articulating axes* with *limb segments* making up the axes (Fig. 6.4, Video E). A limb segment has a constant length, is any part of the body between two joints or a free extremity extending from one joint, and can move with one end fixed to the center of a sphere.

Notated scores often provide details on how each limb segment (including the head, torso, and pelvis) move over time by specifying the *types of movement*

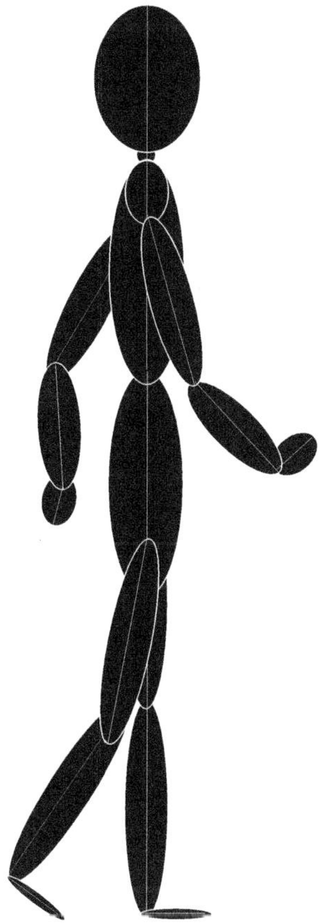

FIGURE 6.4 Limb segments: proximal and distal.

The major limb segments of the human body are shown, with their axes drawn with gray lines. A limb segment has a constant length, is any part of the body between two joints or a free extremity extending from one joint, and can move with one end fixed to the center of a sphere. The head, upper torso, and lower torso, as well as each section between the joints on the arms, legs, hands, and feet, are considered limb segments.

performed (Video F) (Figs. 6.5 and 6.6), their magnitude of size, spatial orientation, and direction (Video D). Notations offer a range of detail. Comprehensive notations, ones that may include many limb segments, are preferred for writing specific choreographies for dancers to study (similar to the way music scores are used by composers and musicians). In animal behavior, EWMN often zooms in on specific aspects of the animal's movement, for example, each limb segment on the fingers, or the relationship between animals, for example, the relative positions among a group of animals, in order to acquire more information on a puzzling behavior.

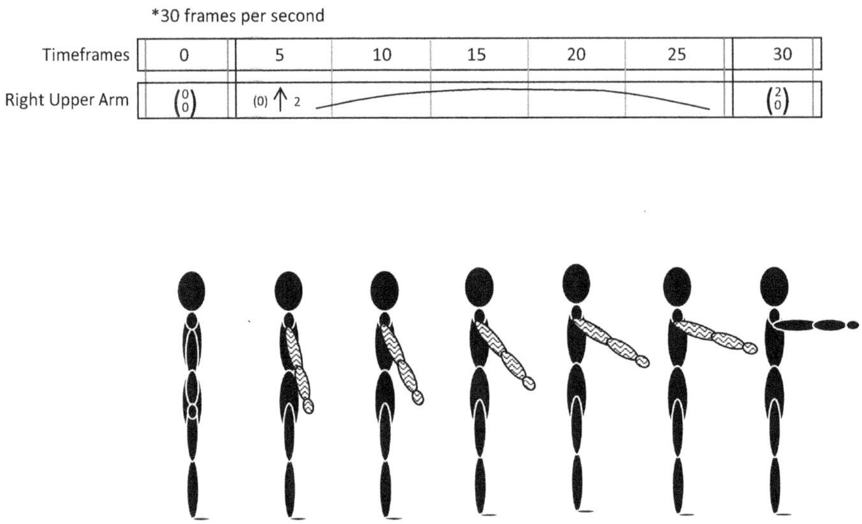

FIGURE 6.5 Limb movement and notation I.

A notated score of the right forearm being lifted from a neutral position to become parallel with the ground (learning exercises 1.1a and 1.3a) is shown. In the starting position the forearm is in the neutral position. The movement begins in frame 5 when the arm lifts upwards in the (0) vertical plane; the number 2 indicates the arm will travel two spatial units of 45 degrees. The bow from frame 5 to frame 25 describes how many units of time the forearm moves in this direction. The ending position is written in the final frame. Assuming this notation was made from watching a video of the performer, and the video was recorded at a speed of 30 frames per second (as noted above the notation), then the notation shows it took the forearm nearly 1 s to complete the movement. Below the notation is a sketch of a character performing the movement described in the notated score. For the purpose of this figure, limb segments that are not moving are depicted in black. Limb segments with a pattern help illustrate when the arm is in motion.

Learning exercise 2: partnerwise orientation

Partnerwise orientation describes the longitudinal axis of one animal in relation to the other.

1. Label the horizontal plane of the EWMN sphere onto the human body.
   a. Use your pen or marker to number each sticky note with its own number using numbers 0 to 7. Make sure each sticky note only has one number, and each number is written on one sticky note.
   b. Use the sticky notes to label the horizontal plane of the EWMN sphere on yourself. To do this attach the sticky notes onto your clothing (if you like, you may use tape to secure the sticky notes in place).

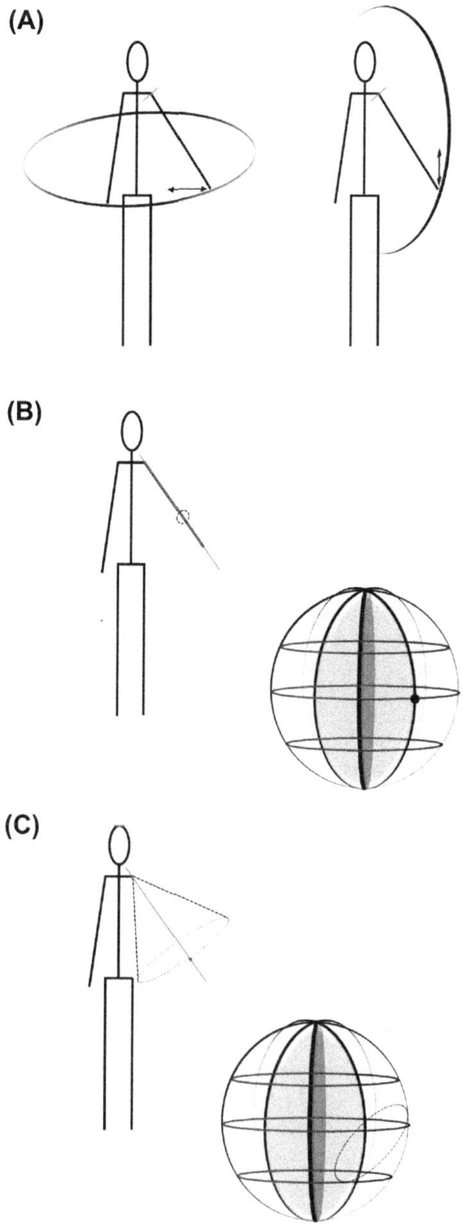

FIGURE 6.6 Types of movement.

For each of the three types of movement shown, (A) planar, (B) rotatory, and (C) conical, a gray line illustrates the axis of movement. (A) Planar movements maintain a right angle between the axis of movement and the axis of the limb. When you move your arm

2. Complete this step with a partner (*we have included the answers to help you work out the rules*):
   a. Select one person to be the *focal participant* (*focal animal*).
   b. The focal participant "carries" the EWMN coordinates with them during this exercise.
   c. Stand facing each other. What numerical direction is the partner facing in relation to the focal participant? *The answer is 4*. Thus the partnerwise orientation is {4}. Curly brackets { } are used for partnerwise reference.
   d. Refer Fig. 6.7A to help identify how partnerwise orientation is determined. Imagine you can draw a straight line from the partner through the focal partner, the line will help you identify the correct numerical direction for partnerwise orientation.
   e. Stand side by side, with the partner facing the right shoulder of the focal participant. What is the partnerwise orientation? *The answer is {6}*.
   f. Stand with the focal participant in front of the partner while you both face the same direction, one in front of the other as though you are waiting in line. What is the partnerwise orientation? *The answer is {0}*.
   g. Stand side by side, with the partner facing the left shoulder of the focal participant. What is the partnerwise orientation? Write your answer on worksheet 3 *Interanimal Relationships in EWMN*. Note that regardless of the orientation of the focal participant, EWMN notes the orientation of the other participant in relation to the focal participant.
   h. Play around with a few more configurations and work out the partnerwise Orientation. Look at Fig. 6.7B for more examples. Notate the partnerwise Orientation shown in the examples on worksheet 3.

sideways, your limb traces part of the path of the horizontal plane in the Eshkol–Wachman Movement Notation (EWMN) sphere. When you move your arm up and/or down, your limb traces one of the vertical planes of the EWMN sphere. Perform sideways and up and down movements with your limbs and compare the circular paths you trace with those in this figure. (B) Rotatory movements do not trace a path. The axis of movement is the same as the long axis of the limb. As the limb rotates, the distal end of the limb appears to spin on a point in space. Try rotating your arm or leg without allowing the distal end of your limb trace a path in space. Stand next to a desk and hold a pencil in your hand. Keep your arm straight and the end point of your pencil touching a sheet of paper. Rotate your arm; there should be a point forming on the paper. If you have drawn a line, it means you added a planar movement to the action. (C) Conical movements vary the angle between the axis of movement and the axis of the limb (between 0 and 90 degrees). Instead of tracing a path along a plane, the limbs pass through the planes of the sphere. Conical movements combine aspects of planar and rotatory movements.

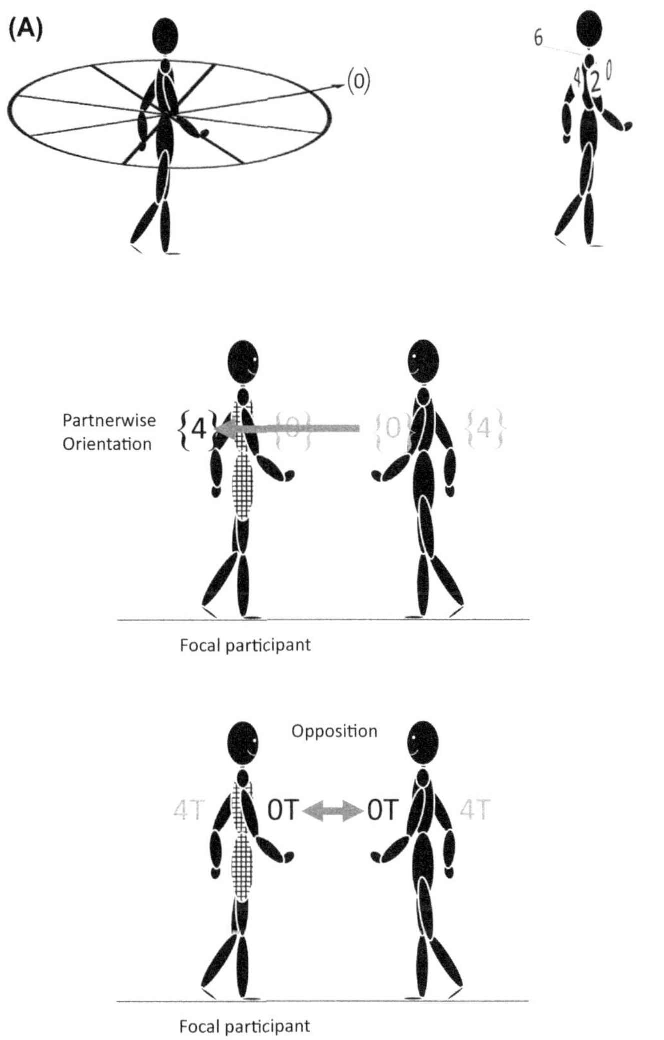

FIGURE 6.7 Examples of partnerwise orientation, opposition, and relative distance.

(A) To use the partnerwise frame of reference, the numerical values from the horizontal plane are assigned to specific locations on the body: {0} = front, {2} = right side, {4} = back, and {6} = left side. Partnerwise frame of reference uses curly brackets and is made in relation to the partner (instead of to the absolute space or bodywise frames of reference). In the first example of the two bodies facing each other, the nonfocal participant's front {0} is facing toward the direction of the focal participant's back {4}, thus {4} is the partnerwise orientation. For opposition, the two values facing are referenced along with the area of the body. Opposition is 0T/0T, where T = torso, as shown in the second example of the two bodies facing each other. Other possible areas for opposition may be H = head, S = shoulder, L = leg, etc. (B) Six images show various spatial relationships among two bodies. For each drawing a notation of partnerwise orientation, opposition, and relative distance is shown. Partnerwise values are made from the surface of the focal participant (indicated by the pattern on the torso) that the partner is facing toward. Notice that for opposition, numerical values used to represent each subject are taken from the body surface that is facing the partner. T = torso.

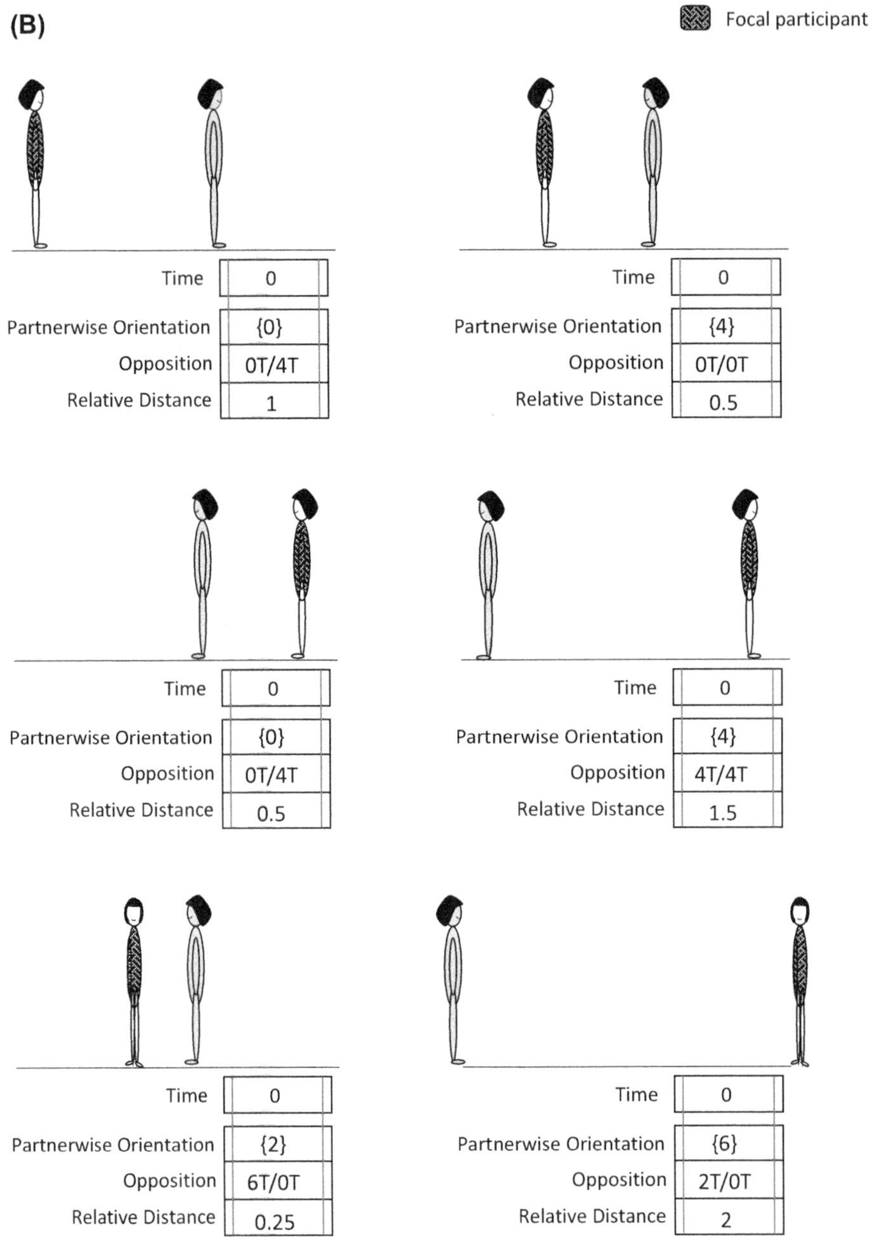

FIGURE 6.7 Continued

Learning exercise 3: opposition

Opposition describes which part of the body of one animal is closest to the body part of the other animal.

1. Label the horizontal plane of the EWMN sphere onto the human body (repeat step 1 in learning exercise 2).
2. Complete this step with a partner:
   a. Just as with partnerwise orientation ensure the front of the sphere (numerical value of 0) is attached to the front of the body and the other end of the sphere (numerical value of 4) is attached to the back of the body.
   b. This time, instead of using all 0–7 values label the right side of the body as 2 and the left side of the body as 6. For opposition, values 0, 2, 4, and 6 are sufficient.
   c. In addition label each side of the body with categorical values; for example, H for head and S for shoulder.
   d. Now you can track the body parts opposed by you and your partner. For example, when standing face to face (as you did for learning exercise 2.2c), what is the numerical value from each partner that is closest to the partner? *The answer is 0T/0T.* Look at Fig. 6.7A to help identify how the opposition is determined.
   e. Now imagine when the two of you are positioned in such a way that the left side of your head is opposing the left side of your partner's head, opposition would be marked as 6H/6H. Move into this opposition with your partner. Are you both facing the same direction? Explain. Fig. 6.7B provides a few examples.
   f. Sit in a chair and have your partner stand next to you on your left. If you are both facing the same direction, how do you notate your opposition? The answer is either 6H/2T (T for torso) or (depending on the height of your partner) 6H/2S.
   g. Play around with a few more configurations and work out the opposition. Share your configurations with the rest of your group/class. Notate the opposition shown in the examples on worksheet 3.

The Eshkol–Wachman Movement Notation activity

Use EWMN to describe interanimal relationships of cricket combat in Video A and Video B.

Step 1. Create EWMN sheets

In EWMN, movements are described with notations similar to musical notations in a music sheet. Body parts are represented in horizontal rows, with columns marking units of time (e.g., video frames). The notation is read from left to right and from bottom to top (Figs. 6.1, 6.5, and 6.8). Refer to Fig. 6.8 to create your own notation sheet for each video.

- Use *frame numbers* from the videos to indicate time units.
- Remember to use double bar lines to depict the beginning and ending of the notation and single bar lines to depict each time unit (as shown in Figs. 6.1, 6.5, and 6.8).

Step 2. Using the video A (aerial view) notate interanimal dynamics

1. Identify the EWMN numerical values for a cricket body.
   a. On a blank sheet of graph paper make a simple drawing of a cricket with the head of the cricket facing the top of the page. The drawing does not need to be detailed but make sure that both the location of the head and the long axis are clearly depicted (e.g., for the body draw a rectangle instead of a square, for the head draw a small triangle at one end of the rectangle).
   b. Draw a circle to surround the drawing of the cricket. Make sure that the line of the circle does not intersect the drawing of the cricket. The cricket should be centered inside the circle.
   c. Draw one line along the longitudinal axis of the cricket, with each end of the line meeting a point in the circle. This line should run through the entire midline of the cricket.
   d. Draw a second line along the cross-sectional axis of the cricket, with each end of the line meeting a point in the circle. This line should cross the center of the cricket body. The two lines should form a plus (+) sign.
   e. Draw two more lines that form an "x" with the center of the "x" matching the center of the plus sign. Each end of these two lines should meet a point in the circle.
   f. Number each end of each line, starting with 0 for the line of the + sign that intersects the head of the cricket. Moving in a clockwise direction number each remaining point until you reach 7.
   g. Now you have virtually shrunk the horizontal plane of the EWMN sphere onto the body of the cricket. The numbers 0–7 are set at 45 degree units around the longitudinal axis of the cricket body.
   h. Complete the worksheet 4 *EWMN Numerical Values for a Cricket Body.*
   i. After you complete worksheet 4, you will use the numbers 0–7 to establish a numerical value for *partnerwise orientation* (the orientation of one cricket in relation to the other).

*30 frames per second

Timeframes	0	1	2	3	4	5
Partnerwise Orientation	{4}				{6}	
Opposition	0H/0H				2T/0H	
Relative Distance	1				0.5	

Timeframes	6	7	8	9	10	11
PW Orientation						{6}
Opposition						2T/0H
Rel. Distance						0.5

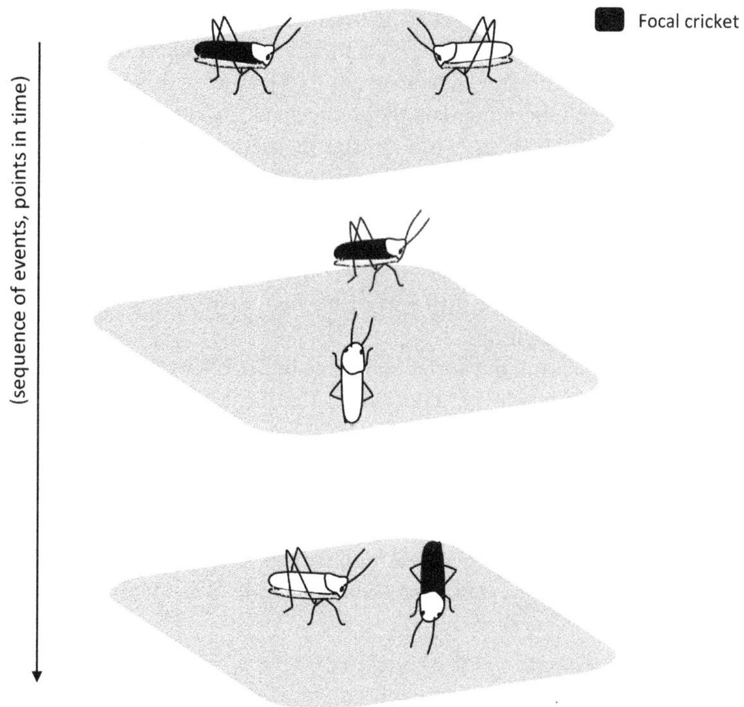

FIGURE 6.8 Partnerwise orientation, opposition, and relative distance in a notated score.

A notated score that focuses on the relationship between two crickets by scoring *partnerwise orientation*, *opposition*, and *relative distance* is shown. Double bar lines indicate the beginning and the end of the movement sequence. Below the notation is a sketch of two crickets performing the movement described in the notated score. The focal cricket is depicted with black wings. The notation reads: when the two crickets face each other head to head, the partnerwise orientation is {4}, the opposition is 0H/0H, and they are approximately one body length apart. Next, one cricket is positioned to the right side of

2. Describe partnerwise orientation during cricket combat.
   Partnerwise orientation is a frame of reference that describes the longitudinal axis of one animal in relation to the other.
   a. Select one cricket to be the focal cricket.
   b. Regardless of the orientation of the focal cricket, you will notate the orientation of the other animal in relation to the focal cricket on the notation sheet. For example, in your drawing of the focal cricket facing the top of your page, if the other cricket is facing the bottom of the page the partnerwise angle of the pair would be {4}. If the focal cricket moves to face the right side of the page and the other cricket moves to face the left side of the page, the partnerwise angle of {4} would be maintained. Use worksheet 5 *Partnerwise Orientation, Opposition, Relative Distance* to work out a few other examples.
   c. Curly brackets { } around the number indicate a frame of reference that is referred to as partnerwise. This means that you are referencing a point on the focal animal rather than a sphere that moves with an animal (bodywise) or the larger sphere that is fixed in the space around the animal (absolute). On your notation sheet, write the starting partnerwise angle in the first column (time frame) of the row for partnerwise. Watch the video and write the partnerwise angle in the proper column each time the angle changes. Remember to use curly brackets { } for partnerwise frame of reference.
3. Describe opposition during cricket combat.
   Opposition describes which part of the body of one animal is closest to the body part of the other animal.
   a. Virtually shrink the horizontal plane of the EWMN sphere onto each cricket's body (similar to what you did for the focal animal in partnerwise orientation).
   b. Just as with partnerwise orientation ensure the front of the sphere (numerical value of 0) is attached to the tip of the head of the cricket and the other end of the sphere (numerical value of 4) is attached to the other end of the cricket's body. Remember, for opposition, values 0, 2, 4, and 6 are sufficient. Instead of using all 0–7 values label the right side of the body as 2 and the left side of the body as 6.

the focal cricket, with its head facing the thoracic part of the focal cricket's body. In this second configuration, they are one half of a body length apart. Moving from the first configuration to the second took 4 units of time. After a brief pause (1 unit of time), it took 5 units of time for the crickets to move into the final position where both crickets are in a new location, having maintained their second partnerwise orientation; that is, both crickets moved to new area in space but maintained their spatial relationship. This notation reveals this spatial relationship during this behavior may be important to the cricket. Now we know to ask the question: what strategies are required for the crickets to maintain this spatial relationship with one another even as they move throughout the space? If the crickets are modifying their movements to maintain a specific partnerwise orientation then this suggests that, for the cricket, there is value in maintaining that spatial relationship during the behavior. What drives the crickets to keep that orientation? Advanced movement notation can reveal the strategies crickets use, how they modify these strategies, and, ultimately, how the behavior is organized.

c. In addition label each side of the body with categorical values. For example, H for head, T for thorax, and A for abdomen.
d. Now you can track the body parts opposed by the two animals. For example, when the two animals are positioned in such a way that the left side of the head of one animal is opposing the left shoulder of the other, opposition would be marked as 6H/6T. Refer to Figs. 6.7 and 6.8 for a few examples. Use worksheet 5 to practice notating a few other examples.
e. On your notation sheet write the starting opposition values in the first column (time frame) of the row for opposition. Watch the video frame by frame and write the opposition value in the proper column each time there is a change. Brackets are not used for opposition values.

4. Describe relative distance during cricket combat.
   Relative distance describes the distance in terms of animal lengths. This is useful when the actual distance (e.g., centimeters) cannot be accurately measured in video recordings.
   a. Pause the video in a section when one cricket is standing in a relaxed position and you are able to see the entire length of its body.
   b. Place one transparent sheet on the screen and draw a line down the center of the longitudinal axis of its body, starting from the tip of its head and ending at the base of its rear. You will use the length of this cricket as 1 unit of measure for *relative distance.* Half this length is 0.5, two times this length is 2, etc.
   c. On your notation sheet write the starting relative distance in the first column (time frame) of the row for distance.
   d. Watch the video frame by frame to identify the first frame in which the relative distance between the crickets begins to change.
   e. Draw a horizontal bow (Figs. 6.5 and 6.8) starting from the first frame in which the relative distance begins to change and ending in the last frame of the change. Include the new relative distance in this last frame. Brackets are not used for relative distance values.
   f. Continue to watch the video, notating the changes in the relative distance until the end of the video.

Results/discussion

1. What did the movement analysis reveal that you missed from your initial description?
   a. What was different from what you originally perceived?
   b. What information did you gain from looking at both aerial and lateral perspectives?
   c. What information did you gain from watching the videos repeatedly?
2. During the interaction, what matters to the cricket?
   a. What position are the crickets working to stay relative to one another?
   b. What are the crickets working to keep constant?

Connections with current literature

Readily identifiable behavior patterns, such as *antennal fencing*, *mandible spreading*, *mandible engagement*, and *tactical combat*, are used to identify levels of aggression and stereotyped escalation in male cricket behavior (Stevenson et al., 2000). For example, if neither cricket retreats after mandible spreading, both crickets lock mandibles in head-to-head battle while pushing against each other (Killian & Allen, 2008). During head-to-head battle, one may flip the other or use its head to butt the other backward (Alexander, 1961).

Although it is clear that such behaviors occur during cricket aggression, less is known on *how* they arise. Is head-butting the result of one animal engaging in aggression against the other or does it stem from a combination of actions from both animals (i.e., one cricket blocking access to the abdomen)? This type of question is crucial for understanding a behavior.

Consider the combat of giant Madagascar hissing cockroaches. Males will head-butt their opponents on the flank or the head; both locations of head-butting have been traditionally classified as one action. However, movement analysis reveals a deeper tale. Specifically, combat in these animals involves competitive access to the opponent's flank that enables one to unbalance the other. During this interaction, butts to the opponent's head is the result of foiling the other's strategies for gaining access to one's own flank. This suggests head-butting to the flank is an aggressive attack, whereas head-butting to the head is a defensive maneuver; these are two different behaviors (Pellis & Bell, 2020) and lumping them as one behavior will likely lead to inaccurate results and conclusions.

Acknowledgment

We thank Nikolas Gom for assistance with videos A-B, and Zahra Foroud for assistance with videos C-F.

References

Alexander, R. D. (1961). Aggressiveness, territoriality, and sexual behavior in field crickets (Orthoptera: Gryllidae). *Behaviour, 17*(2/3), 130–223.

Eshkol, N., & Wachmann, A. (1958). *Movement notation*. London, UK: Weidenfeld & Nicholson.

Fentress, J. C. (2008). Stepping outside the traditional "science" box. In *Proceedings of measuring behavior 2008, 6th international conference on methods and techniques in behavioral research* (p. 6).

Fentress, J. C. (2009). Streams and patterns in behavior as challenges for future technologies. *Behavior Research Methods, 41*(3), 765–771.

Foroud, A., & Pellis, S. M. (2002). The development of "anchoring" in the play fighting of rats: Evidence for an adaptive age-reversal in the juvenile phase. *International Journal of Comparative Psychology, 15*(1), 11–20.

Foroud, A., & Pellis, S. M. (2003). The development of "roughness" in the play fighting of rats: A laban movement analysis perspective. *Developmental Psychobiology, 41*(1), 35–43.

Foroud, A., & Whishaw, I. Q. (2006). Changes in the kinematic structure and non-kinematic features of movements during skilled reaching after stroke: A laban movement analysis in two case studies. *Journal of Neuroscience Methods, 158*(1), 137–149.

Foroud, A., & Whishaw, I. Q. (2012). The consummatory origins of visually guided reaching in human infants: A dynamic integration of whole-body and upper-limb movements. *Behavioural Brain Research, 231*(2), 343–355.

Golani, I., & Fentress, J. C. (1985). Early ontogeny of face grooming in mice. *Developmental Psychobiology, 18*(6), 529–544.

Himmler, S. M., Himmler, B. T., Stryjek, R., Modlińska, K., Pisula, W., & Pellis, S. M. (2016). Pinning in the play fighting of rats: A comparative perspective with some methodological recommendations. *International Journal of Comparative Psychology, 29*, 1–14.

Himmler, B. T., Pellis, V. C., & Pellis, S. M. (2013). Peering into the dynamics of social interactions: Measuring play fighting in rats. *Journal of Visualized Experiments, 71*, e4288. https://doi.org/10.3791/4288.

Killian, K. A., & Allen, J. R. (2008). Mating resets male cricket aggression. *Journal of Insect Behaviour, 21*, 535–548.

Laban, R. (1960). In L. Ullman (Ed.), *The mastery of movement.* (Revised 3rd ed.). London: Macdonald & Evans.

Lourens, T., van Berkel, R., & Barakova, E. (2010). Communicating emotions and mental states to robots in a real time parallel framework using laban movement analysis. *Robotics and Autonomous Systems, 58*, 1256–1265.

Moran, G., Fentress, J. C., & Golani, I. (1981). A description of relational patterns of movement during 'ritualized fighting' in wolves. *Animal Behaviour, 29*, 1146–1165.

Panskepp, J. (1981). The ontogeny of play in rats. *Developmental Psychobiology, 14*, 327–332.

Pellis, S. M., & Bell, H. C. (2020). Unraveling the dynamics of dyadic interactions: Perceptual control in animal contests. In W. Mansell (Ed.), *The interdisciplinary handbook of perceptual control theory: Living control systems IV* (pp. 75–97). Academic Press. https://doi.org/10.1016/B978-0-12-818948-1.00005-8.

Pellis, S. M., Blundell, M. A., Bell, H. C., Pellis, V. C., Krakauer, A. H., & Patricelli, G. L. (2013). Drawn into the vortex: The facing-past encounter and combat in lekking male greater sage-grouse (*Centrocercus urophasianus*). *Behaviour, 150*, 1567–1599.

Pellis, S. M., & Pellis, V. C. (1987). Play-fighting differs from serious fighting in both target of attack and tactics of fighting in the laboratory rat *Rattus norvegicus*. *Aggressive Behavior, 13*, 227–242.

Pellis, S. M., & Pellis, V. C. (1998). The play fighting of rats in comparative perspective: A schema for neurobehavioural analyses. *Neuroscience and Biobehavioral Reviews, 23*, 87–101.

Pellis, S. M., & Pellis, V. C. (2017). What is play fighting and what is it good for? *Learning & Behavior, 45*, 355–366.

Pellis, S. M., Pellis, V. C., & Foroud, A. (2005). Play fighting: Aggression, affiliation and the development of nuanced social skills. In R. Tremblay, W. W. Hartup, & J. Archer (Eds.), *Developmental origins of aggression* (pp. 47–62). New York, NY: Guildford Press.

Stevenson, P. A., Hofmann, H. A., Schoch, K., & Schildberger, K. (2000). The fight and flight responses of crickets depleted of biogenic amines. *Journal of Neurobiology, 43*, 107–112.

Szechtman, H., Ahmari, S. E., Beninger, R. J., Eilam, D., Harvey, B. H., Edemann-Callesen, H., & Winter, C. (2017). Obsessive-compulsive disorder: Insights from animal models. *Neuroscience Biobehavioral Reviews, 76*(Pt B), 254–279.

Part II. Instructor notes

Classroom management/blocks of analysis

The activity in this chapter can be completed in two standard laboratory periods or over four class sessions. Several organizational options are suggested in the following. It is recommended that students read the sections on Learning Goals, Background Information, and the Methods section in this chapter before laboratory periods or classes. Students may complete all the learning exercises in one laboratory. The next laboratory session may be used for completing all the EWMN activities (analysis of crickets). Additionally, the directed questions in the Results/Discussion section may be assigned as a written assignment or an in-class presentation. Either way, a class discussion is highly recommended.

The worksheets are designed to direct focus on essential aspects of the learning material. Learning exercises prepare students for completing the worksheets. The figures and videos provide a great deal of information, both have been formulated to coach readers to become movers! Use the figures and videos to help embody the lessons. Learners who perform the movements described in the figures, and who copy the performer in the videos, will gain a more concrete understanding of EWMN and its applications.

Teaching the activity

It is not expected that Institutional Animal Care and Use Committee (IACUC) approval be required for this chapter. Videos for analysis of cricket combat are provided for use in this chapter.

If you plan to use your own videos, in order to avoid the need of IACUC, video recordings should be made of spontaneous interactions of one's own pets, freely moving animals in the wild, or animals in a pet store or a public zoo, provided permission has been requested and granted ahead of time.

Teaching movement analysis

In cases where instructors are not yet experienced with movement analysis, it is recommended they practice the exercises on their own or with a colleague or teaching assistant as part of their own preparation. Additional information on the EWMN sphere, axis of movement, and frame of reference may be found at http://noaeshkol.org/wp-content/uploads/2016/11/The-System-of-Reference.pdf and http://noaeshkol.org/about-eshkol-Wachmann-movement-notation/basic-principals-of-ewmn/.

Recommendations regarding selection of species and/or setting for exercise

The videos provided in this chapter display approximately 2–3 s (80 frames) of male–male combat in *G. sigillatus*. Videos should be made accessible to students in all learning environments associated with this chapter. Effective use of EWMN requires the ability to look at the same video(s) repeatedly, including *frame by frame*. Students will need controls that enable rewinding and reviewing shorter sections of the video(s) repeatedly. It is recommended to view the video(s) on a computer monitor.

Ideas for in-class or online discussion

1. What types of behaviors can EWMN be applied to?
2. Why is it important to look at the same behavior from different angles?
3. How can you choose the best set of angles?
4. Why is it important to use movement analysis to study behavior?
5. How does movement analysis help the researcher understand what the animal experiences, expects, and seeks?
6. What does your analysis from the activities in this chapter add to the current understanding of cricket combat? (Suggestion: review the videos of cricket combat and your movement analysis in order to identify how head-butting arises in the crickets in your videos.)

Modifications to the activity

Activities from this chapter can be applied to most social encounters in a variety of other species. For example, social encounters in cockroaches, fish, birds, squirrels, hamsters, cats, dogs, etc. Pigeons or gulls scavenging in a parking lot, on a dock, or in a beach often engage in social interactions that are suitable for use in the activities outlined in this chapter.

Video duration and quality

It is important that the analysis be made of video recordings so that students can rewind and have frame-by-frame access. Video length should be limited to 1–3 s. Low-quality videos, videos with low resolution, or videos recorded at speeds lower than 29 frames per second will not meet the standards required for successful applications of movement analysis. Similarly, unsteady videos will not be suitable because it can be difficult to identify movements of the animals when the frame of reference is constantly shifting. The use of a tripod during video capture is highly recommended. Finally, video recordings should capture the entire body of each animal to be studied. When parts of the body, or the entire body, move out of frame, movement notation cannot be completed.

FIGURE 6.9 Pipe cleaner model of the Eshkol–Wachman Movement Notation (EWMN) sphere.

Seven pipe cleaners were used to model the EWMN sphere. A different color was used for each of the vertical planes and one color was used for the three levels of the horizontal plane. Make your own pipe cleaner model and label each coordinate.

Options to lengthen or shorten learning activities

Lengthen learning exercise 1 by asking students to create an EWMN sphere using pipe cleaners (Fig. 6.9) and to label the coordinates. Ask them to complete the worksheet 6 *Coordinates of the EWMN Sphere*.

To lengthen learning exercises 2 and 3 ask students to find new configurations for partnerwise orientation and opposition that they have not yet tried and that have not been provided in the examples or worksheets. Challenge students to notate configurations for partnerwise orientation and opposition that involve different levels, such as using stairs, sitting down, or lying down.

To shorten the chapter, first ask students to complete step 1 of the EWMN activity. Collect and save their descriptions for comparison after they complete the learning exercises. Next, focus on the learning exercises to help students gain a solid introduction to EWMN. Finally have students repeat step 1 of the EWMN activity and compare with their initial written observations.

Areas of potential confusion or difficulty for students

A concrete understanding of the system of reference is foundational for using EWMN. Learners who dedicate a part of their time to perform the instructions as described in the figures and learning exercises, and who copy the movements while watching the videos, will gain tangible knowledge of the system of reference.

The learning exercises provided are designed to help learners gain a concrete, experience-based understanding of the movement analysis applications described in this chapter. The steps in the activity are made possible by the experiences gained from completing the learning exercises while studying the figures and videos.

Students generally find it both helpful and encouraging to see the instructor model a willingness to act out the examples.

Encourage students to move and act out/perform the actions in the learning exercises. It is essential for the learner to both experience and observe the performance of the exercises. Encourage students to apply the aspects of the notation that they are learning on the observations they are making during the learning exercises. For example, ask them to notate their own movements and those of their class/laboratory partner.

When working with the videos of the crickets (during the EWMN activities), it will be both valuable and insightful to try to act out parts of the behavior that are more challenging to notate.

Recommendations for extensions or continuations for more advanced classes

The applications of activities from this chapter to other species may be used for extensions or the continuation of more advanced classes. Refer to the section Modifications to the activity in Part II Instructor notes for guidelines.

Extensions/continuations involving class discussion and/or a written assignment may center on (1) integrating information from student notations on cricket combat and current understanding of the behavior (refer to item 6 in the section Ideas for in-class or online discussion in Part II Instructor notes) or (2) reviewing published research using similar methods. The study by Pellis et al. (2013) is highly recommended for this type of extension.

Continuation/advanced Eshkol–Wachman Movement Notation

Learning exercises 4 and 5 and the related activity steps 3 and 4 offer a continuation and more advanced learning of EWMN. It is important students complete the learning exercises before working on the activity steps. Depending on time, and student interest, the two learning exercises may be a sufficient continuation.

Learning exercise 4: types of movement

Learning to identify different types of movement and how to notate limb segments over time helps reliably identify the time points for when one movement begins and ends. Consider the simple action of curling your fingers into a fist. When does the movement begin? When does it end? The action may be perceived as one movement; however, it is made up of a bunch of movements, occurring sometimes in synchrony and at other times in a sequential pattern (one leading into the other). Objectively notating when the movement of each limb segment begins, when it ends, and what type of movement it employs provides greater information than lumping the series of movements into one.

EWMN defines three types of movement, namely, planar, rotatory, and conical (Fig. 6.6). In this exercise, we will explore planar and rotatory movements. *Planar* movements occur when "*the angle between the axis of movement and the axis of the limb is 90 degrees*" (Eshkol & Wachmann, 1958). Imagine your fingertips have been dipped in paint and can trace a colorful path in the space around you. You would be painting plane brushstrokes on the surface of the air every time your arm makes planar movements. *Rotatory* movements occur when "*the angle of movement is 0 degrees ... When the angle between the axes is zero, no surface is created by movement of the axis of limb, which simply rotates about itself*" (Eshkol & Wachmann, 1958). Instead of making brushstrokes, you would be painting one point in the air during rotatory movements made by your arm.

In groups of 3 to 5 people practice making planar and rotatory movements with different limb segments.

Perform them for your group and ask them to identify the type of movement.

Together discuss whether the following examples are planar or rotatory movements. As part of your discussion perform each of these examples. Use information from your sense of your own movements, observations from your reflection in a large mirror, and observations of your peers' movements to help you discern the type of movements used in these examples:

a. turning your head to look to your side,
b. tilting your head to look down,
c. clenching your fingers,
d. lifting your knee up,
e. waving hello with your hand,
f. bending forward at the waist,
g. twisting at the waist,
h. spinning an imaginary skewer while roasting marshmallows,
i. turning a page in a book.

Can each of these examples be performed using a different type of movement? Experiment with this idea and discuss.

Movements made by the body as a whole, or by any limb segment, are described as the *distal* end moves on (i.e., rotatory), or across (i.e., planar), the surface of the sphere, while the *proximal* end is anchored to the center of the sphere (e.g., in the center of its proximal joint or the larger sphere fixed in space). In cases where multiple connected limb segments are moving at the same time, the proximal end of each limb segment may be the distal end of the preceding limb segment (Fig. 6.10).

Learning exercise 5: notating movement of limb segments

Begin by standing in a neutral position (refer to learning exercise 1 for instructions).

Lift up one leg while simultaneously bending your knee to keep your foot parallel to the ground (Fig. 6.11). Try to keep the rest of your body still. For the sake of this exercise, we will ignore what your foot and the rest of your body is doing. To

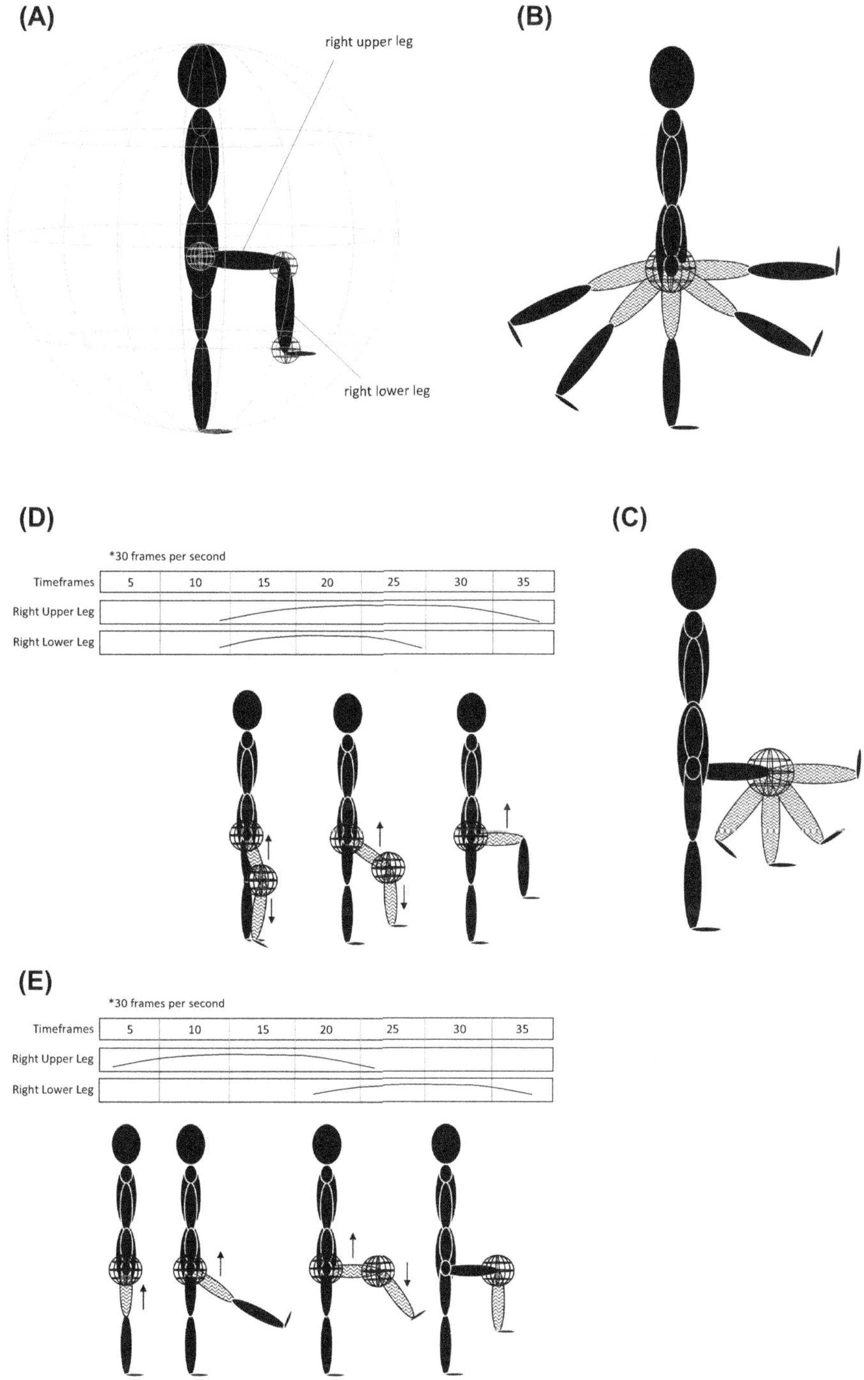

FIGURE 6.10 Proximal and distal limb segments.

help you balance, you may rest one hand on a table or against the wall and/or slightly bend the knee on your standing leg.

Here, we describe the location of the smaller spheres for two of the *limb segments* in your raised leg: your upper leg and your lower leg.

1. First, your upper leg is anchored in the center of the sphere at your hip joint (its proximal end).
2. Your knee is at the distal end of the upper leg, and it is the knee that appears to move across the surface of the sphere.
3. Second, your lower leg is anchored in the center of the sphere at your knee joint (its proximal end)!
4. Your ankle joint is at the distal end of your lower leg, and for your lower leg, it is the ankle that appears to move across the surface of the sphere (Fig. 6.11).

Activity step 3. Notate movements over time during cricket combat

1. *The beginning and ending of a movement* is defined by any time a movement (1) changes in direction, (2) changes from one type of movement to another, and (3) pauses or stops and then begins again.
2. Use Video B (*lateral view*) to notate torso movements of the focal cricket over time.
   Use the same notation sheet in which you have notated interanimal dynamics (from step 2).
3. For each movement of the torso draw a horizontal bow (Figs. 6.5 and 6.8) starting from the first frame in which the torso begins to move and ending in the last frame in which the torso moved. (Optional: in the first frame [for each movement of the torso] indicate the type of movement. Use P for planar, R for rotatory, and C for conical movements.)
4. Using a separate notation sheet repeat the above-outlined steps (in step 3) using Video A (aerial view).

(A) The hip joint is the proximal (closer to the center of the body) end of the leg and the foot is at the distal (further from center) end. The lower leg is both distal to the upper leg and proximal to the foot. For example, (B) the hip joint is anchored to the center of the sphere, while the upper leg moves along the surface of the sphere. (C) The knee joint is anchored to the center of the sphere, while the lower leg moves along the surface. As shown by the placement of the bows in notation examples (D) and (E), both limb segments and the upper and lower legs may begin and end moving at the same time (D), or in sequence one after another (E). Below the notation is a sketch of a character performing the movement described in the notated score. Limb segments that are not moving are depicted in black. Limb segments with a pattern help illustrate when the leg is in motion. The arrows point to the direction of movement in the sphere.

*30 frames per second

Timeframes	0	5	10	15	20	25	30
Right Upper Leg	$\binom{0}{0}$	(0) ↑ 2					$\binom{2}{0}$
Right Lower Leg	$\binom{0}{0}$	(0) ↓ 2					$\binom{0}{0}$
Right Foot	$_T\binom{2}{0}$	=					

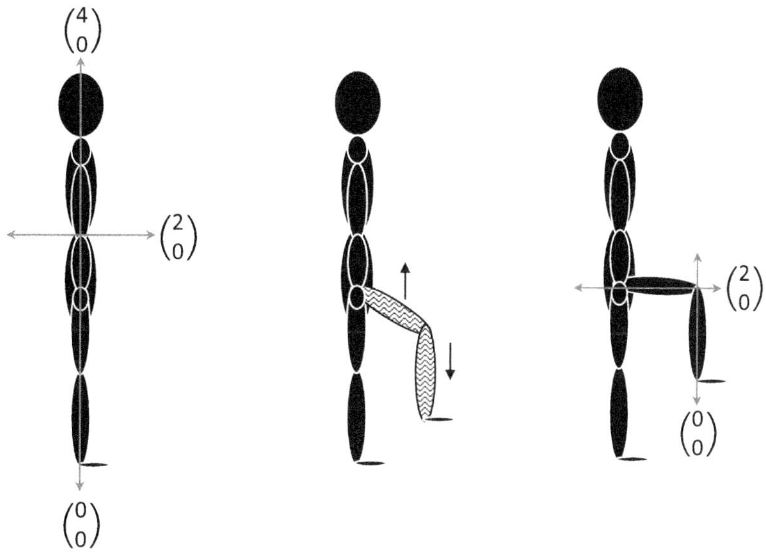

FIGURE 6.11 Limb movement and notation II.

Lifting the leg involves stabilizing the torso and head, shifting balance to the standing leg, and planar movements made by the upper and lower legs, all of which can be notated in the Eshkol–Wachman Movement Notation. This example shows a detailed notation of the right leg only. In the beginning the leg is in a neutral position (0,0), with the foot bearing weight to the ground (T). The foot releases contact (=) as the leg begins to move. The upper leg performs a planar movement, in the (0) vertical plane, with an upward direction for two spatial units in the sphere. The lower leg makes a planar movement, also in the (0) vertical plane, with a backward direction for two spatial units in the sphere. At the end, the upper leg is at 0,2 both for bodywise and in absolute space (it is only necessary to indicate either the bodywise or the absolute coordinate in the notation). The notation reveals something more interesting about the movement of the lower leg. Notice that both starting and ending positions of the lower leg are the same for absolute space, 0,0. Yet the lower leg moved! It moved away from the body as it was carried by the upper leg. When reading the notation, information about the lower leg is gained by looking at the notation of the upper leg. Furthermore, the lower leg performed its own movement. By bending at the knee, the lower leg moved back toward the body. The reader gains this information by reading the notation row for the lower leg. Additionally, its bodywise coordinate can be written instead of the absolute coordinate. In this case, bodywise for the end position of the lower leg is 4,2 (a chapter for another time). Below the notation is a sketch of a character performing the movement described in the notated score. Limb segments that are not moving are depicted in black. Limb segments with a pattern help illustrate when the leg is in motion. Small black arrows point to the direction of movement in the sphere. The long gray arrows highlight limb axes.

Activity step 4. Notate limb movements over time during cricket combat

1. Using the same notation sheet from step 2 repeat step 3 (notation of the torso) for each limb in the focal cricket using Video B (lateral view).
2. Complete the notation for one limb for the entire length of the video before notating the next limb.
3. Using the same notation sheet that was used in step 3 for Video A (aerial view)
4. Repeat step 3 for each limb in the focal cricket.

Answer key

1. Learning exercise 1.3a. For example, when your arm was held straight in front of you and parallel to the ground, you were at (0) in the horizontal plane and (2) in the vertical plane.
2. Learning exercise 1.3b. There was one more point at which the horizontal and vertical planes intersect in this exercise, what is the name of that coordinate? *(0,6).*
3. Learning exercise 1.3c. If someone says "move your arm to zero two," you will understand that to mean directly in front of you and parallel to the ground. *A sketch is provided in* Fig. 6.12.
4. Learning exercise 1.3d. If you are asked to "move your arm to two zero," where should you place your arm? *The arm would be in a neutral position. When the vertical value is 0 the limb is in a neutral position. A sketch is provided in* Fig. 6.12.
5. Learning exercise 1.3e. If you are asked to "move your arm to two three," where should you place your arm? *The arm would be extended on the right side of the body and pointing up diagonally. A sketch is provided in* Fig. 6.12.
6. Learning exercise 2.2g. Stand side by side, with the partner facing the left shoulder of the focal participant. What is the partnerwise orientation? *{2}.*
7. Learning exercise 3.2e. Move into this opposition with your partner. Are you both facing the same direction? Explain. *No. As 6H is on the left side of the head for each body, they would have been facing opposite directions.*
8. Learning exercise 4: types of movement
   a. turning your head to look to your side, *rotatory*
   b. tilting your head to look down, *planar*
   c. clenching your fingers, *planar*
   d. lifting your knee up, *planar*
   e. waving hello with your hand, *planar*
   f. bending forward at the waist, *planar*
   g. twisting at the waist, *rotatory*
   h. spinning an imaginary skewer while roasting marshmallows, *rotatory*
   i. turning a page in a book, *a combination of planar and rotatory (or conical)*

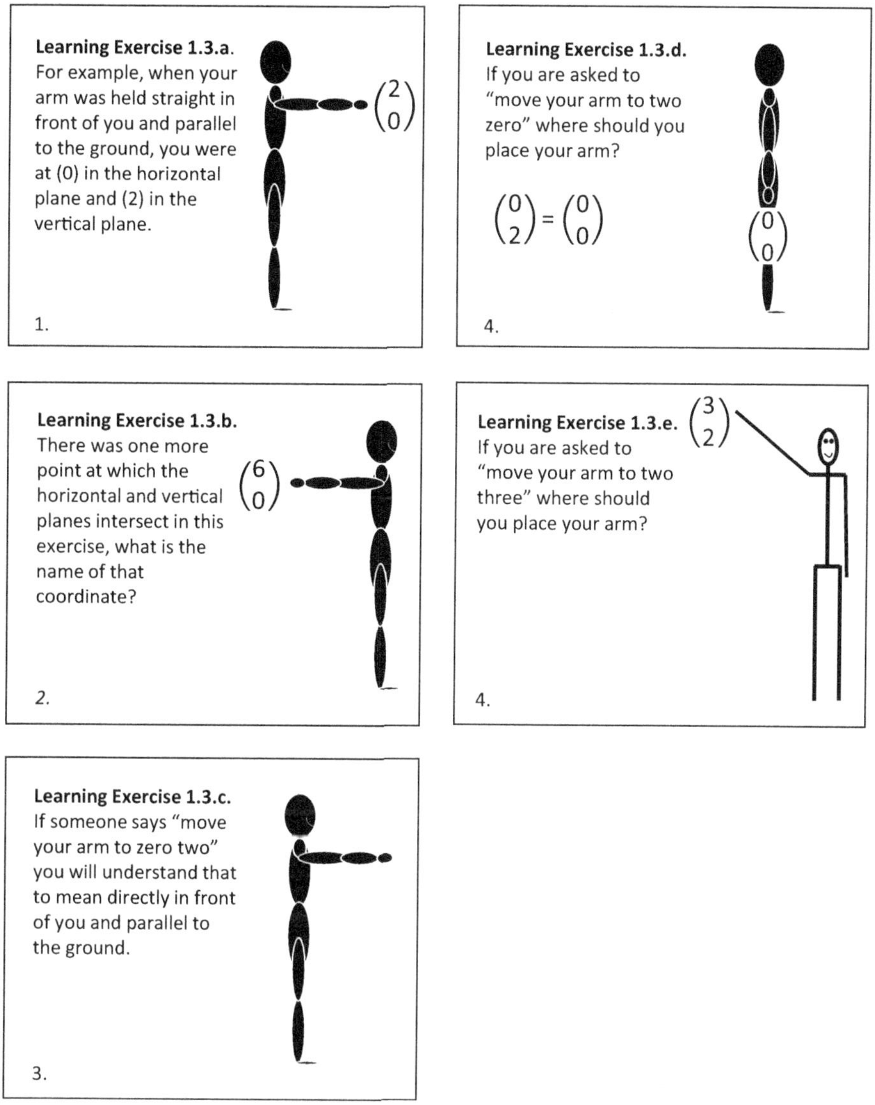

FIGURE 6.12 Illustrations for the answer key.

9. Learning exercise 4. Can each of the abovementioned examples be performed using a different type of movement? Experiment with this idea and discuss. *Yes. (e) The hand can be waved using a planar movement, a rotatory movement, or a combination of both. (g) One may use rotatory movements, or a combination of planar and rotatory movements (or conical movements), to turn the marshmallow during roasting.*

Part III. Supplementary material

Supplementary data related to this chapter can be found online at https://doi.org/10.1016/B978-0-12-821410-7.00008-X.

PART 2

Theory of behavior

CHAPTER

The evolution of behavior: a phylogenetic approach

7

J. Jordan Price[1], Ken Yasukawa[2]

[1]*Department of Biology, St. Mary's College of Maryland, St. Mary's City, MD, United States;*
[2]*Department of Biology, Beloit College, Beloit, WI, United States*

Chapter outline

Exploring Animal Behavior in Laboratory and Field. https://doi.org/10.1016/B978-0-12-821410-7.00013-3

Part I. Student instructions

Learning goals, objectives, and key concepts

- To understand how phylogenetic trees are constructed and interpreted in evolutionary studies.
- To gain experience scoring behavioral characters and mapping them onto a phylogenetic tree.
- To learn how to infer evolutionary histories of behavior using phylogenetic comparative methods.

Background

Evolution is the underlying principle of biology and, starting with Darwin, animal behavior rested on a firm foundation of evolutionary principles (e.g., Chapin, 1917; Friedmann, 1929; Wheeler, 1919; Whitman, 1899). For some anatomical traits, evolutionary history can be followed in the fossil record, especially when a relatively complete fossil series of species is available. Unfortunately, behavior does not fossilize, so until time travel becomes a reality or the behavior of extinct animals can be studied using the methods described in the *Jurassic Park* movies, evolutionary studies of behavior must rely on **comparative methods**, by which ancestral states and the pattern of evolutionary changes in behavior can be inferred (Martins, 1996). Comparative studies of behavior gained considerable momentum in the 1940 and 1950s as influential ethologists such as Konrad Lorenz and Niko Tinbergen championed the view that behavior could be used to infer evolutionary relationships (**phylogeny**) among species and that behavior must evolve in a phylogenetic context. For example, in his study of the evolution of courtship displays of "dabbling" ducks (those that feed from the surface), Lorenz (1958) said, "every time a biologist seeks to know why an organism looks and acts as it does, he must resort to the comparative method." Tinbergen (1963) included evolution as one of the now famous four areas of animal behavior, and he described the comparative method as follows:

> *Through comparison [the naturalist] notices both similarities between species and differences between them. Either of these can be due to one of two sources.* Similarity *can be due to affinity, to common descent; or it can be due to convergent evolution.... The* differences *between species can be due to lack of affinity, or they can be found in closely related species. The student of survival value concentrates on the latter differences, because they must be due to recent adaptive radiation.*
>
> **(Tinbergen, 1964).**

Tinbergen's characterization of comparative biology foreshadowed the basic concepts of modern phylogenetic comparative methods.

Curiously, from this high point in the evolutionary analyses of behavior, phylogeny was increasingly ignored by students of animal behavior, and by the 1970s, most biologists believed that morphological characteristics were much more valuable for comparative analyses than behavioral characteristics (Brooks & McLennan, 1991; Harvey & Pagel, 1991). With the rise of the "new" comparative biology, behavior has once again been thrust into the evolutionary spotlight (Owens, 2006).

Building and interpreting phylogenetic trees

Comparative studies attempt to identify evolutionary patterns by examining traits of different **extant** (i.e., currently living) organisms. The traits can be molecular, morphological, or behavioral characteristics. All modern comparative methods center on building and interpreting "trees," or phylogenies, which represent the pattern of shared ancestry among organisms. In other words, phylogenies are like family trees, except that instead of showing the ancestry of individuals in a family, they are "evolutionary genealogies" (Rubenstein & Alcock, 2018). The term *tree* describes the physical appearance of these proposed phylogenetic relationships, which are depicted as branching patterns of evolutionary events. An example of such a tree (or phylogeny) is shown in Fig. 7.1. Fig. 7.1 shows seven hypothetical organisms (A–G) and their phylogenetic relationships. These organisms could represent different species, subspecies, genera, or any other taxonomic group, so in the language of comparative biology, we refer to them as different **taxa**, with each referred

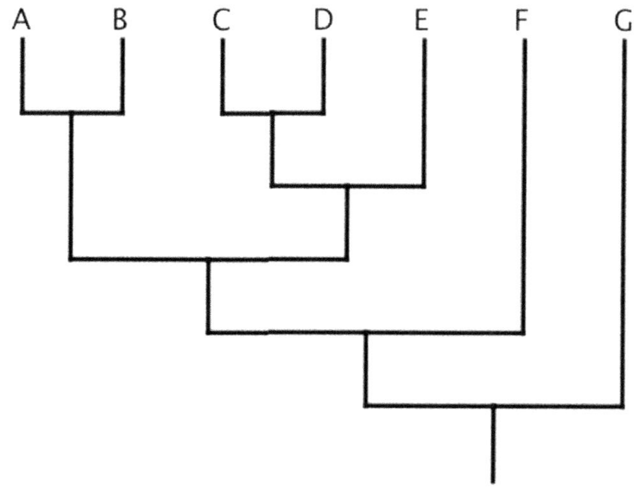

FIGURE 7.1

A phylogenetic tree for seven hypothetical taxa, labeled A–G.

to individually as a **taxon**. Extant taxa occur at the **tips** of branches, while all other parts of the tree represent evolutionary ancestors.

What can we infer from this tree? Well, for one thing, the tree shows us that taxon A and taxon B are each other's closest evolutionary relatives (comparative biologists call them **sister taxa**) because they both diverge from the same **node** (the point at which the **branches** leading to A and B emerge). Taxa C and D are likewise sister taxa, whereas taxon E is more distantly but equally related to both C and D. Together these taxa form a **clade** (a group of taxa sharing a common ancestor, with closer relationships to each other than to members of any other group). In tree terminology, a clade comprises all the branch tips that come from the same node. In Fig. 7.1, A and B are considered a clade, as are C-D, C-D-E, A-B-C-D-E, A-B-C-D-E-F, and A-B-C-D-E-F-G. A clade can also be called a **monophyletic group** (the terms are synonymous), which refers to a group of taxa descended from a single ancestral taxon and including the ancestral species and all descendant species. Note that the common ancestor is included in a clade (or monophyletic group). In tree terminology, taxonomic groups can include ancestors as well as extant taxa. Two other terms that are important in discussing trees are the **root** (base) of the tree and the **internodes** (branches between nodes).

How do we determine relationships among taxa to build a phylogenetic tree? These analyses are based primarily on a careful examination of **characters** across taxa. A phylogenetic character includes a set of possible states or conditions that are thought to evolve from one to the other, with each alternative condition called a **character state** or **trait**. We group organisms together based on **shared derived traits**, which are novel character states that are shared by two or more taxa because they were inherited from a common ancestor in which the trait originated. For example, in songbirds, we can divide the species into those that build open-cup nests and those that build domed nests with roofs (Price & Griffith, 2017). Suppose that open-cup nest building behavior evolved in an ancient species of bird (indicated by *X* on the tree shown in Fig. 7.2) and that the ancestors of that species built domed nests. Domed nests were thus the **ancestral** state and cup nests the relatively **derived** trait (note that it is also possible for domed nest building behavior to reappear through the loss of the derived trait). This transition from domed to cup nests could have happened anywhere along the internode indicated by *X* (even though the whole internode is coded as cup nest building in Fig. 7.2). Suppose further that two species, A and B, evolved from that cup nest building ancestor. If species A and B also build cup nests, and if they inherited this behavior from their common ancestor, then cup nest building is a shared derived trait in species A and B and is said to be **homologous** (it was inherited by both species from a common ancestor). The goal of modern taxonomy (called "phylogenetic systematics") is to organize taxa into clades based on homologous, shared derived traits.

Of course, there are many ways to sort the taxa in Fig. 7.2 into groups that are not monophyletic (not clades) and thus do not reflect evolutionary relationships. For example, a taxonomic group including species A-B-C, based on the presence of cup nests, would be called a **polyphyletic group** because the trait the group is based

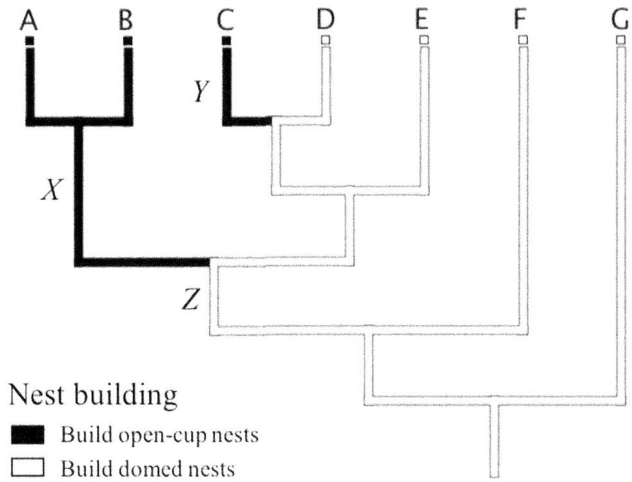

FIGURE 7.2

A phylogeny showing the same relationships as Fig. 7.1, but with two character states mapped onto the tree: open-cup nest building (indicated by black) and domed nest building (indicated by white). Cup nest building presumably evolved from domed nest building twice on this tree, along branches X and Y. If cup nest building evolved earlier on branch Z, two more transitions back to dome building would have had to occur on the branches leading to taxa D and E, which would be a less parsimonious explanation of evolutionary history.

on (cup nest building) is shared due to the independent evolution of this trait on branches *X* and *Y* rather than common ancestry. Cup nest building behavior in this case is said to be a **homoplasy** (it was not inherited from a common ancestor). Polyphyletic groups are based on homoplasies and do not include a common ancestor. Alternatively, a group including just D-E-F-G, based on having domed nests, would be called a **paraphyletic group** because it includes the common ancestor but not all the descendants of that ancestor. Paraphyletic groups are based on homologous, shared ancestral traits rather than on shared derived traits. The easiest way to distinguish polyphyletic and paraphyletic groups is by whether or not they include the group's most recent common ancestor. Both groupings include taxa that are more closely related to nongroup members than to other group members (e.g., D is more closely related to C than to its other group members, and vice versa).

Many well-known animal groups are not monophyletic. Pandas, for example, provide a good example of a polyphyletic group. Despite some striking similarities (including an extra "pseudo thumb" for eating bamboo), giant pandas (*Ailuropoda melanoleuca*) and red pandas (*Ailurus fulgens*) are actually distantly related and did not inherit these traits from a common panda ancestor. Their shared traits are homoplasies that were derived independently through convergent evolution. Reptiles provide a good example of a paraphyletic group because the common ancestor was a reptile but some descendants of that ancestor, such as birds, are not considered reptiles. Reptiles are grouped together based on shared character states that are relatively ancestral in comparison to those of birds (e.g., scales vs. feathers). As a result, interestingly, some reptiles, such as crocodiles, are actually more closely related to birds than to other reptiles.

Although behavioral characters have been used to construct phylogenies, these days, trees are constructed primarily using molecular data, such as DNA sequences. This is not necessarily because DNA nucleotide characters exhibit less homoplasy than do behavioral characters but because DNA provides so *many* characters—thousands or even billions rather than dozens. Phylogenies are now being constructed based on entire genomes. The phylogenetic trees produced from such molecular analyses have been extremely useful in animal behavior studies in recent decades because they provide an independent and relatively reliable way of estimating phylogenetic relationships. Using them, we can infer the sequence of events through which behavior evolved.

Using phylogenies to reconstruct the evolution of behaviors

Rather than using traits to build a phylogeny, in this exercise, we will use independently generated phylogenetic trees to reconstruct the evolution of behavioral traits. How do we use trees to reconstruct behavioral evolution? **Parsimony** (simple explanations are better than complicated ones) is the most commonly used criterion in both building phylogenetic trees and in using these trees to trace past evolutionary changes in a character. In practice, the parsimony criterion means trying to minimize the number of character-state changes (state transitions) needed to explain the current distribution of character states among extant taxa. For example, let us return to Fig. 7.2. Theoretically, cup nest building behavior could have evolved earlier on in the tree, just once in ancestor *Z* rather than twice in ancestors *X* and *Y*. However, given the current distribution of cup and domed nest building taxa today, this would mean that cup nests were lost at least twice in taxa D and E. This alternative scenario would therefore involve three evolutionary changes, one gain and two losses of cup nests, rather than just two gains, so we would say that it is a *less parsimonious* explanation.

We can use this same parsimony criterion to determine which character states are relatively ancestral and which are derived; that is, we can determine the order (evolutionary sequence) of character state changes. If cup nests were ancestral to domed nests, for example, our phylogeny suggests that at least four evolutionary changes would have to have occurred, such as domed nests evolving from cup nests four times independently in taxa D, E, F, and G. Can you come up with some other possible evolutionary scenarios that would have resulted in the distribution of cup nest building and domed nest building taxa today? Any way you look at it, using these character states on this phylogeny, it is more parsimonious to assume that domed nest building was the ancestral behavior in songbirds.

Sometimes there are multiple possible evolutionary scenarios that are equally parsimonious. For example, imagine if taxon G were the only domed nest building taxon and all other taxa (A–F) built cup nests. The ancestor, at the root of the tree, could have built cup nests or domed nests and either possibility would be equally parsimonious. How do we decide which scenario is more likely, and thus which character state is ancestral and which derived? In such cases, we can look to additional taxa, called **outgroups**, that are closely related to, but are outside (i.e., not *as* closely related to), the group of interest, which of course is called the **ingroup**. Including outgroup taxa simply increases the size of our tree so that we can determine the directionality of character state transitions. In cases that cannot be resolved using outgroups, sometimes additional information about the characters can help us decide. For example, character state transitions in one direction might be more probable than changes in the other direction. In reconstructing the evolution of flight in birds, for instance, the ability to fly seems far more likely to be lost than gained during evolutionary history simply because it requires so many unique specializations.

When done carefully, comparative studies can help us to infer **adaptation**, or whether the character of interest has evolved into its current state in response to natural selection. For example, building cheap, relatively disposable open-cup nests could have been an adaptation in response to changes in predation, nest parasites, or perhaps even levels of climatic variability. We would need to compare changes in nest building to changes in other characters to see how well they match up. There are a wide variety of methods for statistically comparing such character changes across taxa, ranging from simple to highly complex. Here we will mostly focus on reconstructing the evolution of individual characters.

Purpose

Our general research question in this exercise is: what was the evolutionary sequence of events (i.e., the order of character state changes) that led to the behavioral characteristics we see in a group of species today? To reconstruct these evolutionary changes, you will need to define a set of characters and character states, map the characters onto a phylogenetic tree, and use the principle of parsimony (minimizing the number of character state transitions) to reconstruct when (on which internodes) each character changed from one state to another.

Methods

This exercise can be done with any clade of related taxa. These can be different species, different taxonomic families, or any other level of taxonomic group, as long as you determine their relationships. Here we will use a group of six species from the New World blackbird family (Icteridae). You can do this exercise with paper and pencil or by using a dry-erase whiteboard and colored dry-erase pens, which you can use to draw the tree, the character states, and evolutionary events with different characters color coded. Alternatively, you can use a free computer program called Mesquite (Maddison & Maddison, 2019; available at http://mesquiteproject.org) to visualize evolutionary changes. Directions for using Mesquite are in the Appendix.

This activity consists of two parts. First, the whole class will work on reconstructing the evolution of some behavioral characteristics using the provided phylogeny of blackbirds. Second, working with a partner or in a small group, you will choose another set of taxa and use a phylogeny and the behaviors exhibited by the taxa to reconstruct how those behaviors evolved.

Form working groups of two to four students. If you are going to use a computer to do your analysis, your instructor will take some time to teach you how to use Mesquite. If you are going to do the analyses by hand using paper or a whiteboard, you can proceed directly to the exercises in the following.

Activity 1: Whole-class exercise

You will use the phylogeny in Fig. 7.3, which shows hypothesized evolutionary relationships among six songbird species from the New World blackbird family (Icteridae), including the red-winged blackbird (*Agelaius phoeniceus*) and its evolutionary relatives. This tree is similar to the ones shown in Figs. 7.1 and 7.2 but is turned 90 degrees clockwise. Note that all six species are extant (currently living) and that we are using this hypothesized phylogenetic tree to infer

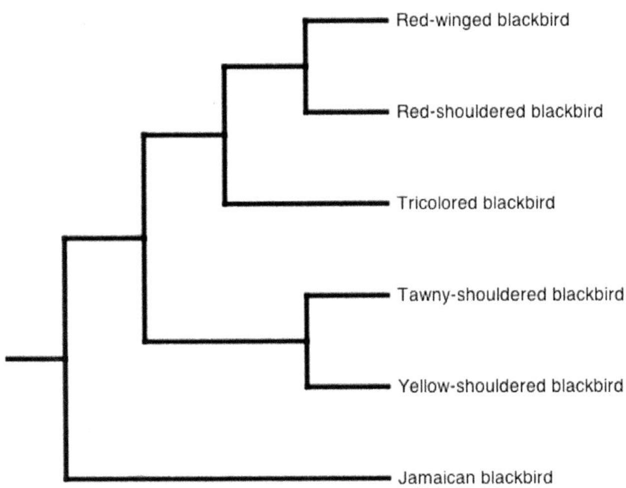

FIGURE 7.3

A phylogenetic tree showing hypothesized evolutionary relationships among six species in the New World blackbird family (Icteridae), based on molecular data (Powell et al., 2014).

evolutionary history from their current character states. This particular phylogeny is based on an extensive comparison of mitochondrial and nuclear DNA sequences (Powell et al., 2014).

A number of studies have been published on the singing and breeding behavior of these species. In fact, the red-winged blackbird is among the most well-studied songbirds in the world (Searcy & Yasukawa, 1995). In some of these species, including red-winged blackbirds, only males produce complex songs, whereas in other species, both males and females produce similar songs at similar rates. In a few species, males and females even coordinate their vocalizations to produce complex duets. In some species, males and females mate as pairs (they are socially monogamous), whereas in others, some males pair with multiple females (they are polygynous). Some breed in the tropics and others breed in the temperate regions of North America. Some migrate annually, whereas others are sedentary and stay in the same areas year-round.

From published studies (Odom et al., 2015; Price, 2009), we can assemble the following descriptions of each species. You will use the following information and Fig. 7.3 to infer the evolutionary history of singing, mating system, breeding latitude, and migratory behavior in this group of animals.

A. Red-winged blackbird (*A. phoeniceus*)
Males defend territories using complex vocal displays (song). Females produce a variety of sounds, some of which are as complex as the songs of males, but they do not sing like males. The mating system is polygyny, with some males pairing with multiple females and other males not pairing at all. Most populations breed in the temperate areas of North America and migrate annually.

B. Red-shouldered blackbird (*Agelaius assimilis*)
Both males and females defend their shared territory using songs, often by combining their vocalizations into a duet. Males and females are monogamous. They breed in the tropics and do not migrate (they are sedentary).

C. Tricolored blackbird (*Agelaius tricolor*)
Males sing and females do not sing. Mating is polygynous. The species breeds in temperate regions and migrates annually.

D. Tawny-shouldered blackbird (*Agelaius humeralis*)
Both males and females sing, often by combining their vocalizations into complex duets. Mating is monogamous. They breed in the tropics and are sedentary.

E. Yellow-shouldered blackbird (*Agelaius xanthomus*)
Both sexes sing, but males and females do not coordinate their vocalizations as duets. The species is monogamous, tropical, and sedentary.

F. Jamaican blackbird (*Nesopsar nigerrimus*)
Both sexes sing, but without producing duets. The species is also monogamous, tropical, and sedentary.

How can we make sense of the diversity of the behavioral and life-history characteristics among these six species? What are the character states, and which ones are ancestral and which ones are derived? What is the evolutionary sequence of events that produced the current distribution of characteristics we see today?

Defining character states

The first and, at this point, most important step in your analysis is to identify the behavioral characters of interest and their possible states. The simplest characters are those with only two possible states (the so-called **binary characters**). Discuss within your group the characters and their states. For example, singing could be scored as a **multistate character** with three states (only males sing, both sexes sing, and both sexes duet). Note, however, that one of our goals in scoring phylogenetic characters is to define character states that are mutually exclusive, either one thing or another. Are "both sexes sing" and "both sexes duet" mutually exclusive character states, given that females must sing in order to combine their songs with male songs? We therefore recommend that singing be scored as two separate binary characters (presence/absence of female song and presence/absence of duetting). Your group will decide which characters to use, and you will assign two possible states for each of them. Once your group has agreed on characters and states, compare them to those of another group. It is possible that the two groups will not agree. See if you can find a common system of characters and states.

For the purposes of this exercise, the authors have defined five binary characters, but your class could choose to use a different set. Use the information on each of the six species to fill in the appropriate spaces in Table 7.1, which lists the authors' five binary characters. For some characters the possible states could be "absent" and

Table 7.1 Five behavioral characters for six species of blackbirds, with states not filled in.

	Characters				
Species	**Female song**	**Vocal duets**	**Mating system**	**Breeding latitude**	**Migratory behavior**
Red-winged blackbird					
Red-shouldered blackbird					
Tricolored blackbird					
Tawny-shouldered blackbird					
Yellow-shouldered blackbird					
Jamaican blackbird					

"present" (or 0 and 1 in Mesquite), whereas for others the possible states could be more descriptive, such as "monogamous" and "polygynous" (likewise scored as 0 and 1 in Mesquite).

Mapping characters onto the tree

Use the information from Table 7.1 and the hypothesized phylogeny of Fig. 7.3 to "map" the five behavioral characters onto the branches of the phylogeny. For each character in turn, using either paper/whiteboard methods or Mesquite, use the current character states of these species to figure out which state is ancestral and where (on which internode or internodes) the derived state evolved. See if you can find the simplest, most parsimonious evolutionary scenario for each character.

Note that it is possible to find multiple evolutionary histories for a character that involve the same minimum number of changes. In the ancestors of red-winged blackbirds, red-shouldered blackbirds, and tricolored blackbirds, for example, you may find that character state changes could have occurred twice independently in red-winged blackbirds and tricolored blackbirds or once in the common ancestor of these three species and then again (in the opposite direction) in red-shouldered blackbirds. Both possibilities involve the same number of evolutionary transitions and so are equally parsimonious. How do we decide which scenario is more likely? As mentioned earlier, sometimes looking at additional information can be helpful. For instance, biogeographic evidence (based on geographic distributions) and genetic data suggest that red-shouldered blackbirds moved to the tropics relatively recently from a temperate breeding ancestor (Barker et al., 2008). Thus it seems likely that temperate breeding was gained earlier on in our tree and then lost in red-shouldered blackbirds rather than gained twice independently in red-winged blackbirds and tricolored blackbirds. Does this help us to resolve changes in other characters, such as migration?

Results/discussion

Once your group has completed its phylogeny, discuss the following questions. How did you resolve characters that have more than one possible most parsimonious evolutionary history? Do any character state changes seem correlated with each other (tending to occur together), and do any characters seem likely to influence each other? Once your group has discussed these and any other questions that occur to you, present your phylogeny of blackbird evolution and discuss it with the rest of the class. It is possible that your group's mapping of behavioral character state transitions will differ from that of another group. Be prepared to defend your proposed evolutionary sequence and to question the sequences of other groups. By what criterion can the class decide which proposed sequence of behavioral state transitions is "best"? It may interest you to know that disagreements in phylogenetics (this kind of analysis) can and do become extremely heated, both in print and in person.

Questions for in-class discussion

If time permits discuss the following questions within your group and then with the whole class. Alternatively, if time is short, each group could be assigned to discuss one question and present its conclusions to the rest of the class.

1. According to some comparative biologists, you should *never use the information you want to study to build your phylogenetic tree.* Why is it important to follow this rule? What happens if you do use the information you want to study to build your phylogenetic tree?
2. Originally, ethologists saw behavior as another set of characters that could be used to construct phylogenies, along with morphological information. These days, however, phylogenies are usually constructed from molecular data. Should we use behavioral data when constructing phylogenies? Why or why not? Which source of information is better? Why?
3. Why is the parsimony criterion used in reconstructing the evolution of character traits? Do you think evolution always necessarily operates according to this principle? Explain.
4. Comparative methods can be used to test whether different characters are correlated with each other across species, such as whether female singing is correlated with monogamy. In making such comparisons, should we always treat each species as an independent "statistical sample" so that, say, five species exhibiting an apparent correlation between female song and monogamy provide more statistical support than just three species sharing these traits? What if these character states are homologous across species, having evolved just once in a common ancestor rather than evolving independently as a result of selection due to similar environments? Does this make a difference?
5. Songbird species that build cup-shaped nests are much more common globally than species that build domed nests, despite the fact that cup nests evolved more recently on the songbird phylogeny (Price & Griffith, 2017). Moreover, most cup-shaped nests are simpler in construction than domed nests. What do these patterns tell us about how the rarity and the complexity of traits relate to the directionality of evolution?
6. In most temperate breeding songbird species, males produce complex songs and females vocalize but do not "sing" like males. Countless studies have investigated why male songbirds sing. Yet, recent phylogenetic studies show that sex differences in singing behavior have generally evolved by females losing song rather than males gaining it (Odom et al., 2014). Likewise, although we often ask why birds fly south in the winter, phylogenetic evidence shows that the ancestors of some migratory songbirds were tropical rather than temperate (e.g., Outlaw et al., 2003; Price, 2009), suggesting that migration evolved as a way to breed in northern regions rather than as a way to winter in southern regions. Do these new findings indicate that previous ideas were wrong?

Activity 2: Small-group projects

Now that you have had some practice, your group can attempt a second analysis with a somewhat more complex example. Our general research question, however, remains the same: what was the evolutionary sequence of events for the behavior or behaviors of interest? You will once again have to define behavioral character states and map character state transitions onto a phylogeny to infer the ancestral states and when each derived state evolved.

Your instructor will provide several data sets for analysis or will tell you how to find appropriate phylogenies and behavioral descriptions. Choose one set and have at it! Devise a table to record the states of the behavioral characters for each taxon. Follow the steps from the whole-class exercise (using either paper/whiteboard methods or Mesquite) to map the behavioral states onto the proposed phylogeny and then to infer the evolutionary history of the behavioral characters of interest.

References

Barker, F. K., Vandergon, A. J., & Lanyon, S. M. (2008). Species status of the red-shouldered blackbird (*Agelaius assimilis*): Implications for ecological, morphological, and behavioral evolution in *Agelaius*. *The Auk: Ornithological Advances, 125*, 87–94.

Brooks, D. R., & McLennan, D. A. (1991). *Phylogeny, ecology, and behavior: A research program in comparative biology*. Chicago: University of Chicago Press.

Chapin, J. P. (1917). The classification of the weaver birds. *Bulletin of the American Museum of Natural History, 37*, 243–280.

Friedmann, H. (1929). *The cowbirds*. Springfield, Illinois: Charles C. Thomas.

Harvey, P. H., & Pagel, M. D. (1991). *The comparative method in evolutionary biology*. Oxford: Oxford University Press.

Lorenz, K. (1958). The evolution of behavior. *Scientific American, 199*, 67–78.

Maddison, W. P., & Maddison, D. R. (2019). *Mesquite: A modular system for evolutionary analysis (v. 3.61)*. See http://mesquiteproject.org.

Martins, E. P. (Ed.). (1996). *Phylogenies and the comparative method in animal behavior*. Oxford: Oxford University Press.

Odom, K. J., Hall, M. L., Riebel, K., Omland, K. E., & Langmore, N. E. (2014). Female song is widespread and ancestral in songbirds. *Nature Communications, 5*, 3379.

Odom, K. J., Omland, K. E., & Price, J. J. (2015). Differentiating the evolution of female song and male–female duets in the new world blackbirds: Can tropical natural history traits explain duet evolution? *Evolution, 69*, 839–847.

Outlaw, D. C., Voelker, G., Mila, B., & Girman, D. J. (2003). Evolution of long-distance migration in and historical biogeography of *Catharus thrushes*: A molecular phylogenetic approach. *The Auk: Ornithological Advances, 120*, 299–310.

Owens, I. P. F. (2006). Where is behavioral ecology going? *Trends in Ecology & Evolution, 21*, 356–361.

Powell, A. F. L. A., Barker, F. K., Lanyon, S. M., Burns, K. J., Klicka, J., & Lovette, I. J. (2014). A comprehensive species-level molecular phylogeny of the new world blackbirds (Icteridae). *M*olecular Phylogenetics and Evolution, *71*, 94–112.

Price, J. J. (2009). Evolution and life history correlates of female song in the New World blackbirds. *Behavioral Ecology, 20*, 967–977.

Price, J. J., & Griffith, S. C. (2017). Open cup nests evolved from roofed nests in the early passerines. *Proceedings of the Royal Society B: Biological Sciences, 284*, 20162708.

Rubenstein, D. R., & Alcock, J. (2018). *Animal behavior: An evolutionary approach* (11th ed.). Oxford: Oxford University Press.

Searcy, W. A., & Yasukawa, K. (1995). *Polygyny and sexual selection in red-winged blackbirds*. Princeton: Princeton University Press.

Tinbergen, N. (1963). *Social behaviour in animals: With special reference to vertebrates*. London: Methuen & Co.

Tinbergen, N. (1964). On aims and methods of ethology. *Zeitschrift für Tierpsychologie, 20*, 410–433.

Wheeler, W. M. (1919). The parasitic *Aculeata*, a study in evolution. *Proceedings of the American Philosophical Society, 58*, 1–40.

Whitman, C. O. (1899). Animal behavior. In C. O Whitman (Ed.), *Biological lectures, woods hole* (pp. 285–338). Boston: Ginn & Co.

Part II. Instructor notes

Classroom management/blocks of analysis

This exercise can be done in two sections. The whole-class exercise can be completed in a single lab period of 2–3 h, with the small-group projects being done next, outside the lab time. For the whole-class exercise, allow about 30 min for students to define the character states and discuss them. If you plan to use Mesquite (see Appendix), then plan to spend about 30 min showing the program to students by using an example data file to walk them through the various functions and to give them some time to "play" with the program on their own. Once they understand how Mesquite works give them 30–60 min to build their character matrix, construct the tree, trace character histories, and discuss their results. Make sure you leave enough time for each group to explain their character state reconstructions to the rest of the class or lab.

Students can do their small-group projects outside the lab or classtime, but they should plan to leave some time at the end of the lab period to answer questions they have about the project. Most students are not familiar with "tree thinking," so the time spent introducing and reinforcing concepts and methods to them is extremely important.

Groups of three to four students work best, although pairs of students can also produce very good work. In some cases, four students may be too many, especially if one student is reticent. In other cases, four students are too few because none of them take to "tree thinking." Careful monitoring during the whole-class exercise and asking for progress reports during the small-group projects will help ensure that each group is making progress.

Teaching the activity

Preclass preparation and potential variations

This exercise could be done by hand with paper and pencils or better yet by giving each group a dry-erase whiteboard and several colors of dry-erase pens. Cheap white laminate wall covering can be purchased in 4 × 8-foot sheets and cut into 2 × 4-foot boards to be given to each group.

Hand methods will be sufficient for simple trees and characters, such as in the whole-class exercise, but may prove to be more difficult for more complex trees and sets of characters, as would be encountered in the small-group projects. Some students may feel more comfortable using a computer, whereas others might be

Table 7.2 States of five behavioral characters for six species of blackbirds.

Species	Characters				
	Female song	**Vocal duets**	**Mating system**	**Breeding latitude**	**Migratory behavior**
Red-winged blackbird	0	0	1	1	1
Red-shouldered blackbird	1	1	0	0	0
Tricolored blackbird	0	0	1	1	1
Tawny-shouldered blackbird	1	1	0	0	0
Yellow-shouldered blackbird	1	0	0	0	0
Jamaican blackbird	1	0	0	0	0

bewildered by all of Mesquite's various functions (in the Appendix, we describe only a very small fraction of the program's features). Decide in advance whether you want to do the exercise by hand, on computers, or as some combination of the two. You may also choose to map other characters onto the blackbird phylogeny, such as plumage colors. If you decide to have the students use Mesquite, you should plan to spend a few hours beforehand learning its ins and outs so you can answer their questions (they will have many).

The small-group projects can be based on trees that are provided by the instructor, or students can be asked to find trees in the published literature or other sources. Some excellent websites for generating evolutionary phylogenies are *TimeTree* (http://timetree.org) and the *Tree of Life Web Project* (http://tolweb.org/tree/). Student groups could build their phylogenies based on those relationships for a wide variety of taxonomic groups, although birds may be your best bet given the many phylogenetic and behavioral studies that have been published.

In-class preparation

Once students define their character states (as in Table 7.2), they can map their behavioral characters onto the tree shown in Fig. 7.3 by hand using paper/whiteboards or on a computer using Mesquite. If they do this by hand, first have the students record the character states of each of the five behavioral characters near the tips of the tree. It is easiest to have them record each character on a separate line, and perhaps using a different color, so they will need five lines for each species (one for female song, one for duetting, etc.). Once the character states are recorded at the tips, the students can focus on each character in turn to try to figure out which is the ancestral state and where (on what branch or internode) each derived state evolved. Typically, these evolutionary events are depicted by drawing a short line across the branch or internode and writing the name of the derived state next to it. Good examples are presented in Brooks and McLennan (1991) and Rubenstein and Alcock (2018).

Areas of potential confusion or difficulty for students

Whether they map behavior by hand or with Mesquite, your students should begin to think about the evolution of behavior in a very different way. Tree thinking will be new for most students, so at each step, it would be a good idea to spend some time talking about what they did, what they found, and what it means.

For both the whole-class exercise and small-group projects, students may have trouble scoring characters into clearly defined character states. In the whole-class exercise, for example, some may want to lump female song and duets into a single multistate character. Emphasize that character states should be mutually exclusive and that using lots of simple binary character states is usually much easier to interpret than using fewer, more complicated characters.

Students might also have trouble when there is more than one possible most parsimonious evolutionary history for a character (as in the ancestors of red-winged blackbirds, red-shouldered blackbirds, and tricolored blackbirds). A trait may have been gained twice independently or gained once and lost once; both possibilities involve the same number of evolutionary transitions. Such issues are common in evolutionary reconstructions. Talk about how we might try to resolve such uncertainties.

In the small-group projects, you may want to discourage students from choosing very complex trees (in excess of 30 taxa) or poorly resolved trees with many **polytomies** (unresolved relationships in which more than two branches come from a single node). For example, you could suggest that they use trees with fewer than 15 taxa or you could provide trees for them to use. If they use Mesquite, the most difficult step your students will face is converting a default random tree into the tree that they have chosen to analyze (see Appendix). Moving branches around in Mesquite is easy, but if students do it without thinking about what they are doing, the result will be a very messy tree and a lot of wasted time. Also, students may not realize at first that trees showing the same relationships can look different. For example, selecting sister taxa and rotating their branches at the node will change the positions of the species but will not change their relationships at all. A quick demonstration on a chalkboard/whiteboard would help show this to your students.

Whatever your students choose to do for their small-group projects, once they find a system to analyze, they will need to do the same three steps as in the whole-class exercise: (1) obtain a tree showing relationships among the taxa of interest, (2) define behavioral character states and enter these data into a table or into Mesquite, and (3) map (trace) the behavioral character transitions on the tree.

Another potential modification to the activity

Early comparative studies of behavior often used behavioral characters to construct trees. There is really no reason why we could not do that as part of this exercise, especially given the tools at our disposal. Students could use Mesquite to construct one tree using a set of behavioral characters (i.e., they could use shared derived character states to define relationships, using only behavioral characters) and then compare that tree to another one constructed from other characters (e.g., molecular sequences or morphological character states). In some cases the behavioral trees could be used to resolve polytomies, or vice versa.

Answers to the questions for in-class discussion

1. The idea that one should *never use the information you want to study to build your phylogenetic tree* is an attempt to avoid circular reasoning. If, for example, you constructed a phylogeny using behavioral characters and then used that phylogeny to reconstruct the evolutionary sequence of events for those same characters, it is pretty obvious that you would not have an independent assessment of evolutionary history. "Phylogenetic systematics" (the field of biology focused on determining evolutionary relationships among organisms) seeks to construct phylogenies that are independent of assumptions about phylogenetic history, so systematists usually construct trees based on molecular data such as DNA sequences. A molecular phylogeny provides an independent hypothesis for phylogenetic relationships among taxa, which can then be used to provide information about the evolution of the behavioral characters.
2. Are behavioral characters useful in constructing phylogenies? For any character, the critical distinction made by comparative biologists is between homologies, which can indicate phylogenetic relationships, and homoplasies, which can be misleading about phylogeny (Brooks & McLennan, 1991). Two species can share a derived behavioral state by homology just as they could share a morphological state or a DNA nucleotide sequence by homology. Although it is perhaps less common these days to use behavioral characters to construct trees, there is no a priori reason to assume that they are less valid than other characters. Some biologists assume that behavioral characters tend to be highly variable and are subject to modification by experience. Yet morphological features are also subject to environmental influences, and DNA sequences can be extremely variable across taxa with high levels of homoplasy. Molecular sequences are so frequently used in building trees because they provide so *many* characters, not necessarily because each character (i.e., each nucleotide) is necessarily more informative than other types of characters.
3. Parsimony is a working assumption used by systematists. Theoretically, when building a phylogenetic tree, if we did not limit the number of times each character transitioned between states, we could generate huge numbers of potential trees from the same set of data. How would we decide which tree reflects the actual pattern of relationships among taxa? Clearly, we need some kind of

rule by which we can choose a tree (remembering that any tree is a hypothesis for the true phylogeny). Choosing the most parsimonious tree, the one with the fewest possible character changes, is simply a rule we can follow to select among all the possibilities. However, the most parsimonious tree may not necessarily be the correct tree. Evolution can proceed randomly or in response to complex and changing environmental conditions, so evolution may not necessarily operate according to the principle of parsimony.

4. When testing whether different characters are correlated with each other across taxa, should each species be considered as an independent sample, even if shared character states among these species are due to shared ancestry (homology) and thus are not really independent? Modern comparative methods use phylogenies to overcome this inherent difficulty, and they have provided lots of valuable information for biologists interested in the evolutionary history of behavior. Taking phylogenetic relationships into account allows us to make comparisons among taxa that are statistically independent. In other words, if three blackbird species exhibit both female song and monogamy because both traits evolved in a common ancestor, we should not consider them as three independent examples showing a statistical correlation between these traits. Each of these traits may have evolved just once in the past, perhaps together but perhaps not. In contrast, if our analysis showed that female song was lost or gained every time the species transitioned between monogamy and other mating systems, then that would provide good evidence for an evolutionary correlation. Understanding phylogeny allows us to see if traits truly have evolved together.
5. Interestingly, although cup-shaped nests evolved more recently than domed nests, most of the world's songbirds (approximately 75%) build cup-shaped nests (Price & Griffith, 2017). Cup nests are also generally simpler to construct. These observations illustrate two important points. First, although biologists have often assumed that the ancestral bird nest was cup shaped, mostly because this design is so widespread, there is no reason to assume that the current prevalence of a trait indicates the order of events in the evolutionary past. Second, evolution does not always proceed from simple to complex. Indeed, there are many examples (including the evolution of female song) in which complex behavioral patterns have become lost or simplified during evolutionary history.
6. Many biologists, including Charles Darwin himself, have explained the complex singing behaviors of male songbirds as having evolved through the influence of sexual selection for elaborate male traits. Recent findings, however, show that singing by both sexes was the ancestral state in songbirds and that female song has been lost repeatedly across the songbird phylogeny (Odom et al., 2014). Thus it might be more appropriate to ask why so many females *do not* produce complex songs, rather than just focusing on why males do. In a similar way, phylogenetic evidence that temperate migratory species evolved from tropical, nonmigratory ancestors suggests that we should be asking why birds migrate north to breed, in addition to asking why they migrate south in the winter. These new findings from phylogenetic studies do not necessarily mean that old evidence is wrong. Rather, they bring up new questions and suggest new ways of thinking about the evolution of animal behavior.

Appendix: Using Mesquite

Mesquite (available at http://www.mesquiteproject.org) is a free computer program that can help students to visualize evolutionary sequences by mapping evolutionary events (changes in character state) onto phylogenetic trees. The following directions are for Mesquite version 3.61 (Maddison & Maddison, 2019).

Open Mesquite and select File > New. Provide a name for the file (e.g., "Blackbird evolution") and a location where it will be saved. When you click OK, a New File Options window will appear. The Make Taxa Block should be checked (check it if it is not checked), enter the number of taxa (6 in our example), check Make Character Matrix, and hit OK. In the New Character Matrix window enter the number of characters (5 in our example). There are several options in the Type of Data menu; for this exercise select Standard Categorical Data. Click OK.

A Project Window will open showing an empty Character Matrix (Fig. 7.4). Here you can enter the names of taxa, characters, and character states. There are several tools along the left side that allow you to manipulate the matrix. When you hold the cursor over each of the buttons, a description of what it does appears at the bottom of the window.

Click on the Edit tool and use it to enter each taxon name along the left column and character name along the top as shown earlier in Table 7.1. Then enter your

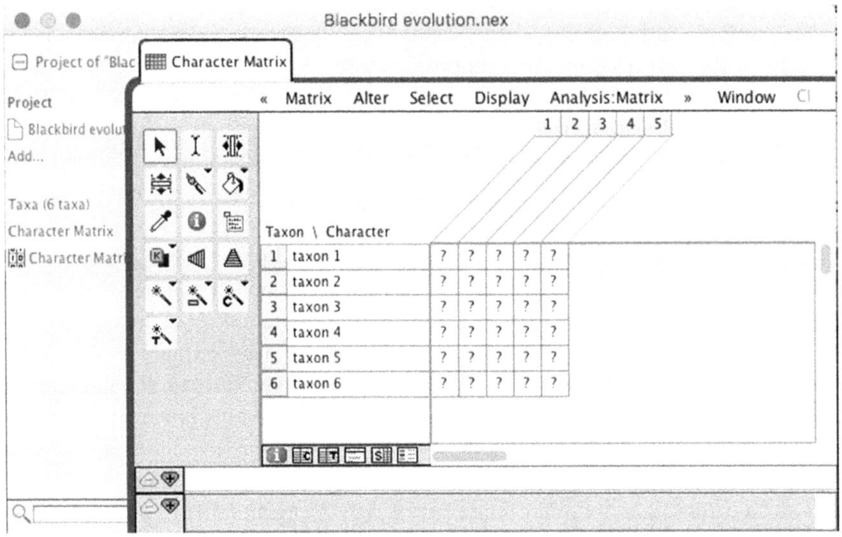

FIGURE 7.4

Project window with empty character matrix.

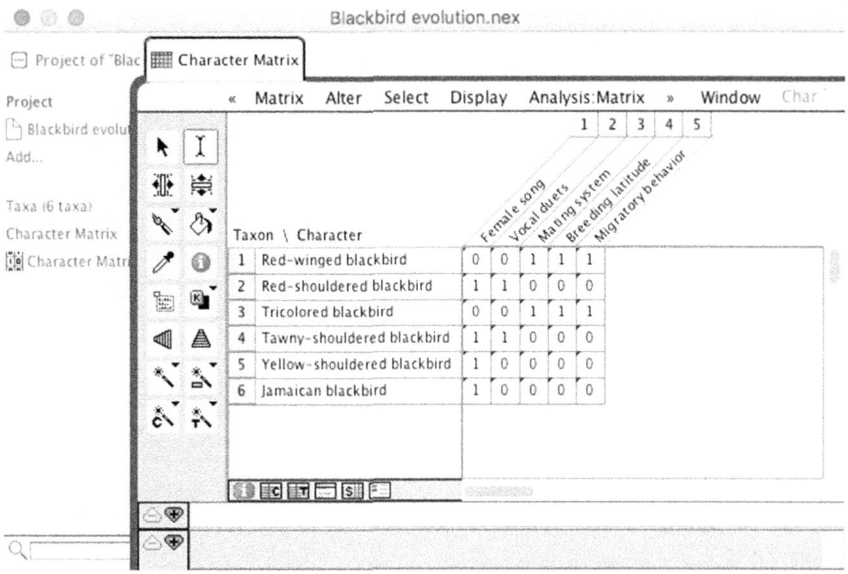

FIGURE 7.5

Character matrix with taxa, characters, and states entered.

previously defined binary character states in each cell as either a 0 or 1, as in Fig. 7.5. For example, you could score species without female song as 0 and with female song as 1, and you could score monogamous species as 0 and polygynous species as 1. In this simple example, we have only five characters with just two states each, but in other cases, there could be more characters, which could have two or more discrete states. Alternatively, our characters could be continuous rather than discrete or could be nucleotides rather than behavioral categories. Each has a specific option in the Type of Data menu (see earlier discussion).

A matrix filled with 0s and 1s is often not the easiest thing to interpret; we would be better served if the character states were more descriptive. You can name the character states as follows. At the bottom left are five small buttons next to a small blue "i"; each looks like a little window. Select the one that is second from the right, the Show State Names Editor Window (this is also available in a drop-down menu as Matrix > Edit State Names). Select the Edit tool and use it to give names to our characters and their states (Fig. 7.6). Now go back to the Character Matrix by clicking the tab at the top of the window. As you can see, all the 0s and 1s have been replaced with character state names. One thing to note is that when adding or editing character states in the matrix you still need to type a 0 or 1, even if the states now have more descriptive names.

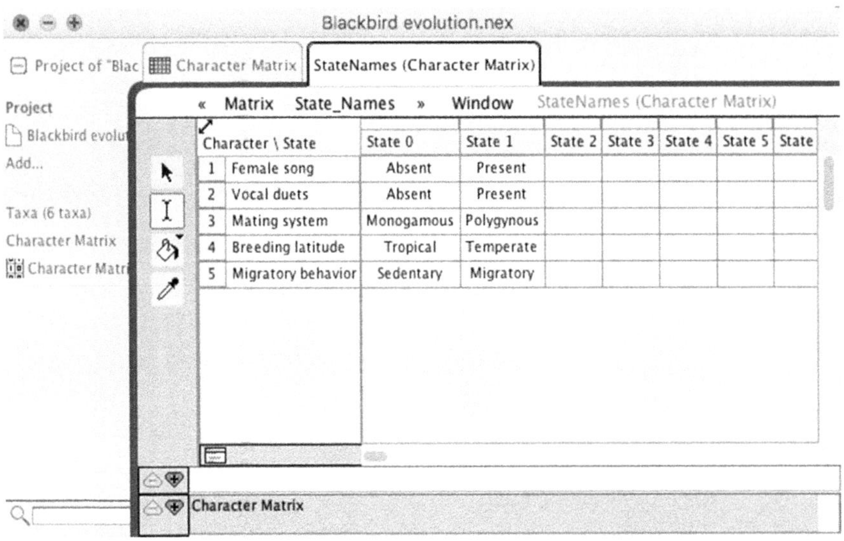

FIGURE 7.6

Entering names for each character state.

Creating and editing trees

There are ways to import previously constructed phylogenies into Mesquite, but this is a little beyond the scope of our exercise. Here we will manually recreate the tree shown in Fig. 7.3. Select Taxa & Trees > New Tree Window > With Tree To Edit By Hand. A window tab will appear with a randomly organized tree of our six taxa (Fig. 7.7), which we will need to modify to match the relationships in Fig. 7.3. Click on the Select tool (the arrow) and use it to click on the various branches and drag them over to connect with others. We suggest starting with pairs of sister species (e.g., red-winged blackbird and red-shouldered blackbird). Note that for sister taxa, it does not matter which one is to the right and which one is to the left—they just have to branch from the same node. The order in which you click and drag determines left and right positioning, so if you want a topology that is identical to that shown in Fig. 7.3, you may need to use Edit > Undo a few times (or the Interchange branches tool) to get the particular branching pattern you want. By default, Mesquite will draw a Square Tree, but you can change it to a different form using Display > Tree Form. You can rotate the tree using Display -> Orientation. You can also change other aspects of the tree, such as the thickness of the lines (Display > Line Width) and the angle of the taxon names (Display > Names > Taxon Name Angle). When you are happy with how your tree looks save it using Tree > Store Tree in Tree Block As ....

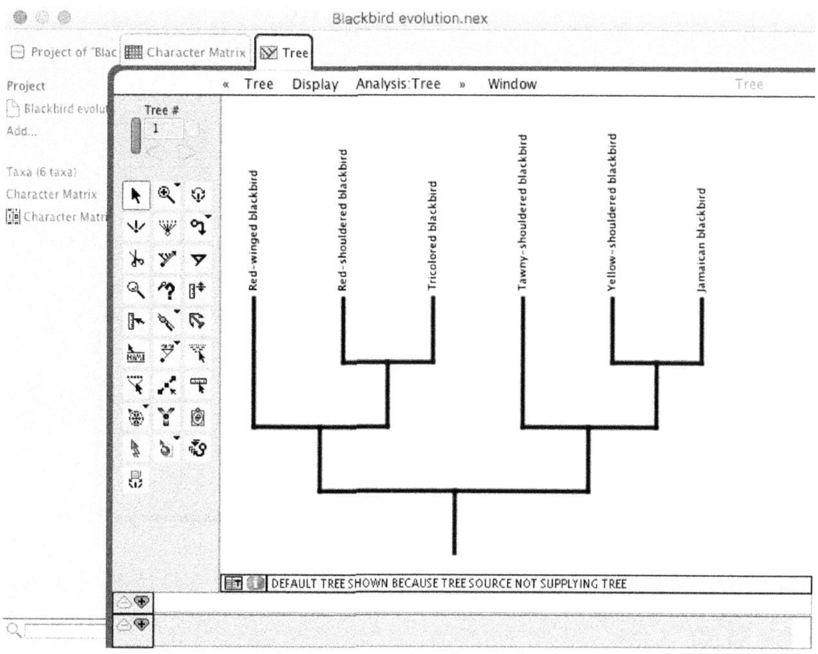

FIGURE 7.7

Default tree before moving branches.

Discrete character state reconstruction using parsimony

With the Tree Window open select Analysis:Tree > Trace Character History. Select Parsimony Ancestral States and hit OK. Ancestral states for our characters will appear on the branches (Fig. 7.8), similar to how they were shown in Fig. 7.2. You can look at the reconstructions of each character by clicking on the arrows in the Trace Character box (which is usually at the bottom left of your Tree Window). In Mesquite, branches are shown with the colors of all the possible states that they could have under parsimony. Thus if there are two most parsimonious scenarios on a branch, it will be shown as unresolved with both character state colors (e.g., white and black). When you move the cursor over a branch, you will see the character state or states at the bottom of the Trace Character box. You may have to move this box up to see its bottom, or so it does not obscure parts of your tree.

Now we are ready to interpret the evolution of our characters on the blackbird phylogeny. For each character, which state is ancestral and which is more recently derived? Do any character state changes seem correlated with each other, and do any characters seem likely to influence each other? Do any characters exhibit multiple possible evolutionary scenarios that are equally parsimonious? How will we

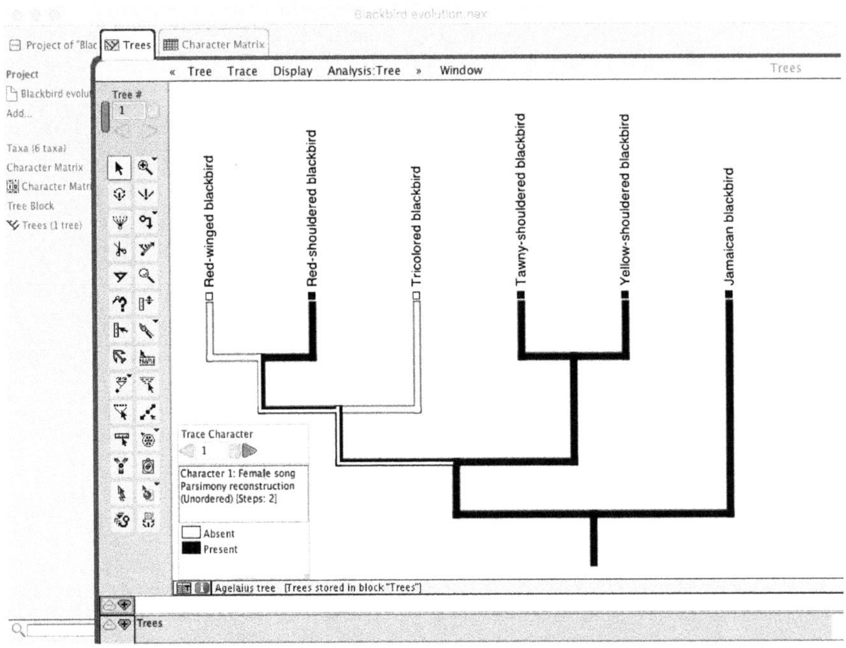

FIGURE 7.8

Ancestral states for female song reconstructed on the blackbird phylogeny.

resolve these? Note that the tree itself is a hypothesized phylogeny, so these evolutionary changes are also hypotheses.

If you change a character state in your matrix, the results will automatically appear in your tree (i.e., the tree is "hot linked" to the character matrix). Go to the Character Matrix window and change the character state of the character you are currently tracing in your tree window for one of the taxa. The character state will automatically change in the tree window. This is a general feature of Mesquite; whenever you change something in one window, it will affect what is happening in all the windows that are related to it.

The abovementioned description of Mesquite covers only a small fraction of the functions it offers for evolutionary studies. The best way to learn Mesquite is to play with it. Students may also want to spend some time going through the Mesquite manual at http://www.mesquiteproject.org/.

CHAPTER

Examining variability in the song of the white-crowned sparrow (*Zonotrichia leucophrys*)

8

Douglas W. Wacker

Division of Biological Sciences, School of STEM, University of Washington, Bothell, WA, United States

Chapter outline

Exploring Animal Behavior in Laboratory and Field. https://doi.org/10.1016/B978-0-12-821410-7.00022-4

Part I. Student instructions

Learning goals, objectives, and key concepts

- To identify and quantify characteristics of animal vocalizations using spectrograms
- To quantify and ruminate on individual and inter-individual variation in animal behavior
- To consider inter-observer variation in behavioral analyses
- To address proximate and ultimate questions about behavioral variation

Background

Behavior varies in nature, across taxonomic groups, between individuals, and even within a single animal over time. Bird song is a behavior that varies at all these levels. Bird song is a learned collection of vocal phrases, composed of elements or syllables separated by silence (Marler & Tamura, 1962; Thompson et al., 1994) (Fig. 8.1). Only a subset of bird taxa learn and produce songs, including the oscines

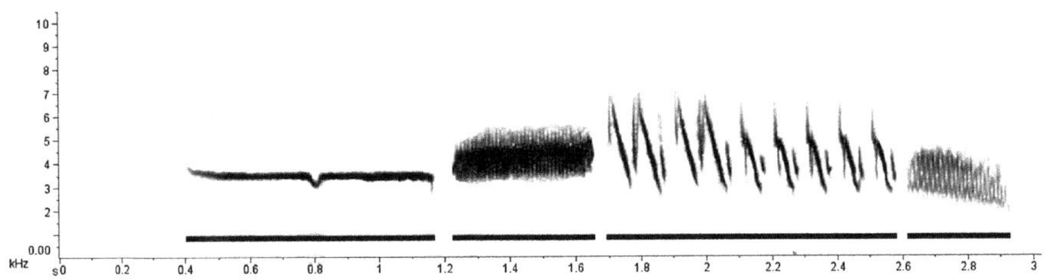

FIGURE 8.1 A bird song and its components.

This is a spectrogram of the song of a white-crowned sparrow (*Z. l. nuttalli*) recorded at Bodega Bay, California by Dr. John Wingfield. A **spectrogram** is a visual representation of sound, with time on the x-axis and frequency, perceived by our brain as pitch, on the y-axis. Differences in loudness, or intensity, are represented by different intensities of shading, or different colors in black and white or color spectrograms, respectively. This spectrogram has been cleaned up for clarity, so some of the harmonic structure is no longer visible. The literature is not always consistent in how components within an individual song are named. Some researchers use the terms syllable and phrase interchangeably, while others introduce different terms, such as motif. For the purposes of this activity, the term **phrase** will be used to denote any collection of similarly sounding song components that occur together (Wada, 2010). In this example, one can see four phrases, all delineated by lines at the bottom of the spectrogram. Each phrase has a certain **phrase duration** in seconds.

or songbirds, parrots, and hummingbirds (Päckert, 2018). Classically, songs were considered to be primarily the domain of male birds, who sing to attract mates and repel rival conspecifics (Eriksson & Wallin, 1986; Searcy et al., 1998). More recent work suggests that female song is not as rare as previously thought (Riebel et al., 2019). Although defining what constitutes a *song type* is not always straightforward, it is clear that some bird species sing a single song type, whereas others sing a repertoire of different types (Krebs & Kroodsma, 1980; Macdougall-Shackleton, 1997, pp. 81–124; Nelson & Poesel, 2011). In species that sing only a single song type, there is still variation in the presentation of their songs. For example, some species, such as the white-crowned sparrow (*Zonotrichia leucophrys*), develop regional dialects of their singular song type (Baptista, 1977; Marler & Tamura, 1962). Individual variation in songs has also been reported in free-living white-crowned sparrows, but at lower levels than inter-individual variation (Marler & Tamura, 1962; Orejuela & Morton, 1975). Understanding how a behavior, such as bird song, varies within and across individuals and populations can help researchers better understand its proximate and ultimate underpinnings.

Purpose

This activity will introduce you to the quantitative assessment of bird song. You will examine songs of the white-crowned sparrow, a species that sings a single song type. You will assess phrase duration utilizing a commonly used audio analysis program, Raven Lite (Center for Conservation Bioacoustics, 2016), and using your resultant dataset, you will compare individual and inter-individual variation in this variable.

Methods

Species selection

This activity uses recorded white-crowned sparrow songs, available from the online repository xeno-canto (https://www.xeno-canto.org/). The activity can be modified to use recordings of other bird species and/or other taxonomic groups with individuals that vocalize.

Materials needed

You will need computers that have Raven Lite (http://ravensoundsoftware.com/software/raven-lite/) installed and access to spreadsheet software (e.g., Microsoft Excel, Google Sheets) to complete this exercise. Speakers or headphones are also required.

Step-by-step instructions

1. Ask your instructor for your *group number* and write this and the names of your group members in the results table provided in the supplementary material.
2. Visit https://www.xeno-canto.org/ and search for "white-crowned sparrow." You'll be presented with a list of recordings made of this species, which can be sorted by different variables. Sort the recording list by catalog number by clicking the *Cat.nr.* (catalog number) heading at the top of the list of recordings. Clicking *Cat.nr.* more than once will resort the list in descending or ascending order. Please note you cannot currently search xeno-canto by entering a specific catalog number in the search box, rather you *must* search by another variable, such as the common or scientific name of the bird of interest.
3. Once you are viewing the list of white-crowned sparrow recordings find our example song by advancing through the pages, listed in the burgundy boxes at the bottom and top of the list of recordings, until you see the Cat.nr. XC353606 (Lambert, 2015). If you have trouble locating this song, here is a direct link, https://www.xeno-canto.org/353606. Click the play button on the left of the recording to listen. Note the common and scientific names, date, and location for this recording in the results table in this packet. Download this audio file by clicking the download icon under *Actions.*
4. Open Raven Lite, click File/Open Sound Files.../, select the downloaded file (XC353606), and click *Open.* When you see the Configure New Sound Window click *OK.* This will open a window with visual representations of the audio file (Fig. 8.2).
5. For this activity, you won't need to use the waveform, so you can close it by unclicking the box next to *Waveform 1* on the upper left of your screen. XC353606 is a stereo file with two channels; close one channel by unclicking the box for Channel 1 on the left under *Channels.*
6. Expand the Raven Lite window so that the spectrogram fills most of your computer screen. The entire recording should appear upon opening, but if it doesn't, then you can use the zoom buttons to adjust this (Fig. 8.3). Select the first song on the recording by clicking and holding your mouse button and highlighting the area where the song appears in the spectrogram. Use the *Zoom to Selection* button at the top of the screen to zoom in on that particular song (Fig. 8.3). The complete first song from Cat.nr. XC353606 is provided in Fig. 8.4 for your reference.
7. Click the *Loop Play Selection* button to listen to your selected song (Fig. 8.3). Please note that you'll need to hit the *Stop Playback* versus the *Pause* button if you want to switch between playback types. Close your eyes and listen as the song repeats. Try to identify the number of different phrases in your selected song. Now look at the spectrogram as the file plays and watch the bar as it moves across the selected song. How many phrases can you identify now? Come to an agreement with your group members on how many phrases are in this song and note this in the results table. Ask a few other groups how many phrases they identified in the first song of XC353606 and note their findings in your results table as well.

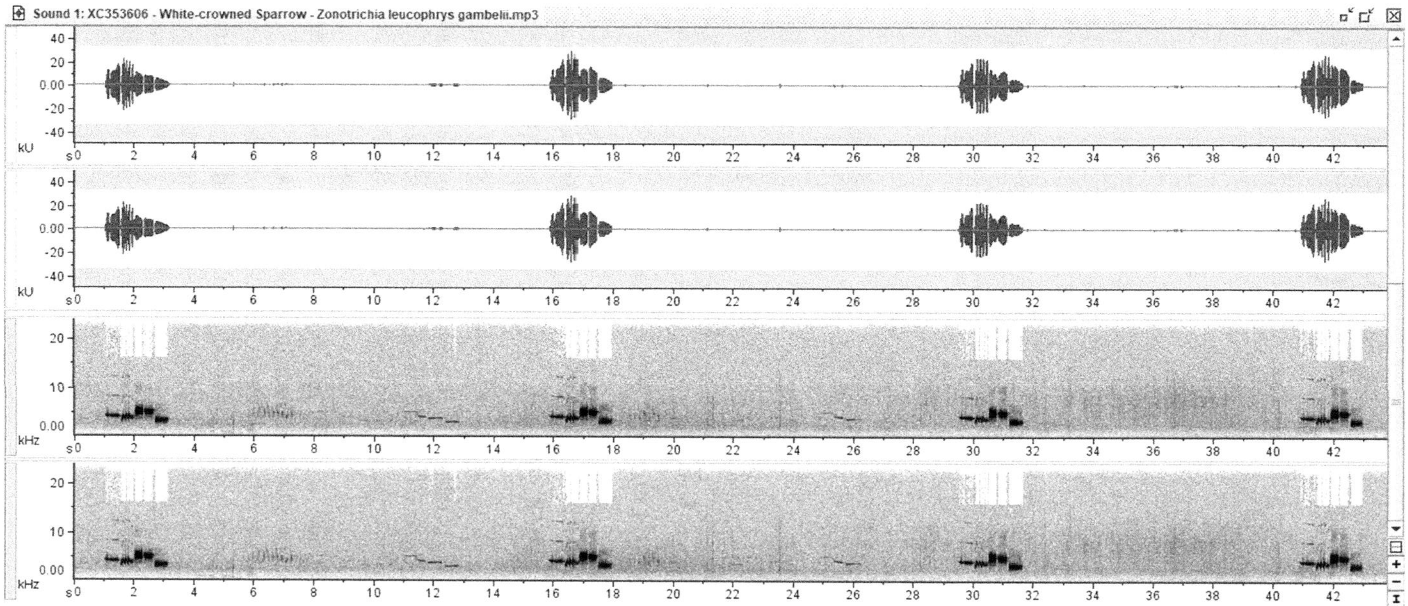

FIGURE 8.2 Screenshot of Cat.nr. XC353606 in Raven Lite.

The bottom channels show spectrograms, while the top channels show waveforms, different visual representations of, in this case, four songs made by a white-crowned sparrow (Lambert, 2015). Note, your spectrogram may be in color; you can change the palette for your spectrograms by clicking the drop-down menu on the left of your screen above *Views*. What we perceive as sound is actually a wave of compressed and rarified air molecules. Imagine a drum head. As you strike the drum, the head moves inward and pulls air molecule backward with it, creating a pocket of less dense (i.e., rarified) air in front of the drum head. As the drum head moves outward, it pushes air molecules forward and together. This back-and-forth movement creates waves of compressed and rarified air that project away from the drum. This is what is represented in the waveform.

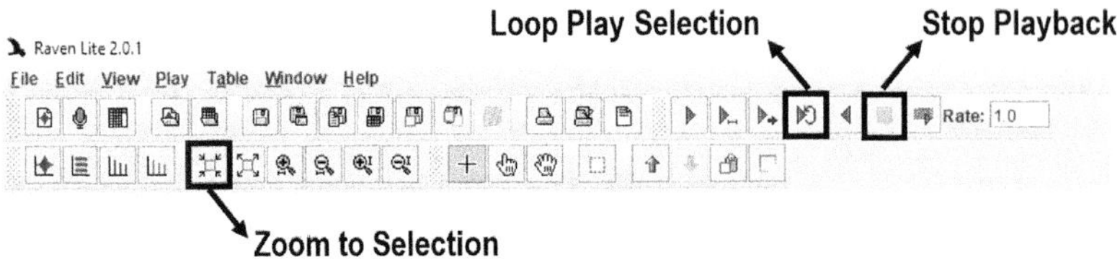

FIGURE 8.3

Zoom and play buttons in Raven Lite.

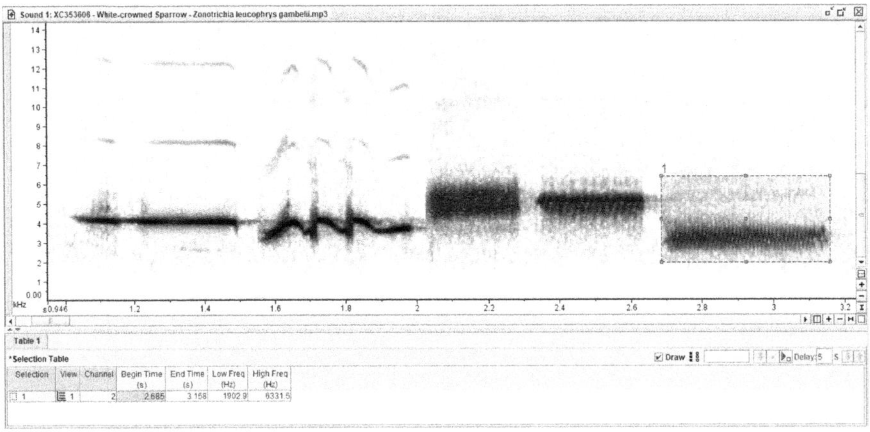

FIGURE 8.4 Phrase selection in Cat.nr. XC353606 in Raven Lite.

You can alter the brightness and contrast of the spectrogram to better resolve the different phrases. Note that the last phrase of the song is highlighted and represented in the *Selection Table* located below the spectrogram.

8. Select the first phrase of the selected song by holding down the mouse button and highlighting it on the spectrogram. Adjust your selection using the small boxes on the selection window. Do your best to select the full frequency (y-axis) and duration (x-axis) of the phrase; however, we'll primarily be concerned with phrase duration for this exercise. Once you've successfully highlighted the first phrase hit *Enter* to commit the selection. A box labeled *Annotate Selection 1* may appear when you hit *Enter.* If so type "Song1-Phrase1" into the box and hit *Enter.* Notice the selection table at the bottom of the screen. If you can't see it, you can reveal this table by holding your mouse button and pulling up on the small dotted bar below the spectrogram window. Your selection's *Begin Time (s)*, *End Time (s)*, *Low Frequency (Hz)*, and *High Frequency (Hz)* are listed (Fig. 8.4). You can click on a selection in this table and then readjust your selection boxes on the spectrogram as necessary.
9. Repeat these steps for all phrases of the song. Be sure to hit *Enter* to commit each selection once you've highlighted it. Once you've finished committing all

your selections, you can save your selection table by clicking File/Save Selection Table "Table 1" As.../, choosing a location to save your file, and hitting *Save*. **If you do not save your selection table, it will be lost**. If you close your sound file, you'll need to do the following to reaccess your selections: open the sound file (as mentioned earlier) in Raven Lite and then, separately, open your selections by clicking File/Open Selection Table.../ and clicking on the selection table text file (.txt) that you created when you saved your table.

10. In Cat.nr. XC353606, you'll find four white-crowned sparrow songs, all likely from the same individual sparrow (Fig. 8.2). There are a few other bird sounds on that recording, but the white-crowned sparrow songs all look relatively similar and should be easy to identify; remember, this species only sings one song type. Conduct the same measurements (as mentioned earlier) for all phrases of the three remaining songs on this recording, trying to be consistent in how you define each phrase across song iterations. Save each entry to your selection table, using *Annotations* such as "Song2Phrase1". Once you've saved your completed selection table, you can close the viewing window for Cat.nr. XC353606.
11. Download recordings from three additional white-crowned sparrows. On xeno-canto.org, the following recordings are good candidates: XC251254, XC253835, and XC335466 (Cruickshank, 2015; Hoyer, 2015; Lagerquist, 2016). Alternatively, with instructor approval, you can find different recordings, but just be sure they're of white-crowned sparrow song, weren't recorded at the same locations, and each have at least four iterations of a song that are (likely) from the same individual. Remember, you can easily listen to each recording by clicking the play button on the far left of the list to ensure your chosen recordings meet these requirements. Be sure to note the common and scientific names, date, and location for all new recordings in the results table in this packet.
12. Analyze four songs on each of your chosen recordings as described earlier, creating and saving new selection tables with different filenames to avoid copying over existing datasets. Once you've finished the abovementioned analyses and saved your data, proceed to step 13.
13. Open/import your first selection table (for Cat.nr. XC353606), saved as a text file, into MS Excel, Google Sheets, or a similar spreadsheet program. Each datapoint should be imported into a separate cell. Add these four columns to your spreadsheet, "CatalogNumber", "SongNumber", "PhraseNumber", and "PhraseDuration". Using your annotations enter the correct catalog number, song number, and phrase number for each entry. Then calculate phrase durations by subtracting the *Begin Time (s)* from the *End Time (s)* for each phrase. In MS Excel and Google Sheets, you can enter this simple calculation in one cell and then drag copy the formula down to other cells in a column.
14. Now, you need to add the selection table information for the other recordings into your spreadsheet. The easiest way to do this is to import each new selection table (e.g., for Cat.nr. XC251254) into its own spreadsheet, append each table with the additional information as described earlier (i.e.,

"CatalogNumber", "SongNumber", "PhraseNumber", and "PhraseDuration"), delete the column headers, and then paste your cells into your first spreadsheet. Do this for all your selection tables from your other analyzed recordings.

15. Now, you will calculate the **individual variation** in song within white-crowned sparrows.
    a. Calculate the **mean** and **standard deviation** for each phrase duration across songs for each individual. A mean is a measure of central tendency in a dataset, while the standard deviation is a measure of variation. In MS Excel and Google Sheets, the mean and standard deviation of a group of values can be calculated with the following formulas, *average(RANGE)* and *stdev(RANGE)*, respectively, where *RANGE* is the range of cells whose values are to be included in the calculation. First calculate the mean of all phrase 1 durations across all four songs in Cat.nr. XC353606. Then do this for each additional phrase in XC353606. Then conduct the same calculations for each phrase for (separately) XC251254, XC253835, and XC335466.
    b. Now you'll compare the variation across all phrases for each individual. Using standard deviations to do this is problematic, as some phrases are longer than others. Although these longer phrases may have larger standard deviations, this does not necessarily mean there is higher *relative* variation. For example, the standard deviation of bat ears would certainly be significantly lower than that of elephant ears, simply because elephant ears are larger. However, that does not mean that the relative variation in elephant ears is larger; perhaps, bat ears are more variable in size, once you correct for their relative smallness! In order to standardize individual variation across all phrases, you need to compute the **coefficient of variation** for each phrase number across songs for each individual. You can do this by dividing each standard deviation by the mean value for each phrase number and then multiplying the result by 100. For example, let's say that you calculated the following phrase durations:

CatalogNumber	SongNumber	PhraseNumber	PhraseDuration
XC353606	One	One	0.54
XC353606	Two	One	0.54
XC353606	Three	One	0.53
XC353606	Four	One	0.55

The mean phrase1 duration for XC353606 is (0.54 + 0.54 + 0.53 + 0.53)/ 4 = 0.54. The standard deviation is 0.008. The coefficient of variation for phrase 1 for XC353606 is (0.008/0.54) * 100 = 1.48.

    c. Calculate the mean coefficient of variation across all phrases for each individual. These values represent the variation across song repetitions *within* each individual. Now, take the mean of these average coefficient of variation values across individuals. This is a measure of **individual**

variation in song production within white-crowned sparrows. Write this value in the results table.

16. Now you will calculate the **inter-individual variation** in songs *across* individual white-crowned sparrows.
    a. First find the mean phrase durations that you calculated for step 15. For example, for bird XC353606, you calculated the mean phrase 1 duration across all four repetitions of its song.
    b. Using these means calculate the coefficient of variation for each phrase *across* individuals. For example, take average phrase 1 durations for XC353606, XC251254, XC253835, and XC335466 and calculate a mean of these four averages. Then calculate a standard deviation and coefficient of variation. Repeat this for the first four phrases across individuals. This will leave you with four interindividual coefficient of variation values: one for phrase 1, one for phrase 2, one for phrase 3, and one for phrase 4.
    c. Take the mean of these four coefficient of variation values. This is a measure of **inter-individual variation** in songs *across* individual white-crowned sparrows. Write this value in the results table.

Results/discussion

Using your results table answer the following questions.

1. Did all groups in your class identify the same number of phrases in the first song of XC353606? If not, how did other groups differ in how they defined each phrase? How might one reduce **inter-observer variation**?
2. Researchers have defined and named the various phrases in white-crowned sparrow songs. Use the existing research literature (Google image search is a good place to start) to find a spectrogram outlining these phrase types. Did you define your phrases in the same way? If not, how did your phrases differ?
3. You have generated coefficient of variation values for phrase durations both *within* and *across* individual white-crowned sparrows. Which was greater, **individual variation** or **inter-individual variation**? Why might you have observed this pattern?
4. Cat.nr. XC353606 and Cat.nr. XC253835 represent two different subspecies of white-crowned sparrow, *Z. l. gambelii* and *Z. l. pugetensis*, respectively. Try to find some information about these subspecies online. Distribution maps might be a good place to start. Why might these two different subspecies of white-crowned sparrow have different sounding songs?
5. Cat.nr XC 353606 and Cat.nr. 335466 also sound quite different. Look at the *order* of the different phrases in these songs. Does it make sense to compare phrase 2 from Cat.nr XC 353606 to phrase 2 from Cat.nr. 335466. Why/why not? If not, how might you alter your analyses to more accurately compare these songs?

6. Despite the fact that white-crowned sparrows learn songs from those around them, there is still some inter-individual variation in songs within the same geographical location. Considering the adaptive function of bird song, why might a white-crowned sparrow male sing a slightly different song than its neighbor?
7. In Cat.nr. XC253835, the last phrase differs across song iterations. What's different? Why might this be? White-crowned sparrows are thought to sing only one song type, do you still agree with this assessment? Go to xeno-canto and listen to a few songs from a related sparrow species known to sing multiple song types, the song sparrow, *Melospiza melodia*. Does this change your mind about the number of song types sung by the white-crowned sparrow? Why/why not?

References

Baptista, L. F. (1977). Geographic variation in song and dialects of the puget sound white-crowned sparrow. *The Condor: Ornithological Applications, 79*(3), 356–370.

Bioacoustics. (2016). *Raven lite: Interactive sound analysis software (version version 2.0) [computer software].* Ithaca, NY: The Cornell Lab of Ornithology. Retrieved from http://ravensoundsoftware.com/.

Cruickshank, I. (2015). *XC251254 white-crowned sparrow* Zonotrichia leucophrys oriantha, *2015-05-10, Waterton Lakes National Park - Lonesome Lake, Alberta.* Retrieved from https://www.xeno-canto.org/251254.

Eriksson, D., & Wallin, L. (1986). Male bird song attracts females—a field experiment. *Behavioral Ecology and Sociobiology, 19*(4), 297–299.

Hoyer, R. (2015). *XC253835 white-crowned sparrow* Zonotrichia leucophrys pugetensis, *2015-06-23*. Curry County, Oregon: Brookings.

Krebs, J. R., & Kroodsma, D. E. (1980). *Repertoires and geographical variation in bird song advances in the study of behavior* (Vol. 11, pp. 143–177). Elsevier.

Lagerquist, B. (2016). *XC335466 white-crowned sparrow* Zonotrichia leucophrys, *2016-04-29*. Washington: Carlson Canyon, Kittitas County.

Lambert, F. (2015). *XC353606 white-crowned sparrow* Zonotrichia leucophrys gambelii, *2015-05-29, Denali wilderness within Denali National Park, Denali, Alaska (approx location).* Retrieved from https://www.xeno-canto.org/353606.

Macdougall-Shackleton, S. A. (1997). *Sexual selection and the evolution of song repertoires current ornithology.* Springer.

Marler, P., & Tamura, M. (1962). Song "dialects" in three populations of white-crowned sparrows. *The Condor: Ornithological Applications, 64*(5), 368–377.

Meitzen, J., Thompson, C. K., Choi, H., Perkel, D. J., & Brenowitz, E. A. (2009). Time course of changes in Gambel's white-crowned sparrow song behavior following transitions in breeding condition. *Hormones and Behavior, 55*(1), 217–227.

Nelson, D. A., & Poesel, A. (2011). Song length variation serves multiple functions in the white-crowned sparrow. *Behavioral Ecology and Sociobiology, 65*(5), 1103–1111.

Orejuela, J. E., & Morton, M. L. (1975). Song dialects in several populations of mountain white-crowned sparrows (Zonotrichia leucophrys oriantha) in the Sierra Nevada. *The Condor: Ornithological Applications, 77*(2), 145–153.

Päckert, M. (2018). *Song: The learned language of three major bird clades bird species* (pp. 75–94). Springer.

Riebel, K., Odom, K. J., Langmore, N. E., & Hall, M. L. (2019). New insights from female bird song: Towards an integrated approach to studying male and female communication roles. *Biology Letters, 15*(4), 20190059.

Searcy, W., Nowicki, S., & Hughes, M. (1998). The territory defense function of song in song sparrows: A test with the speaker occupation design. *Behaviour, 135*(5), 615–628.

Thompson, N. S., LeDoux, K., & Moody, K. (1994). A system for describing bird song units. *Bioacoustics, 5*(4), 267–279.

Wada, H. (2010). The development of birdsong. *Nature Education Knowledge, 2*(2), 16.

Part II. Instructor notes

Classroom management/blocks of analysis

Assuming Raven Lite is preinstalled on computers, this activity can be completed in a single 3-h laboratory period or during several shorter classroom sessions. There will be some variation in the amount of time it takes students to get used to Raven Lite. The latter part of this activity will be completed more quickly by students with a greater familiarity with MS Excel or Google Sheets.

Teaching the activity

Preclass preparation

Institutional Animal Care and Use Committee (IACUC) approval is not required for any aspect of this activity. As written, this activity can be completed entirely in a classroom or laboratory outfitted with computers preloaded with Raven Lite (FREE; http://ravensoundsoftware.com/software/raven-lite/) and/or access to a spreadsheet program. Students may wish to download Raven Lite, a free program, onto their own laptops. The activity uses recorded white-crowned sparrow songs, available from the online repository xeno-canto (http://https://www.xeno-canto.org/), and candidate files are suggested. As Raven Lite is available as a free download, most aspects of this activity could be completed outside class if students have access to a computer with the aforementioned programs and an internet connection. For more help or to learn more functions of Raven Lite or Raven Pro, click on Help/Raven Lite User's Manual. Chapter 1 *Getting Started* has lots of helpful information for first-time users.

One could alter the protocol to use/compare other white-crowned sparrow recordings, recordings from individuals of other songbird species, and/or append this activity to include a field component where students record live birds and then compare their vocalizations. If the latter option is chosen, then adequate microphones and audio equipment will be required. The Cornell Lab of Ornithology, Macaulay Library website has good resources for procuring the best recording equipment based on your needs (current website: https://www.macaulaylibrary.org/how-to/audio-recording-gear/). It is important to ensure that each recording includes at least four iterations of each individual's vocalization. This activity could also be altered to use vocalizations from other taxa, such as insects, anurans, fish, etc. For more advanced classes, the analysis portion of this activity could be completed using R (FREE; https://www.r-project.org/), with additional time required depending on student and instructor familiarity with R programming.

In-class preparation

It is recommended that students break up into groups of no more than two or three to complete this assignment, and it could easily be converted to an activity completed by students individually and/or independently. If you use groups, then assign each group a unique *group number*. Measurements in Raven Lite can be largely completed by one person, so it is important to make sure that all students get a chance to actively use the program. Still, consistency of measurement is also an important consideration for this activity, and thus it provides ample opportunity to discuss inter-observer variation. For example, the instructor could ask their students how they plan to split up their analysis to ensure adequate precision.

Data analysis is relatively straightforward, including the calculation of means, standard deviations, and coefficient of variation values. In the supplementary materials linked at the end of this chapter, you'll find a sample spreadsheet with calculated phrase durations for four songs of XC353606, XC251254, XC253835, and XC335466. What constitutes a phrase and where a phrase starts and ends is open to interpretation so student values may differ. Variation in student measurements can be used to discuss inter-observer variation, and one could expand this activity by further integrating the research literature to explore how others have previously assessed phrases, syllables, and notes in white-crowned sparrow songs.

Recommendations for extensions or continuations for more advanced classes

One could easily alter the activity to include a larger sample of birds, thereby creating a dataset large enough for statistical comparisons. In more advanced classes, one could utilize **cross-correlation comparisons** to analyze variations in bird song within and across individuals. This type of analysis is more consistent with how vocalizations are compared in the current research literature. Raven Pro, a paid version of Raven Lite, has these capabilities.

Once students learn how to use audio files from xeno-canto to quantify differences in bird vocalizations in Raven Lite, you may want them to apply their new skills with a more independent activity. Here is a suggestion. Have students choose a local bird species. If they are unsure about which birds occur in your area, The Cornell Lab of Ornithology's All About Birds website is a free online reference they can explore, https://www.allaboutbirds.org/. Once students have found a candidate species, they can search for relevant audio recordings on xeno-canto. Remind students that not all birds sing and a bird's song may differ considerably from its other vocalizations. Students could then use techniques learned from the first exercise to compare songs or other vocalizations both within and across individuals and then calculate both individual and inter-individual variations. In addition to having students examine phrase durations, you might have them compare frequency (pitch) differences. If you wish, you could then have your students explain their findings based on the natural history of their chosen species or, if they compare birds over larger geographical areas, any potential subspecies or population differences.

Answer key

Answers to student questions may vary.

1. Did all groups in your class identify the same number of phrases in the first song of XC353606? If not, how did the other groups differ in how they defined each phrase? How might one reduce **inter-observer variation**?
Nontrained observers will often define what constitutes a phrase differently. Some will highlight individual syllables or notes, whereas others will highlight phrases more consistent with what is typically seen in the scientific literature. For example, in the selected Z. l. gambelii recording (Cat.nr. XC353606; Fig. 8.4), *one might parse the first component into two whistle phrases, whereas others might consider this a single phrase. Inter-observer variation can be reduced by using the existing literature to establish rules for defining phrases before analysis and by conducting trial analyses by observers, comparing their output, and only initiating final analyses once some level of consistency is reached.*
2. Researchers have defined and named the various phrases in white-crowned sparrow songs. Use the existing research literature (Google image search is a good place to start) to find a spectrogram outlining these phrase types. Did you define your phrases in the same way? If not, how did your phrases differ?
There are a variety of scientific articles that have defined the components of white-crowned sparrow songs. These definitions can differ by author, subspecies, and dialect. See Baptista (1977), Marler and Tamura (1962), and Wada (2010) *for some examples.*
3. You have generated coefficient of variation values for phrase durations both *within* and *across* individual white-crowned sparrows. Which was greater, **individual variation** or **inter-individual variation**? Why might you have observed this pattern?
Typically, students will detect more variation across versus within individuals. Interestingly, consistency within an individual increases with the hormonal and neural changes associated with breeding (Meitzen et al., 2009), *which suggests that song stereotypy might be an important adaptive signal in white-crowned sparrows. More simply, the recorded birds are from different subspecies and geographical regions, so higher inter-individual variation reflects different dialects learned by different individuals.*
4. Cat.nr. XC353606 and Cat.nr. XC253835 represent two different subspecies of white-crowned sparrow, *Z. l. gambelii* and *Z. l. pugetensis*, respectively. Try to find some information about these subspecies online. Distribution maps might be a good place to start. Why might these two different subspecies of white-crowned sparrow have different sounding songs?
A little sleuthing will show that Z. l. gambelii is a long-distance migrant, which breeds in Alaska and Canada and winters in the continental United States and Mexico, whereas Z. l. pugetensis populations may either be sedentary or shorter

distance migrants that move along the west coast of British Columbia and the continental United States. Thinking about the development of behavior, birds learn *songs. White-crowned sparrows learn and form a crystalized, or final, unchanging version of their song, early in life. Therefore they acquire dialects from individuals they hear during this developmental period, rather than dialects of birds at more distant locations. It is possible that variations in songs across these subspecies are due to chance, but they could also relate to different selective pressures due to differing environments and/or life history requirements.*

5. Cat.nr XC 353606 and Cat.nr. 335466 also sound quite different. Look at the *order* of the different phrases in these songs. Does it make sense to compare phrase 2 from Cat.nr XC 353606 to phrase 2 from Cat.nr. 335466? Why/why not? If not, how might you alter your analyses to more accurately compare these songs?
Phrase 2 sounds different in Cat.nr XC 353606 versus Cat.nr. 335466. Listening to both songs and looking at the spectrograms, phrase 2 in Cat.nr XC 353606 is more comparable to phrase 3 in Cat.nr. 335466 and phrase 3 in Cat.nr XC 353606 is more comparable to phrase 2 in Cat.nr. 335466. It would probably be most appropriate to make some determination about the order of phrases, such as which is a buzz, which is a whistle, etc., before determining which phrases to compare across different birds/populations.
6. Despite the fact that white-crowned sparrows learn songs from those around them, there is still some inter-individual variation in the songs within the same geographical location. Considering the adaptive function of bird song, why might a male white-crowned sparrow sing a slightly different song than its neighbor?
Differences in male songs may relate to its functions—to repel rival males and attract females. Sexual selection acts on the ability of male birds to produce songs attractive to females. Subtle differences to our ears may represent important cues to females about which males to choose as mates. Fitter males may produce more attractive songs.
7. In Cat.nr. XC253835 the last phrase differs across song iterations. What's different? Why might this be? White-crowned sparrows are thought to sing only one song type, do you still agree with this assessment? Go to xeno-canto and listen to a few songs from a related sparrow species known to sing multiple song types, the song sparrow, *M. melodia*. Does this change your mind about the number of song types sung by the white-crowned sparrow? Why/why not?
Across song iterations, one or Across song iterations in Cat.nr. XC253835, one or two syllables were dropped from the last phrase. Without more context, it isn't clear why this occurred. The singing male may have simply become distracted and arrested its song prematurely. Alternatively, such variation could be a signal to females or rival males (Nelson & Poesel, 2011). *Listening to the variety of song sparrow song types, it should become clear that white-crowned sparrows sing a single song type with regional dialects and only occasional*

individual changes. Compared with songbird species that sing multiple song types, these differences (e.g., dropping a syllable or two at the end) are small. Still, what constitutes a song type is qualitative, so one typically bases his/her opinion on the predominant view in the existing scientific literature.

Part III. Supplementary material data sheets

Supplementary data related to this article can be found online at https://doi.org/10.1016/B978-0-12-821410-7.00022-4

CHAPTER

Learning to be winners and losers: agonistic behavior in crayfish

9

Elizabeth M. Jakob[1], Chad D. Hoefler[2]

[1]*Biology Department, University of Massachusetts Amherst, Amherst, MA, United States;*
[2]*Biology Department, Arcadia University, Glenside, PA, United States*

Chapter outline

Exploring Animal Behavior in Laboratory and Field. https://doi.org/10.1016/B978-0-12-821410-7.00020-0

Part I. Student instructions

Learning goals, objectives, and key concepts

- To observe crayfish and to construct an ethogram of their behaviors
- To develop hypotheses about the role of experience in the aggressive interactions of crayfish and to test these hypotheses with staged fights
- To analyze data and to draw conclusions about whether the data support your hypothesis

Background

Many animals engage in aggressive, or agonistic, interactions with conspecifics. Animals may compete over food, territories, mates, or any other limited resource. Agonistic interactions may range in intensity from threat displays in which no animal is physically harmed to physical combat that may even result in death. How animals decide whether to escalate fights or to give up has been the subject of a great deal of experimental and theoretical research (e.g., Green & Patek, 2018). The outcome of agonistic contests depends on many variables, including the fighting ability of the contestants, their size and strength, and how much each contestant values the resource.

Here you will focus on the role of experience in determining contest outcome. Animals that have recently won aggressive interactions may often be more likely to win when they face new opponents—that is, winners keep winning. This phenomenon has been found in a variety of taxa, including juncos, chickens, paradise fish, red deer, and spiders (Jackson, 1991). Although this phenomenon is widespread, we focus on crayfish because they are large enough to observe easily, have interesting behaviors, and are easily kept in tanks of fresh water at room temperature. Much is known about what affects their fighting behavior: reproductive status (Figler et al., 1995), light intensity (Bruski & Dunham, 1987), body size (Pavey & Fielder, 1996), and previous contest experience (e.g., Goessmann et al., 2000; Momohara et al., 2013). Crayfish readily interact with one another, and interactions are generally easy to score. Other species might also be appropriate; your instructor will provide instructions.

Purpose

In this chapter, you will test the role of fighting experience in crayfish. We know that bigger crayfish are more likely to win fights (e.g., Pavey & Fielder, 1996) and that experience plays a role, at least in some species (e.g., Goessman et al., 2000). You will begin with crayfish that are matched in size. During the training phase, you will train one crayfish as a "winner" by allowing it to interact with smaller training partners that it is likely to defeat. The other crayfish will be trained as a "loser" by having interactions with larger training partners. In the test phase, your "winning" and "losing" crayfish will encounter one another.

Methods

Materials needed

You will work in a group of four. Each group needs two medium crayfish (which will be your test subjects) and two small and two large crayfish (which will be the training partners for the test subjects). The training partners can have multiple fights and can be shared with up to two other groups. If possible, every group should also have an extra pair of crayfish for pilot observations and constructing an ethogram.

Ideally, you'll have three types of containers: larger holding tanks or kiddie pools that you share with the class to keep crayfish before the experiment starts, isolation cups to keep crayfish alone before their fights so that they are calm, and two 20-gallon test tanks for developing ethograms and for observing crayfish interactions. Crayfish can live in tap water, but the water needs to be dechlorinated either with special droplets or by leaving it out overnight so don't add water straight from the tap into the containers. Every group should have a small dip net to capture crayfish. You will need a way to measure their size: plastic rulers, calipers, or a balance will all work, and your instructor will tell you which method your class will use. Nail polish is excellent for giving crayfish individual numbers.

Step-by-step instructions

Construct an ethogram, a catalog of species-typical behaviors, and decide on how to score interactions. For this task, you can use your extra pair of crayfish designated for constructing your ethogram if you were given these. If not, you can use one large and one small crayfish.

- Pick up a large crayfish by the posterior end of the carapace, pointing the chelae away from you, or use a net. Examine its ventral side to determine its sex. In adult males, you will see a pair of long claspers at the base of the tail (Fig. 9.1).
- Place it in an empty test tank and observe its motion. How does it move its chelae, antennae, and walking legs as it explores its tank? Does it generally move forward, backward, or sideways?
- Read through the behaviors in the sample data sheet. These provide a starting point for scoring behavioral interactions.
- Add a small crayfish to the tank. How do the crayfish interact initially? How do they use their chelae in interactions? Look for the behaviors listed on the sheet. Do you observe both dramatic behaviors, such as tailflips and chelae strikes, and more subtle ones, such as avoidance?
- In your group, finalize your data sheet. You may decide that some of the behaviors listed in the sample sheet are difficult to see and need to be deleted or

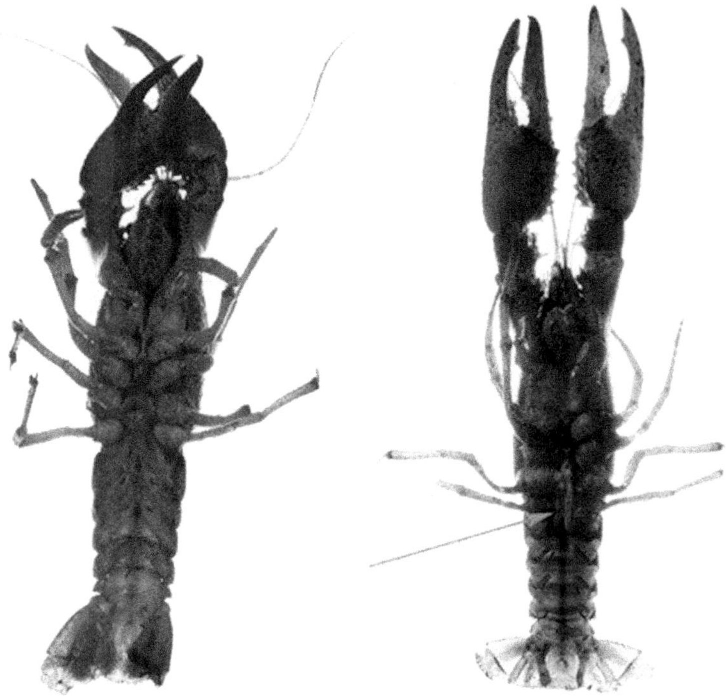

FIGURE 9.1

The dorsal sides of a male and female crayfish. The arrow points to the modified swimmerets used to transfer sperm in males.

modified. You may also observe additional behaviors not listed there. If so write descriptions that are so clear and unambiguous that anyone can use them. Remember that the different behaviors should be mutually exclusive.

- Work together with the rest of the class in order to come up with a common ethogram.
- Decide how you will determine the winner of a fight. Different definitions are possible, as long as you agree on a standard method. One idea is to decide that a crayfish has lost a fight when it performs a tailflip, which previous research has established as a submissive signal (see the sample data sheet for the definition). Often, however, crayfish do not always perform this behavior (e.g., Figler et al., 1995) so you might need another way to score trials. Another suggestion is to determine in advance the duration of each trial (10 min works well) and score all the behaviors that you see, such as avoidance, threats, and tailtucks. You can then establish a winner and loser by adding up the number of times different behaviors were performed. After you discuss the scoring within your groups, you can decide with the rest of your class on a common approach.
- Decide, as a class, what you will do if a "winning" crayfish does not win all of its training trials or if a "losing" crayfish does not lose all of its training trials. You may decide to stop at three trials regardless of their outcome or to add additional

trials until a certain number of successful trials are reached. Both methods have their advantages and disadvantages.
- Finalize the data sheet that you will use to take notes during the interactions.

Next, you can work with your group to develop a secondary hypothesis. The primary hypothesis that your class will test is that crayfish trained to be winners will be likely to win fights against crayfish trained to be losers. The data your class collects can also be used to test additional hypotheses. For example, "winners" may not win all their training fights—if this is the case in your class, what prediction might you make about the relationship between training experience and the likely outcome of the fight? If you measure both the weight and length of your crayfish and find that they are not perfectly correlated, what might you predict? What might you predict about a crayfish that initiates fights? Develop a secondary hypothesis and a prediction before you begin the next step.

Run the experiment as follows (Fig. 9.2).

- Place crayfish in separate containers if they are not already housed individually. If they are not already marked, then dry their carapace with a towel and use nail polish to number them on their backs, being careful not to get polish into their joints. Use different colors for males and females so you can quickly tell them apart.
- If your instructor has not already set up pairs of crayfish, then measure your crayfish using the method(s) your instructor has chosen and match the medium-sized crayfish in pairs. Pair males with males and females with females. If any crayfish are missing an appendage, pair them with similarly impaired animals, if possible. If any of the females are carrying eggs beneath their body, then pair them with other egg-carrying females.
- Your group of four students should get a pair of medium crayfish and, using a coin flip or a random number generator, assign one to be trained as a winner and the other as a loser. Two students will train the winner by giving it three fights against smaller crayfish, and two students will train the loser by giving it three fights against larger crayfish. For best results use crayfish that differ in size by 10% or more.
- Begin the training trials by placing two crayfish simultaneously in the opposite ends of the testing tank, oriented toward one another. Crayfish can be very

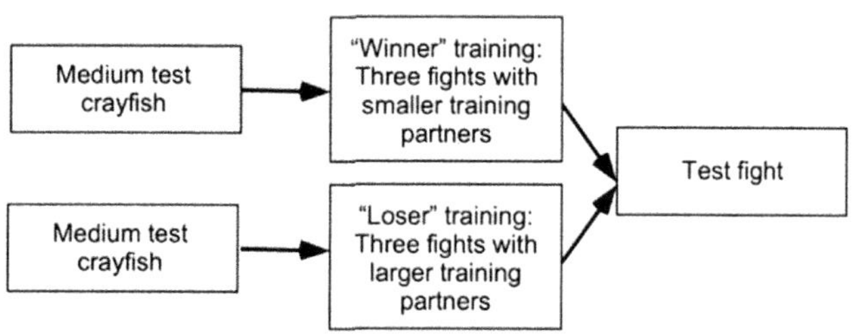

FIGURE 9.2

The design of the experiment. Two crayfish matched in size are given different fighting experiences before encountering each other for the first time.

sensitive to your motion so minimize your movements and avoid causing vibrations on the table top. One member of the pair of observers should focus on the interaction and call out the behaviors as they occur. The other member of the group should take notes using the data sheet your class has developed and keep track of time.

- Finish the training trials according to the plan that the class has developed.
- Run the test trial in your group of four by staging an interaction between the two size-matched crayfish that received different training regimes.

Results/discussion

Consider the principal hypothesis that crayfish trained as winners are more likely to win the test fights against losers. Collect the data from the class on the board.

Number of test fights that "winners" won ______________.

Number of test fights that "losers" won ______________.

Test the class data against the null hypothesis that "winners" and "losers" would each win about half the fights. Appropriate tests include a binomial test or a chi square goodness-of-fit test; calculators are available online.

Did training have a significant effect on the outcome of test fights? Was it in the expected direction? If not, then hypothesize why this might be so.

Next, you can consider the secondary hypothesis that your class developed and collect the data on the board. Do your data support this hypothesis?

For further discussion

1. It may seem maladaptive for the behavior of animals in fights to be influenced by previous experience: animals may withdraw from fights that they could have won. Why do you think this phenomenon exists in such a wide variety of species?
2. In an Australian species of crayfish, winner effects were only seen in repeated fights with the same individual but were not seen when winners fought with new individuals (Seebacher & Wilson, 2007). Why would it be especially valuable to settle fights quickly with familiar individuals?
3. Did your ethogram suffice, or did you see new behaviors as you carried out the trials? If you were advising investigators on how to begin a behavioral project, how would you suggest that they determine how much time to invest in observing the animals before starting data collection?
4. Another well-known influence on the outcome of aggressive interactions is resource ownership. For example, in marbled crayfish, there is a residency effect: crayfish that have had access to a shelter are more likely to win against an intruder and can even defeat larger intruders (Takahashi et al., 2019). Design an experiment that tests the relative importance of *three* variables on the outcome of crayfish fights: body size, residency, and experience. What would be a disadvantage of this approach?

References

Bruski, C. A., & Dunham, D. W. (1987). The importance of vision in agonistic communication of the crayfish *Orconectes rusticus*. I. An analysis of bout dynamics. *Behaviour, 103*, 83–107.

Figler, M. H., Twum, M., Finkelstein, J. E., & Peeke, H. V. S. (1995). Maternal aggression in red swamp crayfish (*Procambaraus clarkii*, Girard): The relation between reproductive status and outcome of aggressive encounters with male and female conspecifics. *Behaviour, 132*, 107–125.

Goessmann, Cl, Hemelrijk, C., & Huber, R. (2000). The formation and maintenance of crayfish hierarchies: Behavioral and self-structuring properties. *Behavioral Ecology and Sociobiology, 48*, 418–428.

Green, P. A., & Patek, S. N. (2018). Mutual assessment during ritualized fighting in mantis shrimp (Stomatopoda). *Proceedings of the Royal Society B, 285*. Article 20172542. https://doi.org/10.1098/rspb.2017.2542.

Jackson, W. M. (1991). Why do winners keep winning? *Behavioral Ecology and Sociobiology, 28*, 3497–3506.

Momohara, Y., Kanai, A., & Nagayama, T. (2013). Aminergic control of social status in crayfish agonistic encounters. *PloS One, 8*(9). Article e74489. https://doi.org/10.1371/journal.pone.0074489.

Pavey, C. R., & Fielder, D. R. (1996). The influence of size differential on agonistic behaviour in the freshwater crayfish, *Cherax cuspidatas* (Decapoda: Parastacidae). *Journal of Zoology, 238*, 445–447.

Seebacher, F., & Wilson, R. S. (2007). Individual recognition in crayfish (*Cherax dispar*): The roles of strength and experience in deciding aggressive encounters. *Biology Letters, 3*, 471–474.

Takahashi, K., Yamaguchi, E., Fujiyama, N., & Nagayama, T. (2019). The effect of shelter quality and prior residence on marmokrebs (marbled crayfish). *Journal of Experimental Biology*. Article 222jeb197301. https://doi.org/10.1242/jeb.197301.

Part II. Instructor notes

Classroom management

This lab can be completed over the course of one or two lab periods. If you choose to do two lab periods, then spend the first lab measuring the crayfish, establishing pairs, and developing hypotheses. For best results, training and testing the crayfish should be done in the same lab.

Teaching the activity

Preclass preparation

Crayfish are not under federal animal care guidelines, so the Institutional Animal Care and Use Committee (IACUC) approval will not be needed.

We have worked mostly with *Procambarus clarkii*, but other species may be used. Crayfish can be captured from local streams with bait or in traps; different methods can be easily found online. Live crayfish can be ordered from several biological supply houses (e.g., Carolina Biological Supply) as well as from restaurant supply houses, although we are not experienced with the latter. Note that crayfish undergo an annual molting cycle, so availability of particular species may vary over the year. Note that many organizations will not ship live animals over the weekend.

Note that crayfish released into nonnative ranges easily become established and wreak havoc with native species. If you order crayfish from a supplier, do not release them into local streams. Crayfish can be kept in the lab room as lab pets or humanely euthanized in the freezer.

Carolina Biological Supply offers a free online guide to crayfish care. Crayfish can be held together in large tanks or kiddie pools prior to the lab. Please note that crayfish do fine in tap water, but the water must be dechlorinated either through the use of tablets or by letting it stand overnight. We recommend supplying air bubblers and a filter to each holding tank, if possible. Crayfish tanks get fouled quickly, and water should be partially drawn off and replaced frequently. In general, we recommend that if no filtration system is being used, 30%–50% of the water should be changed once per week. If a filtration system is being used, we advise that 20%–25% of the water should be changed weekly. Crayfish are omnivorous and will eat fish food, bits of frozen fish, or bits of earthworms, among other things.

If you have a short class period (e.g., 2 h), it will be best to measure and number the crayfish yourself before the class meeting. Crayfish perform best if they have been isolated for at least 4 h prior to the experiment. For isolation, inexpensive large deli cups with lids work well; there is no need for individual tanks.

In-class preparation

Every group of students needs one pair of medium crayfish of the same sex. In addition, each group needs access to two large and two small training partners, preferably of the same sex. Training partners can be shared with one or two other groups. The training partner crayfish can have multiple fights with different test crayfish. Reusing training partners helps keep down the number of crayfish required, and in fact, as the small crayfish consistently lose fights and the large crayfish consistently win fights, they become more and more reliable at providing the desired training outcome.

If time allows, students can measure and pair up crayfish during the lab period or they can do it ahead of time. In case of the latter, for best results, separate the crayfish into individual containers prior to the lab, as described in the Student instructions.

If you don't have enough of one sex or the other to create pairs matched by sex, we have found that males and females will often fight one another.

Data for the primary hypothesis can be easily tested with online statistical calculators, as described earlier. We have had results where the data have both unambiguously supported the hypothesis and unambiguously not supported it. Often, sample size is too limited for a strong test, but this can lead to a valuable discussion about power.

Answer key for discussion questions

1. *It may seem maladaptive for the behavior of animals in fights to be influenced by experience: animals may withdraw from fights that they could have won. Why do you think this phenomenon exists in such a wide variety of species?*
   Possible answers: There are both ultimate and proximate answers to this question. The widespread presence of the effect of experience suggests that it is *generally* predictive of what will happen in the future and that animals that follow this strategy will have greater fitness than animals that do not. At a proximate level, fight experience causes physiological changes in neuronal and hormonal states, and these effects may be lasting and may continue to affect behavior.
2. *In an Australian species of crayfish, winner effects were only seen in repeated fights with the same individual but were not seen when winners fought with new individuals (Seebacher & Wilson, 2007). Why would it be especially valuable to settle fights quickly with familiar individuals?*
   Possible answer: The outcomes of repeated fights with the same individual are likely to be similar, so it is not to the loser's advantage to keep challenging the winner. We see often that when animals encounter each other repeatedly, aggressive behavior is minimized; for example, once dominance hierarchies are established in chicken flocks or once neighboring male birds establish boundaries between their territories, the rate of aggressive interactions tends to decrease. It can spike up again if a new, unfamiliar rival enters the scene.

3. *Did your ethogram suffice, or did you see new behaviors as you carried out the trials? If you were advising investigators on how to begin a behavioral project, how would you suggest that they determine how much time to invest in observing the animals before starting data collection?*
   Possible answer: Usually, when we start to study a species new to us, we see lots of unfamiliar behaviors. One way to know that you are ready to start collecting data is if you see new behaviors only rarely.
4. *Another well-known influence on the outcome of aggressive interactions is resource ownership. For example, in marbled crayfish, there is a residency effect: crayfish that have had access to a shelter are more likely to win against an intruder and can even defeat larger intruders (Takahashi et al., 2019). Design an experiment that tests the relative importance of three variables on the outcome of crayfish fights: body size, residency, and experience. What would be a disadvantage of this approach?*
   Possible answer: An obvious answer would be a balanced design in which every possible combination of variables was tested. A disadvantage is that the sample size needed would increase dramatically. If you have discussed statistical power with your students, this would be a place to reinforce the concept.

Optional extensions

- In the week or two before the lab have the students read the primary literature on aggression, particularly aggression in crustaceans. In particular, much is known about neuronal and hormonal bases of aggression that might provide the foundation for secondary hypotheses.
- Provide ample time to develop one or more secondary hypotheses. More advanced classes should be able to come up with some good ideas and develop robust statistical approaches.

Part III. Supplementary material

Supplementary data related to this chapter can be found online at https://doi.org/10.1016/B978-0-12-821410-7.00020-0.

CHAPTER

Love is blind: investigating the perceptual world of a courting parasitoid

10

Robert W. Matthews, Janice R. Matthews

Department of Entomology, The University of Georgia, Athens, GA, United States

Chapter outline

Exploring Animal Behavior in Laboratory and Field. https://doi.org/10.1016/B978-0-12-821410-7.00007-8

Part I. Student instructions

Learning goals, objectives, and key concepts

- To observe and collect data on insect sexual behaviors using a parasitoid wasp as a model
- To identify, define, and quantify male and female behaviors during attraction, courtship, and mating
- To experimentally assess the nature of communication cues used to initiate and maintain courtship
- To discuss the implications of experimental design in bioassay development

Background information

For nearly all sexually reproducing animals, mating with appropriate members of the opposite sex to produce viable offspring is critical to population survival. In its broadest sense, **courtship** can be defined as all behaviors between a male and a female used to obtain copulation or to maintain reproductive interactions with an existing partner. Typically, courtship includes reciprocal signals that allow the male and female to ensure that they will ultimately mate with an animal of the same species, opposite sex, and suitable physiological condition. Thus courtship could also be defined as the set of the precopulatory behaviors that allow both sexes to make all those critical decisions that are involved in the word "appropriate."

Courtship also provides an opportunity to present information about one's quality as a potential mate. A courtship **display** is a set of more or less stereotyped behaviors in which an animal, usually a male, attempts to attract a mate; the mate exercises choice, so sexual selection acts on the display. This issue is complex. There is no simple answer to the question of which potential mate is "best."

Scientific knowledge of courtship behaviors has implications for conservation, management, and control of animal populations. However, among the Arthropoda, the largest group of animals, courtship is both incompletely studied and undeniably diverse. Some insect species exhibit almost none. Others practice complex and lengthy rituals that involve a mixture of visual, sound, chemical, and motion cues, from the visual cues of firefly flashes and the auditory cues of cricket chirps to the communicatory chemical cues (**pheromones**) by which moths attract one another.

Members of the Hymenoptera—an important group that includes the bees, wasps, and ants—display many of the most complex and highly evolved insect behaviors. Like members of many other insect groups, Hymenoptera males often perform elaborate species-specific courtship displays. Visually evident components of their repertoire generally include repeated, stylized movements of the wings, legs, antennae, and mouthparts.

Little wasps known as *Melittobia digitata* or WOWBugs are **parasitoids**—one of a large group of organisms that live in close association with their host and at the host's expense, which sooner or later kill it, but are capable of independent living at some point in their life cycle. Around the world, *Melittobia* raise their young upon other insect species belonging to several different orders, including bees and wasps, beetles, and flies. Being wasps, female *Melittobia* do have tiny stingers, but they cannot penetrate human flesh. Modified ovipositors, the stingers are used only upon insect larvae and pupae that will serve as hosts for their progeny.

Males make up only a tiny proportion of the *Melittobia* population, but by their behaviors, they quickly reduce their numbers even further. Male *Melittobia* are highly aggressive toward their brothers and will eagerly use their legs and jaws to battle to death with these potential rivals (as male *Melittobia* possess no stingers, they are also harmless to humans.). The survivors of these brother-to-brother battles never leave the darkness of the host cocoon that was their birthplace. For the brief week or so that comprises their entire adult life, they repeatedly court and mate one sister after another.

Unlike their brothers, however, females, which make up over 95% of each generation, disperse after mating to undertake a hazardous journey to search for new hosts. Upon locating a suitable host, a female (now called a **foundress**) punctures it and feeds. She must consume some host fluids from the sting site, almost always the only meal of her adult lifetime, to obtain the nutritional resources to mature her eggs. A short time later, she deposits hundreds of eggs on the host's body surface. Within 17–30 days, these eggs will mature to adults that are able to breed and repeat the cycle of brother–sister mating within the closed environs of an intact host cocoon.

The result of this system is a highly inbred population of closely related individuals, yet courtship within the genus *Melittobia* is surprisingly complex. In some of the 12 recognized species, courtship sequences are simple, composed of a single motor pattern or a series of identical motor patterns that follow each other in rapid succession. In other species, a number of different motor patterns occur in a highly predictable sequence.

Purpose

Using both observation and experimentation, you will observe, describe, and quantify *Melittobia* behavior, paying special attention to male and female interactions. You will begin by establishing the attraction of one sex to the other. Then you will observe and describe behaviors involved in courtship and mating. Finally, you will conduct a bioassay designed to experimentally assess the type of cues in use.

Your observations should also help you generate ideas about what sensory cues male and female *Melittobia* might use to find and identify one another. In the course of these activities, you will also become familiar with the terminology and methodology commonly used by researchers in such studies. You will work in pairs, dividing the tasks of observing, recording, and timing, and swapping roles whenever practical. Each partner is responsible for all aspects of the inquiry.

When all three parts of this activity have been completed, you will analyze your own data and the data obtained by the class as a whole. With guidance from your instructor, you will compile appropriate descriptive statistics, answer discussion questions that place these results in context, and develop a flowchart that visually depicts the numerical outcomes of your class data.

Methods and materials

Species selection

This activity introduces a little parasitoid insect, *M. digitata*, commonly called the WOWBug. It is about the same size as the common fruit fly, *Drosophila*, and it has the same advantages of rapid life cycle and production of large numbers of adults. Unlike *Drosophila*, however, the sexes of *Melittobia* are so different that they are easily recognizable even to the naked eye and WOWBugs rarely fly, preferring instead to slowly crawl. Best of all, their courtship rituals are elaborate and can be easily observed, and if simple procedures are followed, they can be dependably elicited.

The following are the materials required for both activities (per pair of students working together):

- two deep-well slides that contain one male *Melittobia* + one female *Melittobia* pupa or adult;
- two empty deep-well slides;
- two one-dram glass vials—one with 10 unmated female *Melittobia* adults and the other with 20;
- a dissecting microscope with cool fiber-optic light (or a high-quality magnifying glass);
- a timing device such as cell phone, watch, timer, or stopwatch;
- six empty one-dram glass vials (for transfers and recaptures);
- six cotton balls to use as stoppers;
- one pipe cleaner, chenille stick, or cotton swab (to pick up and move *Melittobia*);
- one sheet of unlined white paper (8.5 × 11 inches);

one small piece of graph paper (5 squares/inch), about 3 inches square;
one bioassay choice chamber;
eight gelatin capsules (size 0);
one small ruler;
one ultrafine point black permanent marker (such as Sharpie brand);
one translucent, red 6- to 10-ounce "party" cup;
one quilting straight pin;
one small piece of poster-mounting putty.

Part 1. Observing interactions

Obtain your materials and set up the fiber-optic light so that it illuminates your microscope stage. You will begin by familiarizing yourself with the experimental organisms. Then you will undertake a series of observations to determine the nature of sexual attraction in *Melittobia*.

Sex identification

Place the vial containing 10 unmated female *Melittobia* on the stage, adjust the magnification, and focus and observe for a few moments. Choose one to be your **focal animal** to observe more closely. Describe and record her body color, wing and eye form and size, and antennae structure.

Set this vial aside and place the deep-well projection slide on the stage. It contains either two adult *Melittobia* or an adult with a pupa. Readjust the focus and observe for a few moments. How do the sexes differ? Describe the male's body color, wing and eye length and form, and antennae structure.

Wasp wrangling techniques

Transferring *Melittobia* from one container to another without damage or escape is easy with a few simple techniques. To make the dark-colored insects easier to see, just place the sheet of unlined white paper on your work surface. To gently pick up an individual or move it about, you can use a soft cotton swab or pipe cleaner rather than fingers, forceps, or a pin.

Use the 10-female vial as a source for small numbers of wasps; save the 20-female vial for the bioassays in Part 3. Turn the bottom end of the vial upward and toward the light source. Females will crawl upward, away from the cotton stopper. Carefully remove it. Reach into the vial with the pipe cleaner or swab. With a twisting motion, carefully pick up one female on the cushiony surface. When you have carried the wasp to its new location, allow it to step off, or gently tap or swipe the swab.

Keep the source vial upended to help prevent others from attempting to escape. Replace the cotton stopper in the vial promptly and quite firmly (WOWBug females will burrow out through cork or loose cotton.). Should extra females leave the vial during this transfer or at any other time, then stay calm. They will drop and hop or crawl rather than flying. Simply cover them with an inverted empty vial. They will crawl upward and remain captive until you can pick them up and return them back to the rest.

Initial attraction, baseline activity, and latency

Before observing courtship interactions, we need to determine which sex will be the attractor and which will be the attracted. Because attraction is expressed as a motion, a simple way to find out is to compare the activity levels of each sex, both separately and together.

It would also be helpful to know when to expect to see courtship begin. When they are placed in a new environment, most organisms go through a stage of exploration and adaptation (called a **latency period**) before they will begin sexual activity. In other words, after being moved, it takes them a bit of time to "settle in." As a simple measure of latency, we can record the actions of solo wasps for the first 6 min after they are placed in new surroundings and then compare what happens in the next 6-min period.

Place an empty deep-well slide over your piece of graph paper, aligning it to provide as many full squares as possible under the circular arena. To be able to replace the slide back in the identical position each time you have lifted it up, you can scribe a line on the graph paper along the slide's outer edge.

How many complete and partial squares are covered by the slide's observation arena? For data collection in your notebook draw six circles, each 2–3 inches in diameter (A simple way is to trace around the rim of an inverted bottle or cup; the exact size is not important.). Then using the ruler draw equidistant grid lines in each circle to match the pattern showing through the observation arena (Fig. 10.1). Label one circle "female exploration, initial 6 min." Label the second circle, "female adaptation, min. 6–12." Label the third, "male exploration, initial 6 min." Label the fourth, "male adaptation, min. 6–12." Label the fifth, "two-female trial, focal wasp activity, 6 min." Label the final circle, "male-female trial, focal wasp activity, 6 min."

For the first four trials, you and your partner will be running your own two motion-tracing trials. Decide which of you will watch which sex first. Then quickly swap roles for the next pair of trials so that each of you have data for a male and for a female wasp. For the last two trials, you will work together, each watching one focal insect when a potentially interacting pair occupies the same deep-well slide.

- **Female baseline activity: initial exploration**. Remove the lid from the deep-well slide. Choose the 10-wasp-filled shell vial labeled "unmated females" (Temporarily set the 20-wasp vial aside for later use.). Using a cotton swab or

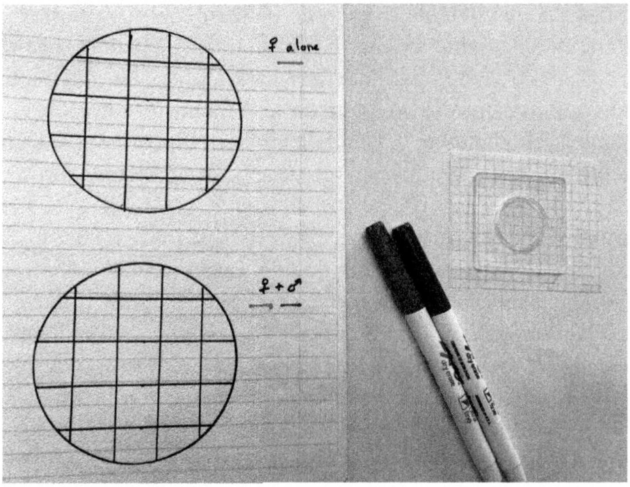

FIGURE 10.1

Tracking arena set up to observe and record activity levels of female and male *Melittobia*.

pipe cleaner, obtain a single unmated female from the stock supply (see the section Wasp wrangling techniques). Place her in the slide and replace the lid. Pick up your pen or marker and start your timer for 6 min. Locate the wasp's position on the grid under the slide and find this same point on your gridded circle 1. As you continuously watch the wasp, simultaneously move your pen or marker to trace her path on your drawn circle. She may crawl on the floor, side, or ceiling; for simplicity we will treat all three substrates as though they were the same.

- **Male baseline activity: initial exploration.** Carefully open the deep-well lid and gently pick up the male wasp on the fuzz of the cotton swab or pipe cleaner (For the moment, leave the female wasp or pupa in place.). Transfer him to the other empty deep-well slide. Run a 6-min trial using the same protocol as for the female trial (abovementioned). Trace his actions on circle 2.
- **Adaptation.** At the end of 6 min, quickly switch places with your partner so that each of you is observing the slide with the opposite *Melittobia* sex from your previous trial. As before, observe and record the wasps' actions in the deep-well slide, tracing their paths for the next 6 min on the appropriately labeled circle 3 or 4.
- **Interactive effects on activity**. Suppose your focal animal were not alone. Intuitively, we would expect that placing a male and female together would change our results. But can we attribute our results solely to sexual attraction? Or would the simple presence of another wasp similarly affect baseline activity

level? Transfer a new female from your 10-wasp vial to join the female from trial 2. Work together with your partner, each tracing a path for one wasp as their focal organism (Following two identical-looking females and tracing two activity paths simultaneously would be quite difficult for one person working alone!). Make a small dot or "x" on the path each time they contact one another. Does the presence of the other female wasp affect the behavior of your focal individual?

- **Sexual attraction.** Return to the deep-well slide that you set aside during the male-only trials. Gently pick up the female wasp or pupa you left inside, place her in an empty shell vial, and close the vial with a tightly packed cotton plug. Label the vial "status?" (Chances are good that if there were already two adults in the slide earlier today, they have already mated. However, you can't be sure.). Into the now-empty deep-well slide transfer in the male from circle 3 and 4 trials and one of the two females used in circle trials 1 and 2. Choose focal individuals. As before trace the path of your wasp over 6 min while your partner traces the path of their wasp. Make a small dot or "x" on the path each time the wasps contact one another.

Because you have placed both wasps into a new environment, their actions will probably reflect a latency period, but you may also see a number of initial tentative antennal contacts. However, if you see the male climb onto the female's back, immediately lift the deep-well slide off the paper grid and place it onto the microscope stage (or under the magnifying class) and move along to the next step (see the following).

Part 2. Observing *Melittobia* sexual behaviors

Begin by watching a short video of *Melittobia* courtship and mating (https://www.youtube.com/watch?v=qDa_C853t34&t=11s). Notice that a central feature of WOWBug courtship is a stereotyped sequence of interactive behaviors (**courtship bout**) that generally proceeds without female interruption until the female signals either willingness to copulate or a refusal to cooperate (Fig. 10.2). If the female does not give the proper signals for mating, the male may leave but often he simply starts over, beginning another bout.

As is true with "nature shows" in general, the video makes it appear as though courtship is a very straightforward affair. Is life really that simple? When male and female *Melittobia* find each other, how often do they actually court? What proportion of courtships ultimately result in mating? Do male and female wasps give equal input into the decisions as to whether to continue?

Take a moment to consider how you will sample *Melittobia* sexual behaviors and what behavioral markers you will use to determine that courtship has begun and has ended. Decide upon a standard format for data collection. If class members are generally familiar with Microsoft Excel spreadsheets, setting one up now will facilitate statistical calculations.

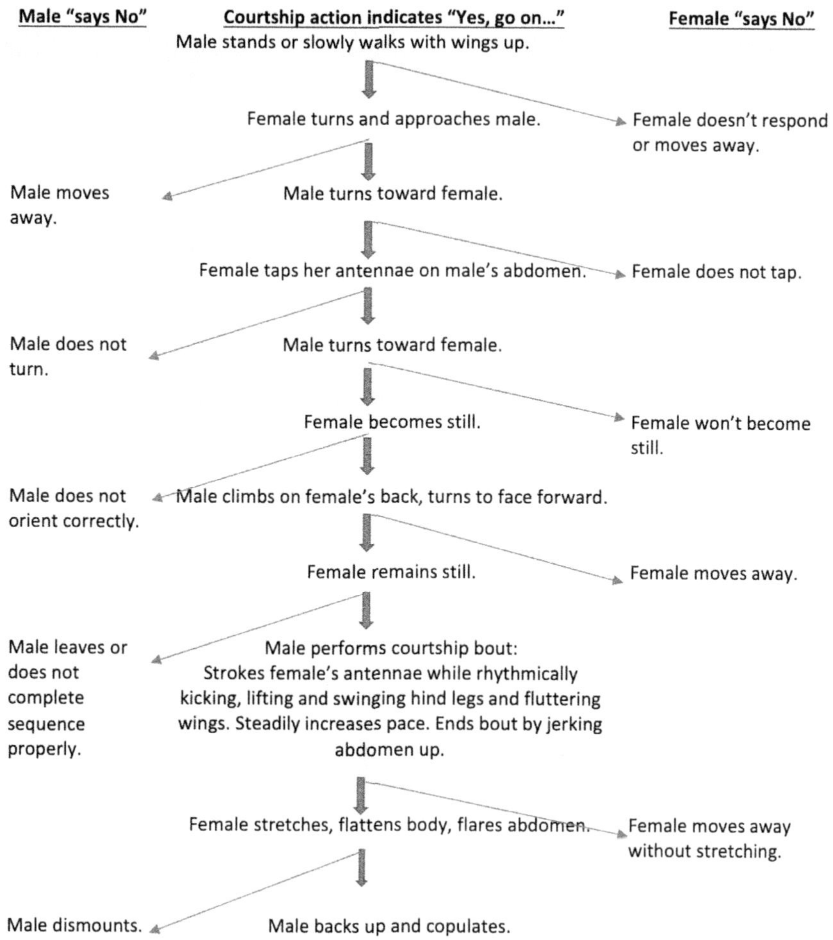

FIGURE 10.2

Reciprocal interactions during *Melittobia* courtship. Arrows link each action with possible responses. Note that unlike behaviors during latency and attraction, courtship bouts are of consistent duration, run a stereotyped uninterrupted course, and may be repeated.

- **Courtship between males and unmated females.** Working over the plain white paper again remove the lid of the deep-well slide and add four more unmated females from the 10-wasp vial to the male–female pair already in the slide. Replace the lid and move the slide onto the microscope stage. Start your timer. While one person watches the wasps nonstop (a method called **ad libitum sampling** or **continuous sampling**) and gives a running commentary on their actions, the other should note the time and record the observed behavior.

Consider a courtship to have begun if the male climbs upon the female's back. When the pair contact one another note the orientation of the male and female, the movements of their body parts, and the sequence in which these movements are performed. Pay particular attention to repetitious patterns of the antennae and legs.

Watch for 10 min. Then remove the male and place him in a gelatin capsule (you will use him in Part 3.). Replace him by transferring the male from the other deep-well slide to the slide with the females. Switch partners to watch for 10 min more. When you have finished, gently transfer these five females into the "status?" vial. Leave the second male in place.

Part 3. Determining courtship attraction cues

Now let's back up a moment. For courtship to even begin, the sexes must find each other. Within the cocoon of their host, *Melittobia* live amidst hundreds of others of their kind. How do the relatively few males and the more numerous unmated females first locate and recognize each other? Are males and females simply bumping into one another at random in the dark? The video mentions a sex pheromone. Is this all that is involved? Or might attraction also involve other cues such as movement, sound, or even a second pheromone? (Given the darkness inside the cocoon, simply seeing one another seems unlikely. Moreover, the males completely lack compound eyes.)

To experimentally investigate these possibilities, you will perform a **bioassay**, an analysis that uses the behavior of living organisms to measure the response to a variable. To conduct it, you will use a **choice chamber**, an experimental apparatus in which a test organism is offered a choice of stimuli that may have significance in its biology, and its reaction to the various stimuli is measured. By its behavior in a bioassay, the insect tells the scientist which of the offered choices, if any, is important and relevant to it.

Prepare the choice chamber

Examine the bioassay choice chamber that you will be using today (Fig. 10.3). Insects to be tested will be placed in the central chamber, from which they are free to move up into four gelatin capsules, allowing simultaneous measurement of the test insect's response to three choice conditions (stimuli) and a control.

- Holding the plastic container in one hand grasp a gelatin capsule between the thumb and index finger of the other hand. Carefully insert the longer, narrower end of the capsule into a hole so that the rounded capsule tip slightly protrudes into the inside of the container. It should fit snugly in its hole and be flush against the sides of the canister. Repeat this for each of the other three holes.
- Pinch off a bit of poster-mounting putty and roll it with your fingers on the palm of your hand to form a short "string" about the thickness of a toothpick.

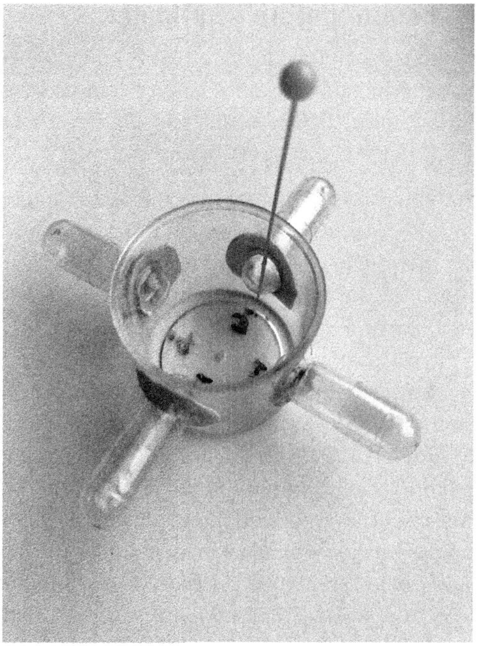

FIGURE 10.3

A bioassay chamber showing the position of a quilting pin to pierce two holes in the gelatin capsule spoke. Note the poster-mounting putty around the base of each capsule to ensure tight seal.

Carefully place the string of putty around the outside of each of the four gelatin capsules where they connect to the bioassay chamber, tamping it in place with your fingernail or pencil tip. You should now have a setup in which the narrow part of the capsule is secure, but the outer part can be easily taken off, filled, and replaced.

- To allow stimuli to pass from the capsules into the chamber pierce a quilting pin once entirely through each capsule, guiding the pin's shaft through the capsule along the **inside** wall of the chamber to produce two holes, one on the top of the capsule and the other on the bottom (see Fig. 10.3).
- The holes in the chamber body are numbered on the container bottom, which will become the top side when the chamber is inverted for use. To minimize confusion during and between trials, whenever you are asked to load or replace capsule ends remove only one at a time. Do not number the capsules themselves, as doing so may obscure your view of any wasps inside.

Before beginning each trial check the bioassay chamber with a magnifying glass or under low power of the dissecting microscope to ensure that everything is in place as prepared.

Identify, test, and control possible variables

One of the marks of good research is the care that is taken to be sure that the results are actually from the experimental variables being examined. Suppose nothing were placed in any of the capsules. What would you predict about the choices that a female *Melittobia* would make? Let's test the prediction.

- Open the lid of the bioassay chamber. Open the vial that has 20 unmated female *Melittobia*, invert them over the chamber, and flick a finger sharply against the side of the vial so that the insects all fall to the bottom of the test chamber at once. Quickly replace the lid and place the chamber, lid side down, on a sheet of white paper. Start your timer.
- While you allow the wasps 6 min to acclimate to their new surroundings, you can decide how you will measure their behavior. One method might be to count each time an insect enters any of the capsules even if it leaves before the end of the time period, recording a tally for each capsule (Noting events each time they happen while scanning an entire group is called **all-occurrence sampling.**). An alternative method might be to divide the observation area in quarters by drawing crossed lines on the white background under the chamber. Once every minute, you should count the number of individuals found anywhere in each section (Data collection at fixed intervals is "on the dot" or **instantaneous sampling.**). Decide which methodology you and your partner will use.
- For the following 6 min use the sampling method you chose. When you are finished, you can add your team's data to a master chart for all class results. How did your results compare with those of other teams? Did the same number of insects enter each capsule (or was the average number present in each quadrat similar)? Determine the means and standard deviations for the class totals. Were class results significantly different from random? If not, you can proceed.
- If they did differ from random, what variable may not have been controlled? Determine a way to control for it (Hint: It might involve the red cup.). To determine whether your control is effective run a second trial using instantaneous sampling. Compile class data and run the same statistical measures. Did you succeed in controlling the variable? If so use the control during the bioassays. If not consider other possible variables and devise tests for them.

Standardize terminology and process

Think about what you and your classmates have already learned about *Melittobia*. The clues are there. Share your ideas. Working together state your prediction(s) in the form of a general hypothesis about the cue(s) thought to be possibly important in initial sexual attraction. Smell, movement, and sound are obvious choices. The independent variable would be the choices presented in the capsules. Because females are so mobile, the dependent variable could logically be the average number of wasps per spoke location. Instantaneous (on the dot) sampling at 1-min intervals would provide straightforward methodology. If you previously identified any uncontrolled variables remember to employ the additional corrective techniques you found were necessary.

Reaching a consensus on sampling, measuring, and reporting methods is also important. For example, it's easy to count the number of insects entering each clear capsule but over what time interval? When will you begin timing? Working together refine your hypothesis in a more specific, measurable form.

Test potential attraction cues

You will use the same choice chamber for these bioassays, but now you will be placing stimuli in the gelatin capsules. What should they be? It would be logical to put a male in one capsule and to leave another one empty as a control, but what about the other two? How about a female? Or how about two females, one unmated and the other previously mated, each in a separate capsule? How might that provide useful information? If you listened closely to the narrative on the video, there were clues … talk with your partner and try to puzzle them out.

The 20 females from the previous trial can remain in the central chamber while you load the capsules. If any wasps are still in a capsule when you go to load it, then return them to the central chamber. We are using a high number of wasps in these bioassays to partially compensate for the relatively short observation periods. If you feel that you have lost a wasp or two, then note that fact but continue the experiment. The actual number used will be confirmed by a recount at the end of the experiment.

- **Bioassay A**. Capsule 1 should remain empty as a control. Transfer the live male *Melittobia* from your deep-well observation slide into capsule 2. In capsule 3 place a live female from a "mated" vial (either your own or one provided by the instructor). In capsule 4 place an unmated female from your "unmated" vial.

Place the bioassay chamber, lid side down, over the white paper on the table. Start your timer, setting it to ring every minute. For the next 12 min, each time the timer beeps record the number of females visible in each of the spokes of the bioassay chamber.

- **Bioassay B**. Remove the contents of each capsule in turn, returning those wasps to their respective containers. Place the now-empty capsules back in their correct color-coded positions on the central chamber. If the central chamber now contains fewer than 10 unmated female wasps, then add the necessary number of replacements from the appropriate shell vial. Record instantaneous (on the dot) samples at 1-min intervals for 12 min, as before.

When both bioassays have been completed transfer all the unmated female responders back to their vial. Open the choice chamber over the piece of white paper. Tap the side and/or shake it gently. As before, you can place the open end of the vial over individuals or groups. Once a covered wasp begins to move up the vial pick up the vial and place it over another wasp. Count the wasps as you transfer them back into their vial. When all have been gathered plug the vial securely with a cotton stopper.

Discard all four gelatin capsules. Return the wasp-filled vials, the males in their deep-well slides, the choice chamber, and all your other supplies to the location from where they came.

Part 4. Results and data analysis

In Part 1, after you learned to recognize *Melittobia* and handle them, you traced their paths to compare male and female activity levels. In Part 2, you watched interactions between males and females to investigate the extent and nature of courtship. In Part 3, you ran bioassays to determine the nature of the cues they might be using. Now turn your attention to the data you and your partner obtained. As you work through the following steps copy your values onto a class master data sheet. Determine means and standard deviations for all class data. Save a copy of the class data sheets to use in your laboratory report.

- To quantify activity levels illustrated in each circle count the number of times the wasp(s) crossed any grid line. A quick way to make these counts is to follow each vertical grid line for its entire length, then each horizontal one. Record the summed total. For the female pair calculate an average activity level for one unmated female *Melittobia* in the presence of another. For the male/female pair, calculate each sex separately. Comparing these totals should give an estimate of the wasps' travel length relative to one another. Then count and record the number of different squares each wasp entered. Comparing these totals should give an estimate of the area they covered relative to one another. Were your class data for the first 6 min of latency significantly different from those for the second 6 min?
- To describe and quantify courtship and mating data work together with your partner to determine the means and standard deviations for the data for all interactions, including the components of both incomplete courtships and courtships that ended in mating. Calculate the percentage of tested pairs that mated for both unmated and previously mated females. With guidance from your instructor construct a flowchart that visually depicts the outcomes of the *Melittobia* courtship interactions observed by your class.
- On your data sheet for courtship attraction cues calculate the average number of wasps per location per minute for both bioassays (For capsules 1–3 be sure to remember to subtract 1 from your total to account for the one wasp placed there prior to the start of the experiment.).

In bioassay A, how did the results for the capsules containing female insects compare with the results for the capsules with the male? How did the results from bioassays A and B compare with each other? What might this indicate? Do you think the attraction cue is a sound or an odor? Which claim does your evidence support? How and why?

Does the variation in class results reflect mere chance or a meaningful association? For this experiment, our null hypothesis has been that when unmated *Melittobia* females are given the choice of an empty capsule, one with a live male, one with a live mated female, and one with an unmated female, the average number of females appearing in each location—as measured at 1-min intervals over a 12-min span—is not statistically different from random. The alternative hypothesis has been that the average number of females appearing in each of these locations—again, as measured at 1-min intervals over a 12-min span—is in fact significantly different from random.

Various statistical analyses have been developed to help guide answers to such questions. With guidance from your instructor choose and conduct an appropriate statistical test to support or reject your hypothesis.

Write your final report

Your instructor may require you to prepare a final report, either a brief summary written with your laboratory partner or a more formal document prepared as though for publication in a scientific journal. Begin with a brief introduction stating what you did and why. Summarize the class data for courtship behavior and the bioassays, with descriptive statistics that include means and standard deviations. Where appropriate state the null and alternative hypotheses and specify the independent and dependent variables used and the controls. Describe any statistical tests employed and give their result. Prepare and include appropriate tables, illustrations, and/or graphs.

Your instructor may also ask you to include your responses to some or all the questions for discussion that follow.

Questions for discussion

1. **Initial attraction**. Compare and contrast the activity levels of male and female *Melittobia* upon encountering new surroundings. Which sex was more active? What differences were better measured by crossed-line counts and by entered-square counts? Did you find evidence of a latency period? Did the female's behavior change after another female was introduced? Overall, what do these trials suggest about which sex is the attractor and which is the attracted? Describe and discuss the evidence in support of your claims. Speculate on the reasons for any differences observed.
2. **Courtship arena interactions**. What types of male–female interactions did you and your classmates observe? What terms did you use to describe them? How long did courtship and copulation usually last? How variable were these behaviors?
3. **Attraction cues**. Which sex approached the other first? What did they do? What does this suggest about the sort of cues the wasps might be using to find each other?

4. **Courtship and male persistence**. Based on the numbers in the class flowchart, when male and female *Melittobia* met, what proportion actually went on to court and mate? Does male persistence appear to be a sound strategy? Why or why not?
5. **Courtship outcomes**. Which sex controls the ultimate outcome of *Melittobia* courtship? In what ways? Recalling the purpose of courtship, is it appropriate to speak of courtship bouts in terms of "success" and "failure" or not? Justify your answer.
6. **Controlling variables.** Before beginning the bioassays, what possibly uncontrolled variable(s) did you consider? How did you test for them? What were your results? How did you deal with these results? How did this influence your sampling protocol?
7. **Choice chamber results**. In bioassays A and B, which capsule(s) contained the most females in your own trials and in the summed trials for the whole class? What types of communication signals were being tested in each spoke of the bioassay? What evidence for chemical communication did these assays provide? If a chemical was implicated, could it clearly be called a pheromone? Why or why not? If a significantly greater number of female wasps appeared in capsule 3 or 4 relative to the empty capsule, what might this suggest about pheromones in *Melittobia*? Name at least two possibilities.
8. **Experimental design and statistical analysis**. These bioassays called for introducing 20 unmated females simultaneously into the testing chamber. What fundamental premise of most statistical tests did this protocol violate? How did pooling the class data help sidestep this issue?
9. **Life history effects**. In the bioassay chamber, you used a single male in captivity to assay the behavior of a group of unmated females. Considering what you know of *Melittobia* life history and behavior predict what would happen if you were to reverse the experiment by using a single captive female and a group of male *Melittobia* together in the central canister of the bioassay chamber.

Part II. Instructor notes

Classroom management

This activity is best done by students working in pairs, taking turns observing and recording the behaviors. As presented, the courtship and bioassay portions of this activity can be completed in a standard 2-to 3-h laboratory period or separately over two 60- to 90-min class sessions.

If two 2- to 3-h sessions are available, the time intervals allotted for each stage of each activity can be extended and/or additional trials can be run. This will increase observational and numerical differences in the data but at the possible disadvantage of student inattention during longer wait times or repetition of activities. Encourage directed multitasking.

The discussion questions could be covered in class, presented as an assignment, or answered in a formal or informal laboratory report. Any detailed data analyses and the final preparation of laboratory reports typically would be completed as homework.

Detailed information on *Melittobia* biology is available in the study by Matthews et al. (2009).

Optional extensions (see later discussion) include a number of individual or small group experiments that could be done for extra credit. These would vary in time and complexity. However, the WOWBug's rapid life cycle means even multigenerational studies can be completed in a single semester or less. Despite a number of published studies, much about *Melittobia* biology and behavior is still unknown, making original research potentially rewarding.

Teaching the activity

Background

There have been many studies on insect mating behavior (Shuker & Simmons, 2014). Traditionally, fruit flies (*Drosophila* spp.) have been used as a model organism for courtship in biology courses, on the basis of ease of culture and availability of numerous individuals. However, in practice, students find it difficult to distinguish *Drosophila* sexes, the adult insects are difficult to handle, and experiments generally require that insects be anesthetized. Furthermore, eliciting *Drosophila* courtship behavior can be unreliable and their behavioral responses are not dependable. These disadvantages can be frustrating for students and teachers alike.

This activity introduces an alternative insect, *Melittobia*, commonly called the WOWBug. The harmless WOWBug is about the same size as *Drosophila* and has the same advantages of rapid life cycle and production of large numbers of adults. Unlike *Drosophila*, the sexes of *Melittobia* are very different and easily recognizable even to the naked eye (Fig. 10.4). WOWBugs rarely fly, preferring instead to slowly crawl, and they never need to be anesthetized. Their courtship is elaborate, easily observed, and both quickly and dependably elicited in the laboratory.

Obtaining and preparing materials

Melittobia can be purchased directly, reared from stock cultures, or collected in nature from mud dauber wasp nests (see Matthews et al. 1996). The Carolina Biological Supply (www.carolina.com) is the sole commercial source for the true WOWBug, *M. digitata* (item # 144570), which they raise on blowfly (*Sarcophaga*) hosts (item #173480).

For immediate use, time your order to have adult *Melittobia* available 7—10 days before you plan to conduct the student activity. Alternatively, if you wish to generate multiples of your initial culture and/or abundant males (see later discussion), then include both WOWBugs and blowflies in your order and time the shipment to arrive 4 weeks before the class activity.

Carolina Biological Supply is also the source for the deep-well slides (item # 603730). These are durable, are reusable, and facilitate easy viewing of a great many types of small living invertebrates. If you don't have glass shell vials of 1-dram size, Carolina Biological Supply carries these (item # 715051), as do many other suppliers.

The reusable bioassay chambers are easily made (see later discussion) from clear, size 0 gelatin capsules and small clear polystyrene plastic craft storage containers (about 30 mm diameter × 25 mm high; exact size is unimportant). These and other materials required for this investigation (listed in the student activity) are readily available from large discount stores and online sources.

Animal care guidelines

Adult *M. digitata* are extremely easy classroom insects. They require no food or water, take up little space, and have no special habitat requirements. Unless subjected to temperature extremes or allowed to desiccate, females will live for up to 2 weeks in the absence of a host; males live slightly shorter lives. Provided with a pupa host, females often live long enough to overlap with their newly emerged adult offspring.

Their respiratory needs are so modest that they can be kept inside a small vial tightly plugged with a cotton ball. The "tight" part is essential—mated females are so eager to disperse that they will chew through cork stoppers and burrow through loose cotton. However, should they escape, they seldom fly. Instead, they tend to persistently crawl and hop toward the nearest directional light source. Being small, escapees generally simply perish unnoticed.

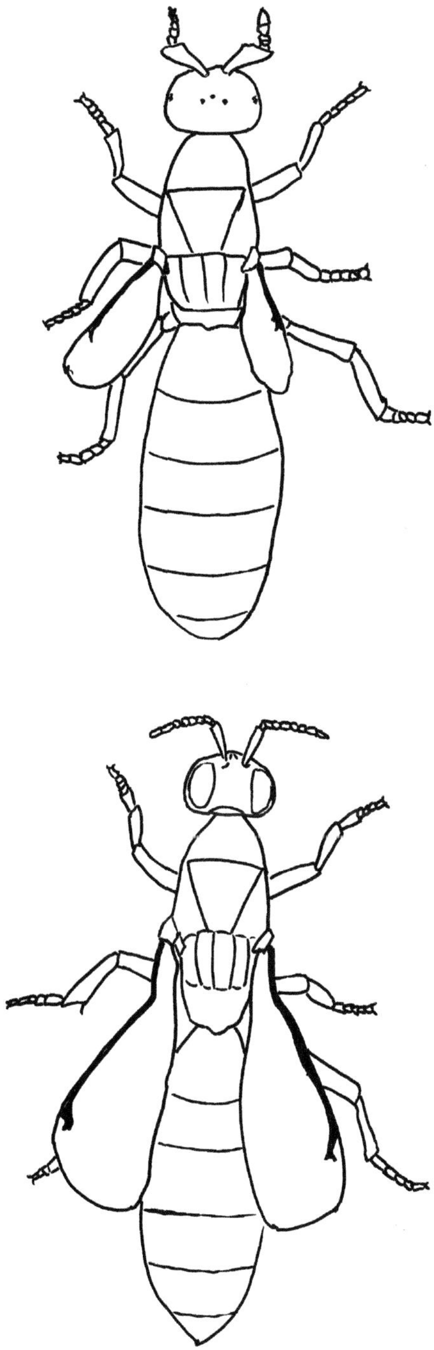

FIGURE 10.4

Student drawing of (A) male and (B) female *Melittobia digitata*. Compare the wings, eyes, and antennae; females are entirely black, whereas males are a caramel brown.

Planning for sufficient experimental organisms

Institutional Animal Care and Use Committee (IACUC) approval is not required but having sufficient numbers of adult unmated female and male wasps is critical for success and the activity requires advance planning and preparation. There are two ways to approach this: one involves money and the other involves a few weeks of lead time.

The money approach: Melittobia have a highly skewed sex ratio, generally producing only 1–5 males per 100 females. A commercially purchased culture will typically contain about 150 wasps, so female numbers will be no concern; however, the culture will usually have only a half-dozen males. Depending upon the class size, the requisite number of males may require ordering a number of cultures.

The time approach: Order an initial culture or find a parasitized mud dauber nest (see Matthews, 1997) and then rear your own *Melittobia.* The haplodiploid sex determination system of Hymenoptera can be easily exploited to provide all-male cultures and/or large numbers of both sexes of *Melittobia* for this activity (see later discussion).

Process to use the initial culture(s) directly

Time your order to have adult *Melittobia* available 7–10 days before you will conduct the student activity. *Melittobia* cultures are shipped as pupae still inside their blowfly host's puparia. These dark brown cases may have been intentionally precracked before shipment, spilling out some of the contents, and one or more living or dead foundress females may also be inside the small plastic shipping box. This is normal. View Carolina's 3-min YouTube video (https://www.youtube.com/watch?v=-dkfYj6x8vs) to see what to expect when the WOWBugs arrive.

At about 26°C, pupae will become adults about a week after reaching the red-eyed pupa stage (Silva-Torres & Matthews, 2003). Like all invertebrates, *Melittobia* development is influenced by temperature. Should it be necessary, their development can be slowed down or sped up by appropriate environmental manipulations, such as placement on a towel-shielded heating pad set on low.

Upon its arrival, immediately open the box and gently tease apart the host pupal skins to reveal the developing *Melittobia* pupae. Shake the pupae out onto a piece of white paper. Sexes are readily recognized (see Fig. 10.4) using a magnifying glass or a dissecting microscope's low power. Female pupae will have visible obvious eyes that become dark red as they progressively develop; males completely lack compound eyes and are functionally blind. This is the easiest stage to sort the sexes and place sets of females in shell vials for student use.

Each student pair will need two filled vials, one with 10 wasps and one with 20 wasps. A quick and easy method is to dump the pupae onto a sheet of paper and then use a cotton swab to brush the appropriate number onto a small paper "dustpan" and "funnel" them into a vial. Label these vials "unmated." Close each vial with a firmly packed cotton ball stopper (Neither *Melittobia* sex requires food, water, or extra ventilation at this time.).

Recall that *Melittobia* have a highly skewed sex ratio and pugnacious males. While they are still pupae, the males must be isolated from one another. As soon as the first male emerges, they will begin fighting, usually to death. Transfer each male pupa to its own deep-well slide and add a single female pupa. You will need one of these slides for each pair or team of students, plus a few spares.

Return the remaining pupae of both sexes to their original culture container. If you have no extra male pupae to add back at this time, then plan to return the males to the container after they have been used in this activity. By doing so, adult females emerging in this container will be mated and can be used for supplemental student research on behaviors such as dispersal and host attraction. They can also be used to set up a continuous laboratory culture (see later discussion), reducing the cost of future investigations.

Process to produce all-male cultures

To ensure having abundant males in time for class activities or to produce additional males for studies of topics such as male aggression (e.g., Matthews et al., 1996), you should start with a *Melittobia* culture and a supply of blowfly puparia about 4 weeks before the class. A standard order of 100–150 blowfly puparia, which arrives in a container roughly the size of a pint ice cream carton, will keep 3 months or more in a refrigerator. They will not develop into adult flies unless they are warmed to room temperature or above for many hours.

From the initial culture, sort a number of red-eyed female pupae into a shell vial labeled "unmated." Return the remaining pupae and all the male pupae back into their original culture container (Adult females taken later from this container will have mated soon after emergence. If desired, they can be used to set up additional new cultures that will produce primarily females. See later discussion.). Within a few days to a week after developing to this red-eyed stage (depending on temperature), most will have emerged as young adults and be crawling about. Use these unmated females to produce all-male cultures.

For each all-male culture, place three to five blowfly puparia into a labeled vial. Lift three to six unmated females carefully with a pipe cleaner and gently tap them into the vial without touching the interior with your fingers. Plug and label the culture vials. The number of cultures needed depends on class size, of course. Cultures derived from unmated females will usually produce about 12–25 males per host. It is a good idea to set a few more cultures than you may actually need. Despite your best efforts, occasionally a culture may fail to develop, usually as a result of introduced mold spores.

Monitor the wasps' development in these cultures. Male *Melittobia* generally mature a few days faster than females do. As before, unless you are specifically interested in the nature of male combat, they must be separated as late pupae. Transfer each male pupa to its own deep-well slide and add a single female pupa.

Process to multiply and/or maintain ongoing mixed-sex Melittobia *cultures*

Continuous culture in the laboratory is quite simple, requiring attention only once a month. Decide how many new cultures you want to set up and label that many new, clean shell vials. Following the procedure outlined earlier for all-male cultures, place three to five blowfly puparia in each vial. Add three to five females, and this time use adult females from the existing culture. Because both sexes are present in that culture and *Melittobia* courtship is both pervasive and reliable, these females will have mated.

Plug the vials firmly with cotton and place them in a box or desk drawer at room temperature. Check them 3–4 weeks later for new mated adult females that have chewed their way out of the puparia. These can then be used to repeat the process.

Older colonies can be allowed to die a natural death, just as they do under natural conditions. It would be rare and highly unlikely for individuals to escape from the laboratory into the outside world. However, it is reassuring to know that these are native insects that are so widespread in occurrence that scientists don't even know their country of origin.

Process to make bioassay chambers

Prepare the reusable central chambers (see Fig. 10.3 in student manual) in advance by piercing the sides of each craft container's sides with a heated tool to make four equidistant holes of appropriate size. A quick and easy method is to heat the tip of a cheap Phillips screwdriver over a gentle flame for 30 s. Holding the polystyrene container in one hand, slowly push the hot tip into and through the plastic at the approximate center of the container's side, rotating the tip slightly as it goes in. Continue pushing straight through until the screwdriver tip pierces the opposite side, then carefully withdraw the screwdriver. Rotate the plastic dish 90 degrees and repeat (reheating should not be necessary if done promptly) to produce a total of four more or less centered and equidistantly spaced holes around the side of the container. Slight variations will not matter; the slight taper of the capsule sides allows a relatively snug fit. Furthermore, when students encircle the juncture between the chamber and capsules with a small string of poster-mounting putty, it will seal any small gaps.

Turn the containers bottom up and number the holes one to four with a permanent marker. During the lab, students will use these numbers as guides when placing gelatin capsules in the holes and conducting their bioassays.

Helpful hints:

- Work under a vented hood either at home or in the laboratory; the heated plastic smells.
- Keep the tool steady; wiggling tends to produce holes that are oval or too large.
- If plastic burrs from around the holes, this is not a concern, but larger ones may be carefully snapped off using one's fingernail, if desired.
- Clean the screwdriver tip by wiping it with a paper towel; alternatively place it back in the flame under the hood to burn off excess plastic stuck on the tip.

In-class preparation

Analytical approach

This two-part activity focuses more on observational skills, experimental design, and control of variables than on precise statistical analyses, but some of the discussion questions (and suggested answers) touch on statistical design.

We suggest using whole-class data for the analyses, which at minimum should include means and ranges. Perhaps the most straightforward way to analyze the bioassay data is by contingency table analysis, using either chi-square or the log-likelihood ratio as the test statistic. Contingency table analysis is included in most introductory statistical texts, incorporated in most commercial statistical programs, and easily calculated by hand. Other analyses are certainly possible. You as the instructor are the person in the best position to determine what level of analytical sophistication your students possess and adjust your requirements accordingly.

We also recommend having students submit a final report following the guidelines given in the student activity. If this is appropriate for your students, then the report should be prepared and formatted as though for publication in a scientific journal. Detailed guidance is available in Matthews and Matthews (2014).

Possible extensions/continuations

Parasitic wasps offer many opportunities for undertaking comparative studies of courtship. An obvious addition to this activity would be to examine the courtship of related species of *Melittobia* (Assem et al., 1982). These can often be obtained from natural hosts (Matthews, 1997; Matthews et al., 1996, 2009) and may be identified using the key presented by Dahms (1984).

Alternatively, a related wasp, *Nasonia vitripennis,* is available from biological supply companies; like *Melittobia,* it can also be reared on blowfly hosts. Barash (1975) outlines a laboratory activity on *Nasonia* courtship. A good overview is provided in Mair and Ruther (2019).

If ongoing *Melittobia* cultures are being maintained, additional experiments with host acceptance requirements are possible. For example, many studies indicate the importance of host-associated chemicals (e.g., Cusumano et al., 2010; Gonzalez et al., 2018) and one appears to show that host shape is a vital cue (Cooperband & Vinson, 2000). Other research studies (e.g., Abe et al., 2014) indicate that females assess whether other females have already laid eggs upon the host and perhaps even adjust the sex ratio of their own offspring in response.

Bioassays are a common tool in the biological and medical sciences, making it important to caution that the choice of which stimuli to use in a bioassay is critical. If all of the choices are biological nonsense, it is possible to get results that look plausible but are nonsense as well (Some young students have run ill-advised *Melittobia* bioassays using choices among soft drinks, types of popular music, and ice cubes vs. a heating pad.). Better choices for male pheromone experiments using the bioassay chamber could include (1) testing the volatility of the purported male pheromone by conducting tests of whole-body squashes on filter paper disks after various intervals (hours to days), (2) investigating the effects on female attraction of such variables as male age or male experience, and (3) isolating the source of the male pheromone by assaying different body regions, such as head, thorax, abdomen initially, carefully separated with a sterile razor blade. Similar experiments could be developed to more fully investigate female aggregation.

Note that in any of these extensions or continuations that might involve repeated trials conducted with the same bioassay chamber, students should use fresh gelatin capsules for each trial. It would also be prudent to wash and rinse the central chamber, then let it dry. Otherwise, lingering pheromones may skew the results.

Sample observational results

Part 1. Observing interactions

Sex identification. Students should easily see that females look like rather ordinary little black wasps. However, males are caramel brown, have short wings, and lack compound eyes. They have enlarged, oddly shaped antennae with thumblike appendages (basis for the species name, *digitata*) that are used to grasp the female's antennae during courtship. If desired, you could require students to draw the insects (Fig. 10.4).

Initial attraction, baseline activity, and latency. Student tracings will all be somewhat different. However, comparison (Fig. 10.5) will show an initial latency period, during which female wasps will explore the perimeter of the arena, then increasingly venture across the middle; the presence of a second individual of the same sex will neither noticeably alter this behavior nor cause a noticeable interaction. Whether measured by lines crossed or by squares entered, males will be much less active than females, both when alone and with females (Table 10.1). As a male releases pheromones, females will approach, but seldom in a straight line. Students may be puzzled when females and males seem to be bumbling around very close to one another without seeing each other. Help them recall that the wasps are zeroing in on an odor. Vision plays little role at this stage; in nature, it would be occurring in total darkness.

Part 2. Observing *Melittobia* sexual behaviors

Melittobia courtship (Fig. 10.6) is mediated by chemical and tactile cues, but in the general form, it is similar to the visual and auditory courtship displays of many birds, in that males perform a stereotyped sequence of behaviors (**courtship bout**) that generally proceeds without female interruption until it ends with the female signaling either willingness to copulate or a refusal to cooperate. If the female does not give the proper signals for a finale, the male may leave but often he simply starts over, beginning another bout.

To introduce this part of the laboratory activity, have students view a brief video (https://www.youtube.com/watch?v=qDa_C853t34&t=11s) that includes courtship interactions in the WOWBug. Then obtain a class consensus as to how they will sample these behaviors and what markers they will use to determine that courtship has begun and has ended. Attraction and courtship operate along a continuum, and even entomologists sometimes give different answers to the question of when courtship begins. For the purposes of this laboratory activity, you may wish to say that a courtship bout begins when the mounted male faces the same direction as the female

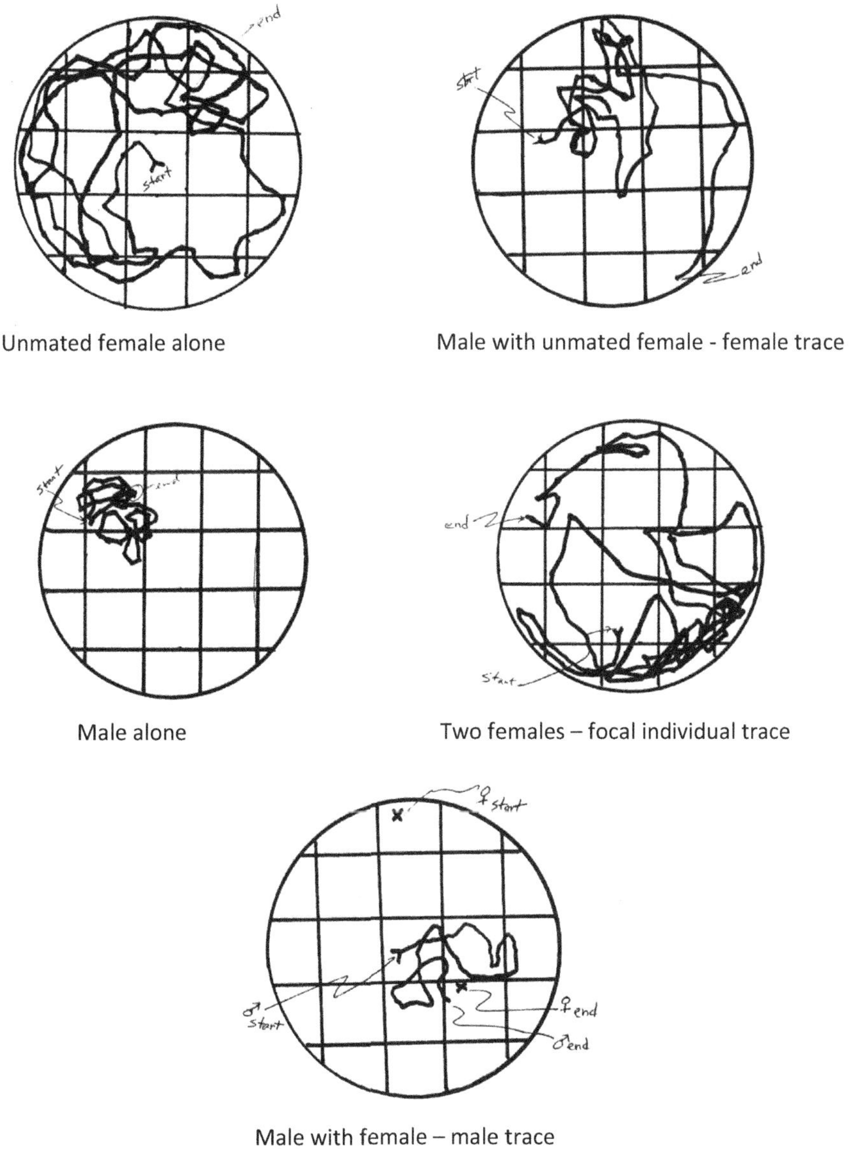

FIGURE 10.5

Examples of students' 6-min baseline activity traces of unmated female, solitary male, male and unmated female together (female trace), one of two unmated females, and male and unmated female together (male trace) placed together in deep-well slides placed on gridded paper. See Table 10.1 for numerical results of similar tracings.

Table 10.1 Typical numerical results (mean, range) obtained by 10 pairs of students tracing paths of focal wasps over the first 6 min after being placed in a new environment.

Focal individual's environment	Total number of gridlines crossed or recrossed (activity level)	Total number of squares entered one or more times (area explored)
Female alone in arena	52 (4–66)	15 (1–22)
Female with another female	53 (5–62)	17 (6–20)
Female with male in arena	59 (11–66)	19 (4–24)
Male alone	6 (4–15)	3 (1–5)
Male with female in arena	9 (4–18)	4 (1–8)

The large ranges reflect the fact that individuals vary widely in their behavior during latency. Overall, females were approximately 10 times as active as males and explored 3–4 times as much area.

and moves forward to initially contact the female's antennae with his (Fig. 10.6A). It could be said to end at the moment when he releases contact with the female's antennae, stretches his body lengthwise, and sticks out his middle legs preparatory to moving backward to attempt copulation (Fig. 10.6B). If the female is receptive, she will lower her body, tilt her head up, and change her abdominal shape to allow genital contact (Fig. 10.6C). If the female is not receptive, she does not lower her body or change her abdominal shape and the male cannot copulate.

Decide upon a standard format for each student pair's data collection. Guide the class to present their pooled results in a table, a Microsoft Excel spreadsheet, and/or an interaction outcome flow chart (Fig. 10.7). Pooled data (e.g., Fig. 10.7) will clearly show that more insects are attracted than courted and more begin courtship than finish it. If a courtship bout ends without the pair mating, then encourage students to keep watching. Some males may just give up and wander off, but many are persistent and will simply begin another bout, upping their chances of eventual success but not assuring it. Help students realize that these numbers do not represent "successes" and "failures." Across the entire animal world, more organisms begin courtship than complete it. This is inherent in the purpose of courtship (see discussion question 5.)

Typically, cannibalism occurs in 2%–10% of the encounters (Gonzalez & Matthews, 2005). Given the highly aggressive nature of males, this may be unsurprising, but researchers disagree as to whether this represents a laboratory artifact, a recognition gone awry, a response to something odd about the female, or an opportunity for nutrition (Deyrup et al., 2006).

If students were to retest a mated female with their male, females invariably would refuse to copulate (as noted in the video), although males typically would court for a single bout of roughly the same duration as they did with unmated females. Most often, mated females brush a male off their back and/or fold their antennae against their body. In nature, mated females exit the host cocoon very soon after their single copulation and thus generally escape the male's continued advances.

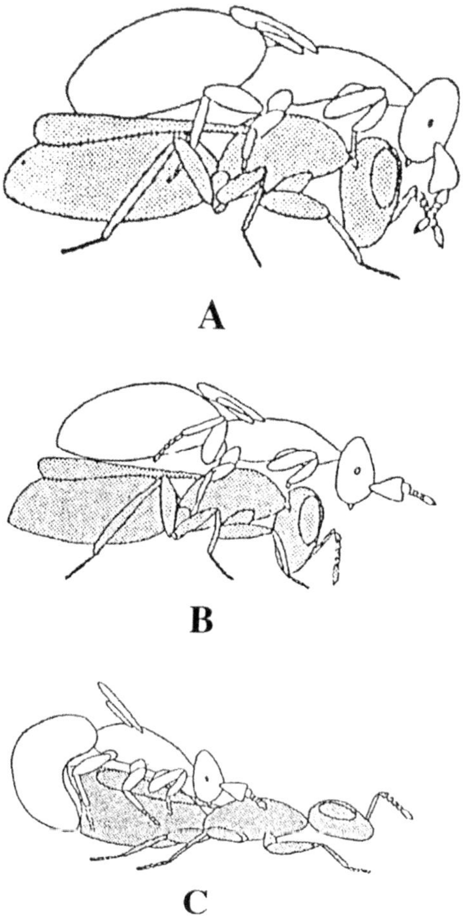

FIGURE 10.6

A male *Melittobia digitata* begins a courtship bout by contacting the female's antennae with his (A) and then repeatedly releases and regrasps her antennae while simultaneously elevating his hind legs in a bicycling motion (B). Courtship concludes when the female stretches and flattens her body and the male moves backward to make genital contact (C).

Part 3. Determining courtship attraction cues

Scientists employ a multitude of designs to determine organisms' responses to various stimuli, and the use of bioassays and choice chambers is widespread. The subjects of Part 3 include not only an examination of pheromone cues but also concepts of appropriate stimulus choices, sampling methods, and well-controlled variables. When students fully appreciate these, the simple chambers introduced here can profitably be used for a great number of additional investigations.

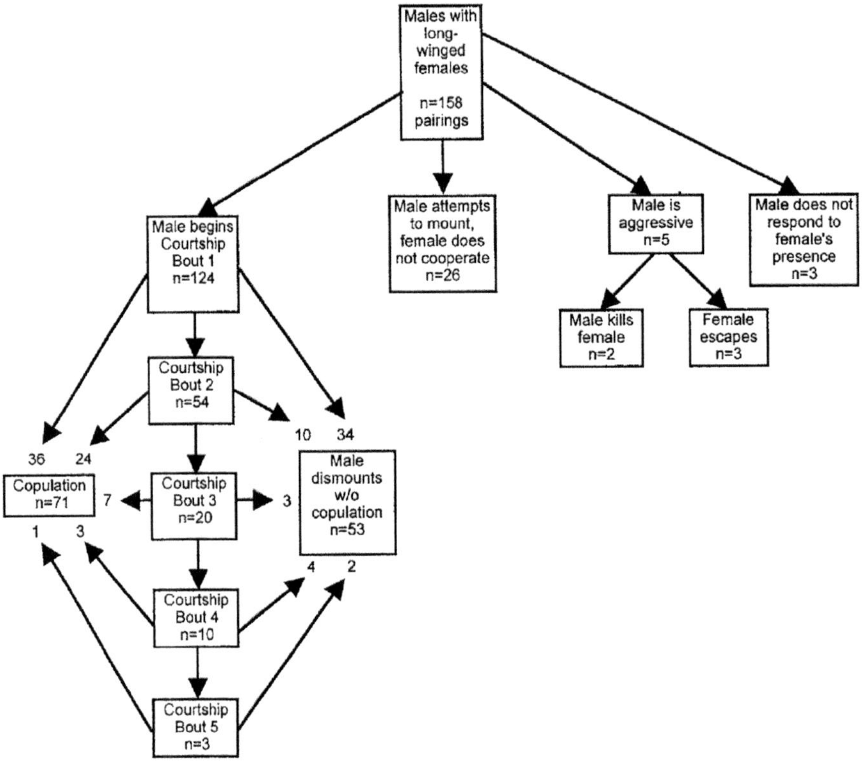

FIGURE 10.7

Flow chart of the courtship interactions between 158 pairs of unmated female *Melittobia digitata* (Gonzalez & Matthews, 2005).

Because time constraints dictate running bioassays A and B for only 12 min each, most female wasps will not make any choice before the experiments conclude. Thus even with 20 potentially responding wasps introduced into the central chamber, student pairs usually tally low numbers for each capsule of the bioassay chamber. Thus to show trends, it will be necessary to pool class data. If classroom mechanics and student schedules allow it, then consider extending the time allotted to the bioassays and/or running duplicate bioassays overnight. Have students compare and contrast the results obtained.

Identifying, testing, and controlling possible variables. The necessity to control directional light as a variable will vary with conditions. In some laboratories and classrooms, conducting an initial trial with an unshielded bioassay chamber and nothing in the capsules will result in the strongly phototropic females congregating in the capsule(s) aligned with the strongest light source. This problem can be easily remedied. One method is to cover the entire bioassay chamber with an inverted red cup, lifting it at 1-min intervals to take on-the-dot samples. Because the vision of insects is generally shifted toward the UV end of the light spectrum, they see red light very poorly. Another method is to pivot the choice chamber a quarter-turn after each sample. Other methods are also possible. Let the students brainstorm and help them realize ways that their experimental design will influence their choice of sampling method (see discussion question 6).

Bioassay A. For this bioassay, students load each of the three capsules with a male, a mated female, and an unmated female, leaving one empty as a control. Combined class data should reveal a preference for capsules containing live insects rather than the empty (control) capsule. A significant attraction to the male's capsule may or may not be evident. In class discussion, have the students consider possible reasons, including both features of experimental design, such as the effects of room temperature and the relatively short observation period, and biological factors, such as age and vigor. Experimental studies (e.g., Consoli et al., 2002) have shown that there is great variability in the release of pheromones both with aging and among males at the same age. Pheromone levels are low soon after emergence, rise to a peak 2 days later, and then decline over the wasp's 5- to 10-day lifespan back to essentially early emergence levels.

Occasionally, when class data are summarized, more female wasps will have moved into the capsules with other female wasps than into the male's capsule. They may be clumped in one female capsule or the other or distributed more or less equally between them. This result will be a "discrepant event"—an unexpected outcome that doesn't seem to agree with what students think they know. At first, they may believe it represents some sort of "mistake." If your instructional setup allows it, then allow a set of choice chambers, loaded as in bioassay A, to sit for 24 h. Then do a recount. Almost invariably, the responding females will all be in with the male; often, even the females initially placed in capsules 3 and 4 will have moved (If you can't manage these as actual trials present this information as a verbal scenario.). Encourage a whole-class discussion of possible explanations for the discrepant event. The reply to discussion question 9 (see later discussion) will provide the guidance you need.

Bioassay B. This time, students expose their responding female wasps to the same capsules after removing the live insects. A high level of attraction to the male capsule without the wasps' actual physical presence would rule out cues such as motion, vibration, or vision (or at least relegate them to a minor role at this stage). If the wasps are producing volatile pheromones, chemical traces might be expected to remain in the capsules, attracting responders. However, the results from bioassay B are usually no different from random. Researchers (e.g., Consoli et al., 2002) believe that the male sex pheromone is short-lived and operates only at a relatively short range. Within the tight confines of a host cocoon crowded with conspecific females, no more than this is necessary or desirable.

Sample numerical results

Table 10.1 shows numerical results for the student pair that obtained the activity tracings shown in Fig. 10.4. By either sampling method, males were less active and did less exploring than females.

Table 10.2 Typical results for class observations of 40 unmated female *Melittobia* placed individually in viewing chambers, each with a male.

Behavior	Mean duration (s)	Range (s)	Times observed (no.)
Latency[a]	255	10–480+	40
Courtship[b]	78	28–303	33
Copulation	3	2–11	27

[a] *Includes all time spent in noncourtship activities such as exploration and grooming.*
[b] *Cannibalism concluded three cases; three others were not completed.*

Table 10.2 presents pooled results for a class of 20 students working in pairs on the courtship observation study. Each watched, timed, and recorded four pairings for a total of 40 trials. The large ranges in duration reflect the fact that the behavior of individual couples vary widely.

Fig. 10.7, taken from a published study of 158 male–female pairings (Gonzalez & Matthews, 2005), shows that successful courtship and mating are far from assured. In this large sample, 22% of the pairs never even began courting; in 28 cases the females did not cooperate and in another 8 cases males did not respond appropriately to the female's initial touch. Although 124 of the 158 pairs did begin courting, only 36 courtship sequences (<30%) were consummated after a single bout. Even after as many as five bouts, a total of 53 (43%) ended with males dismounting without copulation. Thus, overall, less than half (45%) of all potential courtship encounters resulted in mating. You may wish to share these data with your students and have them compare these results with their own.

Answer key to "questions for discussion"

1. **Initial attraction.** *Compare and contrast the activity levels of male and female Melittobia upon encountering new surroundings. Which sex was more active? What differences were better measured by crossed-line counts and by entered-square counts? Did you find evidence of a latency period? Did the female's behavior change after another female was introduced? Overall, what do these trials suggest about which sex is the attractor and which is the attracted? Describe and discuss the evidence in support of your claims. Speculate on the reasons for any differences observed.*

Females are far more active than males. Introducing a second female has little effect upon the first female's actions. Tracing paths (see Fig. 10.4) illustrates that individual behaviors vary widely during the latency period. Females typically walk partially or completely around the perimeter of the arena, then begin exploring the interior. However, some are excitedly "hyper," whereas others simply hunker down and/or spend long periods self-grooming. Counting crossed lines marked during a fixed time period gives a measure of the speed and extent of this roaming.

Counting the number of squares entered shows that the males not only are slower to move but also stay in a much more constrained area. These observations suggest that females are probably being attracted to males rather than vice versa.

2. **Courtship arena interactions.** *What types of male–female interactions did you and your classmates observe? What terms did you use to describe them? How long did each usually last? How variable were these behaviors?*

Answers will vary but typically include terms such as latency, attraction, orientation, antennal stroking, mounting, leg lifting, wing fluttering, backing up, stretching, mating, and grooming. Behaviors during latency usually vary the most; this will often be the longest stage. It may include several contacts between the sexes before the male climbs on top of the female and orients to her head. Copulation is always the briefest stage, and often it occurs so quick making it difficult to time. Courting bout lengths are relatively consistent, with the pace or intensity noticeably increasing at the end. Once mounted, the male's courtship behaviors proceed fairly predictably (see Fig. 10.6) and includes antennal stroking and leg lifting. If copulation doesn't occur at the end, then the male usually pauses for a few seconds and then begins a second bout. Occasionally the male will completely dismount and then climb back on top of the female again to resume courting. After mating, both sexes tend to spend extended periods cleaning or preening their bodies.

3. **Attraction cues.** *Which sex approached the other first? What did they do? What does this suggest about the sort of cues the wasps might be using to find each other?*

In the beginning, the female will usually turn and orient toward the male, approaching him from his side and often tapping him on his abdomen with her antennae. If the female does not move away, the tapped male will usually turn toward her and then attempt to climb onto her body. Once on top, he usually moves first toward her rear end, tapping with his antennae, then reverses and moves toward her head. The female will usually stay more or less motionless once the male is on her back. The antennal tapping by both sexes suggests an odor or chemical cue, as insect antennae are organs of smell equivalent to our noses; however, tactile cues such as body shape can't be excluded. Males also touch with their legs as they climb onto the female, so tactile cues may be involved at this stage as well. Although our ears aren't sensitive enough to detect any sounds without amplification, one can't exclude the possibility that some auditory cues could also be used. That research would certainly be possible, but apparently has yet to be done.

4. **Courtship and male persistence.** *Based on the numbers in the class flow chart, when male and female Melittobia met, what proportion actually went on to court and mate? Does male persistence appear to be a sound strategy? Why or why not?*

Answers will vary. If your students saw very few cases of courtship and mating, you may wish to have them analyze the data shown in Table 10.2. In that sample, courtship concluded with mating after a single bout in about half of all cases. Overall the average was 1.72 bouts, with at most 5 bouts (1 of 71 trials). Persistent males usually did succeed in copulating, but successive bouts often required progressively longer times to complete. Those males that copulated after just one bout required on average less than a minute, whereas those requiring four bouts required a little more than 3 min. Thus the "speedy" males could potentially inseminate three times as many of their sisters as the "slow" males. The benefit of male persistence may relate to the old saying that "a bird in the hand is worth two in the bush" because males probably incur a time cost in waiting for new partners as well as risking discovery by other males.

5. **Courtship outcomes.** *Which sex controls the ultimate outcome of Melittobia courtship? In what ways? Recalling the purpose of courtship, is it appropriate to speak of courtship bouts in terms of "success" and "failure" or not? Justify your answer.*

Although they may appear relatively passive, females are ultimately in control of courtship and mating. If dissatisfied with a male's performance, they will not indicate receptivity by changing their abdomen shape and position in the manner seen in receptive unmated females, so eventually the male climbs off. Moreover, mated females are very resistant to copulation attempts. Males may be able to mount them and will often attempt to perform courting bouts, but already mated females generally will not remain still and often actively try to dislodge the male. Furthermore, it is the unmated females that initially seek out males, not vice versa. Answers regarding "success versus failure" will vary. On the one hand, continuation of the species requires mating; in that sense, expending energy in courtship without obtaining copulation could be viewed as "failure." On the other hand, if the purpose of courtship is discrimination between potential mates of varying quality and appropriateness, then any match that does so is actually successful. Failure would actually be nondiscrimination, leading to inappropriate pairings with the wrong species or wrong sex.

6. **Controlling variables.** *Before beginning the bioassays, what possibly uncontrolled variable(s) did you consider? How did you test for them? What were your results? How did you deal with these finding(s)? How did this influence your choice of sampling method?*

Various answers are possible, but the necessity to control directional light is the most common response. It can be most easily dealt with by covering the bioassay chamber with an inverted red cup. Even with a translucent cup, it is difficult to view the wasps unless the cup is lifted. Thus on-the-dot (instantaneous) sampling is required. If their initial trial with empty capsules did not show any results that differed from random, they might have chosen continuous sampling or all-occurrence sampling. This is an opportunity to engage students in a discussion of the limits, shortcomings, and advantages of various sampling methods.

7. ***Choice chamber results.*** *In bioassays A and B, which capsule(s) contained the most females in your own trials and in the summed trials for the whole class? What types of communication signals were being tested in each spoke of the bioassay? What evidence for chemical communication did these assays provide? If a chemical was implicated, could it be clearly called a pheromone? Why or why not? If a significantly greater number of female wasps appeared in capsule 3 or 4 relative to the empty capsule, what might this suggest about pheromones in Melittobia? Name at least two possibilities.*

Answers will differ. Numbers obtained by individual pairs of students generally are low and variable over the short trial period. However, when data from all the class trials are summed and analyzed, the greatest response usually is to the capsule 2, the one with the live male.

Capsule 1 is the control; it helps us to tell whether females are choosing the spokes randomly or not. Capsule 2 contains a live male; if most females come here, it may be that they are attracted by his smell (either a pheromone or a simple by-product of respiration), some inaudible (to us) sound he produces, or by some movement they can detect. Even though females have functional vision, it is unlikely that it would play a role in their attraction, given that they are normally courted in the dark inside the host cocoon. Capsules 3 and 4 contain females; if movement or presence of a live respiring insect is all that is required for attraction, then one would expect equal numbers of unmated females to choose all three filled capsules in roughly equal numbers. The female wasps in capsules 3 and 4 differ in their mating status.

Because females cannot directly contact the male, the signal must be an airborne chemical. **Note**: the pheromone source appears to be one or more subcuticular gland(s) located in the base of the abdomen (Gonzalez et al., 1985). The most active compounds in the male's pheromone have been identified as alpha- and beta-trans-bergamotenes (Consoli et al., 2002). Interestingly, these are common components of many leaf and flower fragrances.

Can the male's chemical absolutely be labeled a pheromone? Probably yes, but not with absolute certainty. On the one hand, mated females and other males do not respond to it (Consoli et al., 2002), and it does function to attract unmated females to the male. However, on the other hand, it's been found that unmated females will respond to males of other *Melittobia* species (Matthews et al., 1985). Whether they are responding to exactly the same pheromone, or simply to shared characteristics of related ones, has yet to be investigated experimentally.

Because bioassay trials were couched in terms of courtship, students will naturally expect unmated females in the chamber to be attracted only to the male, the insect suspected of producing a sex pheromone. Female *Melittobia* being attracted to one another on occasion suggests several other hypotheses. Students may suggest that the association is pure chance, mistaken identity, an uncontrolled variable, an aberrant outcome produced by the experimental design, or female homosexuality. However, it is far more likely simply that more than one pheromone is involved. Rather than, or in addition to, a male pheromone, wasps of both sexes might produce a generalized aggregation pheromone, or perhaps males are only producing a sex pheromone but females are producing an aggregation pheromone to attract one another.

Some published evidence supports the two-pheromone hypothesis. In addition to the male-produced sex pheromone (Consoli et al., 2002), it appears likely that females produce an aggregation pheromone that attracts one another. They have been observed queueing up around mating pairs and around places where a mated female has begun chewing a hole to escape into the outside world (Deyrup et al., 2005; Deyrup & Matthews, 2007). Mated female *Melittobia* have a strong drive to disperse, and unmated females will follow them.

As an example of a biological variable that is not addressed, students may recognize that the selection of a capsule by a female might be influenced by what other females select. The trials in the deep-well slides failed to show any significant differences in female exploratory behavior with one female or two, but whether this is applicable or not is open to question. You may want to encourage motivated students to undertake single-insect trials as a follow-up or independent investigation.

8. **Experimental design and statistical analysis**. *These bioassays called for introducing 20 unmated females simultaneously into the testing chamber. What fundamental premise of most statistical tests did this protocol violate? How did pooling the class data help sidestep this issue?*

Statistically sophisticated students may recognize that the bioassay protocol violates a fundamental premise of most statistical tests—when the wasps were introduced simultaneously but counted individually, the observations were not independent. Thus instead of considering each wasp in a trial to be an experimental unit, have student pairs each calculate the average number of females in the central chamber and each arm over the trial period. Then set up the class master data set to record each team's overall average for the trial period as a single separate observation (N). Guide students to realize that if wasp distribution were a random affair, they could expect the same average number of wasps in each arm of the bioassay (the null hypothesis). A simple chi-square goodness-of-fit analysis can be used to compare what they might expect to find with what they actually observed.

9. **Life history effects**. *In the bioassay chamber, you used a single male in captivity to assay the behavior of a group of unmated females. Considering what you know of Melittobia life history and behavior predict what would happen if you were to reverse the experiment using a single captive female and a group of male Melittobia together in the central canister of the bioassay chamber.*

Because males are highly aggressive toward one another, they would probably spend the time fighting and perhaps killing one another instead of responding to female cues. Furthermore, such an experiment would be less logical because student observations should reveal that it is usually the female who initiates the preliminary courtship. Finally, males move relatively slower than females, so most males would probably not even arrive at the spokes of the bioassay chamber during the time allotted.

References

Abe, J., Kamimura, Y., & West, S. A. (2014). Inexplicably female-biased sex ratios in *Melittobia* wasps. *Evolution, 68*(9), 2709–2714.

Assem van, J. den, Bosch, H. A. J., & Prooy, E. (1982). *Melittobia* courtship behaviour: A comparative study of evolution of a display. *Netherlands Journal of Zoology, 32*, 427–471.

Barash, D. P. (1975). Reaction chains in insect courtship: *Nasonia vitripennis*. pp. 64–66. In E. O. Price, & A. W. Stokes (Eds.), *Animal behavior in laboratory and field*. San Francisco: W. H. Freeman.

Consoli, F. L., Williams, H. J., Vinson, S. B., Matthews, R. W., & Cooperband, M. F. (2002). *trans*-Bergamotenes—male pheromone of the ectoparasitoid *Melittobia digitata*. *Journal of Chemical Ecology, 28*(8), 1675–1689.

Cooperband, M. F., & Vinson, S. B. (2000). Host acceptance requirements of *Melittobia digitata* (Hymenoptera:Eulophidae), a parasitoid of mud dauber wasps. *Biological Control, 17*, 23–28.

Cusumano, A., Gonzalez, J. M., Colazza, S., & Vinson, S. B. (2010). Behavioral responses of the parasitoid *Melittobia digitata* to volatiles emitted by its natural and laboratory hosts. *Entomologia Experimentalis et Applicata, 136*(3), 301–307.

Dahms, E. C. (1984). Revision of the genus *Melittobia* (Chalcidoidea: Eulophidae) with the description of seven new species. *Memoirs of the Queensland Museum, 21*, 271–336.

Deyrup, L. D., & Matthews, R. W. (2007). The effect of gland secretions on escape chewing in *Melittobia* (Hymenoptera:Eulophidae), including cross-species investigations. *Journal of Hymenoptera Research, 16*(1), 7–10.

Deyrup, L. D., Matthews, R. W., & Deyrup, M. (2006). Feeding and siblicidal cannibalism in a male parasitic wasp (Hymenoptera: Eulophidae). *Florida Entomologist, 89*, 135–140.

Deyrup, L. D., Matthews, R. W., & Gonzalez, J. M. (2005). Cooperative chewing in a gregariously developing parasitoid wasp, *Melittobia digitata* Dahms, is stimulated by structural cues and a pheromone in crude venom extract. *Journal of Insect Behavior, 18*(3), 293–304.

Gonzalez, J. M., Camino, D., Simon, S., & Cusumano, A. (2018). Semiochemical exploitation of host-associated cues by seven *Melittobia* parasitoid species: Behavioral and phylogenetic implications. *Frontiers in Ecology and Evolution*. https://doi.org/10.3389/fevo.2017.00172

Gonzalez, J. M., & Matthews, R. W. (2005). Courtship of the two female morphs of *Melittobia digitata* (Hymenoptera: Eulophidae). *Florida Entomologist, 88*, 258–267.

Gonzalez, J. M., Matthews, R. W., & Matthews, J. R. (1985). A sex pheromone in males of *Melittobia australica* and *M. femorata* (Hymenoptera: Eulophidae). *Florida Entomologist, 68*, 279–286.

Mair, M. M., & Ruther, J. (2019). Chemical ecology of the parasitoid wasp genus *Nasonia* (Hymenoptera: Pteromalidae). *Frontiers in Ecology and Evolution*. https://doi.org/10.3389/fevo.2019.00184

Matthews, J. R., & Matthews, R. W. (2014). *Successful scientific writing. A step-by-step guide for the biological and medical sciences* (4th ed.). Cambridge, UK: Cambridge University Press, 294 pp.

Matthews, R. W. (1997). Teaching ecological interactions with mud dauber nests. *The American Biology Teacher, 59*(3), 152–158.

Matthews, R. W., Gonzalez, J. M., Matthews, J. R., & Deyrup, L. D. (2009). Biology of the parasitoid, *Melittobia* (Hymenoptera: Eulophidae). *Annual Review of Entomology, 54*, 251–266. https://doi.org/10.1146/annurev.ento.54.110807.090440

Matthews, R. W., Koballa, T. R., Flage, L. R., & Pyle, E. J. (1996). *WOWBugs: New life for life science*. Athens, GA: Riverview Press, 319 pp.

Matthews, R. W., Yukawa, J., & Gonzalez, J. M. (1985). Sex pheromones in male *Melittobia* parasitic wasps: Female response to conspecific and congeneric males of three species. *Journal of Ethology, 3*, 59–62.

Shuker, D. M., & Simmons, L. W. (Eds.). (2014). *The evolution of insect mating systems*. Oxford, UK: Oxford University Press.

Silva-Torres, C. S. A., & Matthews, R. W. (2003). Development of *Melittobia australica* Girault and *M. digitata* Dahms (Parker) (Hymenoptera: Eulophidae) parasitizing *Neobellieria bullata* (Parker) (Diptera: Sarcophagidae) puparia. *Neotropical Entomology, 32*(4), 645–651.

Part III. Supplementary material

Supplementary data related to this chapter can be found online at https://doi.org/10.1016/B978-0-12-821410-7.00007-8.

CHAPTER

11 Are squirrels and ants smart shoppers? How foraging choices may meet current and future needs

Sylvia L. Halkin, Alicia M. Bray

Biology Department, Central Connecticut State University, New Britain, Connecticut, United States

Chapter outline

Exploring Animal Behavior in Laboratory and Field. https://doi.org/10.1016/B978-0-12-821410-7.00027-3

Part I. Student instructions

Learning goals, objectives, and key concepts

- To design and conduct a controlled experiment on foraging choices made by tree squirrels or by ants
- To determine the factors that seem to influence foraging choices
- To determine whether different foods are chosen for immediate consumption than for carrying away to be stored at another location(s) and to discuss and attempt to explain such differences, if they are observed
- To think about how to further test for the potential importance of variables that were uncontrolled in the initial experiment

Background

The field of behavioral ecology focuses on how the behavior of animals affects their ability to survive and reproduce. A major focus of the field of behavioral ecology is how efficiently animals use their time and energy in obtaining resources that enhance survival and reproduction. Foraging is the behavior involved in searching for, obtaining, and processing food, including the removal of inedible parts before eating the food, food storage, and retrieval. Animals that minimize the energetic cost of foraging will benefit from a higher net energy gain than animals that do not behave in this way (net energy = energy obtained from food − energetic cost of foraging). The foraging decisions of many species seem to be based on maximizing energy intake per unit of time (Brown et al., 2011; Cuthill & Houston, 1997; Stephens & Krebs, 1986). Mathematical models of the role of energy and time constraints in shaping foraging behavior were first developed by Emlen (1966) and MacArthur and Pianka (1966).

Energy and time are not the only relevant considerations for foraging animals. Acquisition of essential nutrients and/or consumption of a varied diet may also be important. Cooperative communication between foragers may facilitate access to food sources once they have been discovered (e.g., worker ants following chemical trails from the colony to a food source, Hölldobler & Wilson, 1990; Hansen & Klotz, 2005; Ellison et al., 2012). Habitat type can influence food choices made by ants (Castillo-Guevara et al., 2019; Stahlschmidt & Johnson, 2018), as can seasonal changes (Stahlschmidt & Johnson, 2018). The most competitive ant species may monopolize food that less effective competitors would have chosen in the absence of competition (Castillo-Guevara et al., 2019); intraspecific competition may also influence food choices made by squirrels (SLH, personal observation). If predators pose a greater threat to animals using some foraging methods than others, less energy-efficient but safer methods may be used than would be predicted if energy alone were considered. The role of predation risk in shaping foraging behavior of squirrels has been modeled by Lima (1985).

Both tree squirrels and ants eat some food immediately, but both can also store food for later use. Animals that store food for later consumption need to choose food items that are likely to survive the period of food storage in good condition (Steele & Smallwood, 1994). In a typical ant colony, most of the individuals are sterile female workers; often there is just one fertile queen, whose primary activity is laying eggs. The workers collect food for the colony; feed their mother, the queen, and feed and care for their young siblings; enlarge the colony as needed; and defend the colony. Ant colonies generally contain tens to thousands of workers (depending upon the species), so the ants you see above ground foraging for food are a very small proportion of the workers living in their colonies. Some species of ants, such as the desert honeypot ants of the genus *Myrmecocystus*, have workers whose abdomens function to store liquid food that they can distribute to other members of the colony (Duncan & Lighton, 1994). Some ant species store foods within their colonies to eat later, including seeds with nutrient-rich elaiosomes (Fischer et al., 2008). Ants in the species *Pogonomyrmex badius* even store seeds in a damp chamber within the colony and wait for germination to open the hard seed coat, allowing for easier consumption of the contents (Tschinkel & Kwapich, 2016).

Squirrels store nuts that are naturally produced only at certain times of year (mainly in the fall and often in large quantities at that time; think of acorns, a kind of nut produced by oak trees). Rather than storing food in their homes, tree squirrels in the genus *Sciurus* bury (cache) individual nuts in dispersed shallow holes in the ground, a behavior referred to as "scatter hoarding." Squirrels may retrieve the nuts they store within an hour or two (SLH, personal observation), or months later; research has shown that they remember where they have stored nuts (e.g., Jacobs & Liman, 1991). Competitive interactions with other foragers may threaten food stores; for example, nuts stored by squirrels may be stolen (pilfered) by conspecifics (other members of the same species), or by other nut-eating species, including crows and jays, and squirrels may use extra time and energy in multiple kinds of behavior that reduce the probability of pilferage (Steele et al., 2008). Eastern gray squirrels tend to bury nuts with especially high energy content in open areas away from trees; in such areas, both the nut-burying squirrel and potential pilferers searching for nuts to steal are more exposed to predators than they would be under the leafy canopy of a tree with a nearby trunk to climb: this choice of open areas seems to serve as a deterrent to pilferers (Steele et al., 2014). However, apparent competition between conspecifics foraging on the nuts you provide may be tempered by the benefits of keeping one's cold-weather sleeping companions healthy enough to participate in group thermoregulation. Mutual tolerance between related females that share a foraging area may also provide indirect fitness benefits if this tolerance helps relatives to survive and/or produce offspring. Both eastern gray squirrels and fox squirrels may share their sleeping nests with conspecific adults, particularly in cold weather (Steele & Koprowski, 2001). Female eastern gray squirrels (but not males and neither sex of fox squirrels) remain as adults in the areas where they were born, and they sleep in stable groups that include mothers, daughters, and/or sisters (Steele & Koprowski, 2001).

How does foraging efficiency relate to natural selection? It is assumed that more efficient animals will have a higher probability of acquiring sufficient resources to survive and that the time they save and the energy they store can also be eventually used for reproduction (including courtship, gamete and offspring production, and parental care, or, in the case of infertile worker ants, caring for the next generation of young produced by their fertile mother). Thus more efficient individuals will have more descendants (or other relatives that share their genes) in future generations than less efficient individuals (as long as the more efficient animals don't take excessive risks with predators).

Purpose

You will be designing a controlled experiment to observe food choices, to observe what is done with the selected food items, and to determine whether the food choices appear to be adaptive. Although you will attempt to control for all the independent variables that may influence the food choices made by your animals, you will almost surely find that you could not control for every potentially important independent variable. Once you have obtained the results of your initial experiment, you will be asked to design one or two follow-up experiments targeted to test for the potential effect of independent variables not controlled in your initial experiment.

Methods

Species selection

Squirrels in the genus *Sciurus* (such as eastern gray squirrels, *Sciurus carolinensis*, and fox squirrels, *Sciurus niger*) and ants (family: Formicidae) have proven to be willing subjects on many campuses, so we are providing options for both. This exercise is designed for use in the field, but beyond that, the methods are quite different for squirrels and ants, so for the remainder of this exercise, the ant and squirrel instructions will be presented separately, up until the discussion questions and references at the end.

Option 1: Squirrels

Materials needed (for each team of six to eight students)

- Each team will be using one or two types of unsalted nuts provided by your instructor. Peanuts and hazelnuts are both readily taken by squirrels, and they have characteristics that are useful for this experiment. You may also use other species of unsalted commercially available or native nuts.
- One to two nutcrackers per student team to crack hard nut shells (not needed for peanuts).
- Containers to sort nuts and to carry the nuts to your field site.
- Binoculars for each student or pair of students to allow you to make detailed observations at distances that will not disturb the squirrels and to observe squirrels that climb trees or go to other places where it is difficult for you to follow them. If you have local squirrels that are very accustomed to having people nearby, you may be able to get close enough to collect adequate data without binoculars.
- A timing device (stopwatch, cell phone, or wristwatch with a second hand).
- *Four* double-sided data sheets.
- A clipboard.
- A pencil with an eraser.
- A tape measure that is at least 10 m long.
- Large plastic garbage bags for students who would like one to sit on.
- If you will be doing this experiment in cold weather, then be sure to dress very warmly (a hat, gloves, and warm footwear will be useful additions to a warm jacket). Be prepared to stand or sit still for about an hour—this will feel much colder than if you were actively moving around in the same weather conditions.

Step-by-step instructions

Designing your experiment

Your goal is to design and conduct an experiment to test your predictions of what squirrels will do with two options (sets) of nuts you provide. You will choose and prepare your nuts so that you predict squirrels will be more likely to immediately eat nuts from one of the sets and will be more likely to store nuts from the other set. Your challenge is to choose a single property of nuts (independent variable) that you predict would affect a squirrel's decision whether to eat or cache (store) the nuts. Your goal is to design an experiment in which your sets of nuts will differ with respect to this one independent variable, while keeping all other potential independent variables as similar as possible. Such an experiment will help you to determine whether the *independent* variable you chose to vary affected the *dependent* variable of the squirrels' decisions of what to do with the nuts. Squirrels may eat or cache nuts right next to where you provide them or at another location. When food items are taken to a different location, squirrels rarely carry more than one food item at a time, so, for example, transporting two nuts without shells will require more time and energy than if those nuts could be carried together inside a single shell. A squirrel that caches multiple nuts near the source will save travel time and energy, but these nuts may be more easily located by pilferers searching near where they have found other buried nuts.

The preactivity worksheet will guide you through the process of designing your experiment. *It is essential to complete the worksheet before the class activity; it will also be used to assign you to a team.* Your instructor will provide unsalted nuts in shells; either roasted or unroasted nuts may be provided. Nuts are a natural food source for squirrels, although you may be using nut species that are not native to your area. In the week or two prior to this experiment, the instructor will set out nuts of the species you will be using, in their shells, to provide local squirrels with experience with the characteristics of those nuts. You will use just one or two nut species for your experiment, most likely peanuts and/or hazelnuts. Different species vary in the thickness of their shells, the number of nuts inside each shell, and their caloric content, as well as their taste and smell. As you probably know, peanuts have thin, fragile shells that you can easily remove by hand or damage in controlled ways. Peanuts have one to three nuts (or occasionally more) within each shell; two-nut peanuts are the most common. Hazelnuts have hard shells that require a nutcracker to remove; it is very difficult to damage a hazelnut shell in a way that will not cause the shell to fall off of the nut. Hazelnuts almost always have a single nut in each shell. The energy available is approximately 5.7 kcal/nut for each nut inside a peanut's shell and approximately 9 kcal/nut for the nut in a standard one-nut hazelnut.

You will place your sets of nuts only a short distance apart so that a squirrel foraging at either set will also have the other set in view. Hopefully, you will be able to see squirrels taking nuts during the activity, which will allow you to determine which nuts are taken first; to see what the squirrels do with nuts from both sets; and to observe other factors that may affect the squirrels' choices. However, you should design an experiment that will provide meaningful data on overall nut preferences even if squirrels don't show up during your observation time and you return later to count how many nuts have been taken from each set. After you collect the data, you will try to explain your results. For example, did squirrels you observed store only the nuts from the set with characteristics most likely to survive a period of being buried in the ground? Did squirrels eat only the nuts from the set that required the least time and energy to eat? Did squirrels' choices seem to be based on the nut size, caloric content, shell thickness/hardness, or something else?

Preparing for your experiment

Students will work in teams (six to eight students if the class size permits), based on similarities in the experiments proposed in the preactivity worksheets. Within your team discuss and come to an agreement on details of your planned experiment, and then describe the experiment to your instructor, who may have additional considerations or suggestions to bring up before approving your plan.

Once your experiment has been approved, you may prepare your two sets of nuts: each set should contain at least 30 nuts. If the nuts in both of your sets are not the same size, you have a choice between controlling for the total number of nuts (by using the same number of nuts in each set) and controlling for the total volume of nuts (by using the same volume of nuts in each set). Controlling for the total volume may require a lot more work, if you are comparing nuts with and without shells, as you will need to carefully remove more nuts from their shells. Be sure that you take adequate notes on the method you use to remove nut shells (if you do that) and the method and nature of any damage to shells (size and positioning of holes or cracks, etc.). If you remove peanuts from their shells note whether the red integument around each nut is left on the nut (which will help hold the cotyledons together).

Review the data sheet and ask your instructor any questions before you go out to do the experiment. Clearly record the first and last names of each person in your team. Record the number of nuts in each set on your data sheet *before* you set them out for the squirrels (you will be counting the remaining nuts at the end of your experiment to determine whether you missed seeing any nuts taken, so it is important to know how many were there at the start). In each team, at any given time, ONLY one person should record data; this will prevent confusion about double-counting the nuts and will ensure that your team has enough data sheets. The other team members should be watching the squirrels and reporting what they see to the data recorder. At the end of the class, the instructor may provide copies of the data sheets for everyone in the team or you can photograph your team's completed data sheets with your smartphone.

Table 11.1 shows an example of how fairly complex data could be recorded on the data sheet provided. These data are for the comparison of two sets of peanuts, one set with two nuts inside each intact shell and the other set with single peanuts without shells. Each individual food item was given a separate number; if the squirrel divided the food item into portions with separate fates, the fates were recorded on separate lines, with each scored as the appropriate portion of the food item. Information on which squirrel took nuts and which nuts were taken on the same visit is not essential, but it may be interesting when you are able to record it. Note that you are asked to record time taken to eat but NOT time taken to bury nuts.

Table 11.1 Example of a filled-in data sheet.

Order in which nuts are taken	Nut set 1, on LEFT: 2 nuts in shell	Nut set 2, on RIGHT: 1 nut, no shell	Eaten	Stored	*Estimated* distance from nut set to where nut was eaten or stored: *include units*: cm, m, ft, or inches	Other information of interest, including time taken to *eat* nuts
1	✓		✓		5 cm	70 sec to eat both nuts } same squirrel,
2		✓	✓		8 cm	41 sec to eat } same visit
3	✓			✓	8 m	Buried in lawn
4	✓		✓ (.5)		2 m	} from same
4	✓			✓ (.5)	10 m	Buried in lawn, with half of shell around it } shell

Setting up your experiment in the field

Squirrels prefer to forage near the base of trees because they can more easily escape from the reach of ground-based predators if need be, so you will be placing your sets of nuts under a tree. Locate a tree with squirrels foraging below it or in its branches, or a tree with one or more leaf nests made by squirrels, or a tree near a tree with squirrel nests, or a tree near a location where your instructor, you, or another student has often seen squirrels foraging. Each team should choose a different tree, minimizing disturbance to squirrels foraging nearby because you want the squirrels to stay around to be your study subjects. As you set up your experiment and collect data, continue to avoid walking close to squirrels or making sudden large movements, and talk quietly when communicating with other team members.

The next step is to figure out where students will sit or stand to watch squirrels foraging beneath your tree. This location should be as far from the tree as you can go where you can still see which nuts squirrels are taking; it must be at least 6–7 m away from the base of the tree trunk. Have at least one of your team members, with binoculars, walk over to *sit* at this location. Next, you want to set out the nuts as quickly and efficiently as possible to make it more likely that any squirrels that are disturbed will return relatively quickly; send just one or two students to approach the tree, carrying the sets of nuts and a tape measure. With guidance from the seated observer, and before releasing any nuts onto the ground, place the two containers of nuts 20 cm apart and within 1 m of the base of the tree trunk.

Both containers should be the same distance from the main trunk of the tree, and the base of each container should be in clear view of the observer.

Once the observer verifies that the bases of both containers are in view, empty each container onto the ground; each set of nuts should be in its own fairly compact pile (about 10–15 cm in diameter), with the outer edges of the piles about 20 cm apart. Make sure that the observer can still see the nuts (if not, adjust their positions accordingly). Quickly measure the exact distance between the sets of nuts and between each set of nuts and the tree trunk, and record the distances, along with information indicating which set of nuts is on the left, from the observer's perspective (this will be set 1), and which set is on the right (this will be set 2). Once the nuts are in place, everyone in your team should sit near the seated observer, making a fairly compact group to leave more area open for the squirrels to move in freely.

Conducting your observations

Each time a nut is taken by a squirrel, one of the students in the team needs to quickly volunteer to follow that squirrel until the nut is eaten or buried (or meets some other fate). The follower should let the data recorder know the set of nuts from which the nut was taken, and the data recorder should let the follower know the nut number. If a squirrel takes a nut away from the immediate vicinity of the tree, the follower must follow the squirrel (getting no closer than 5 m from the squirrel) to discover where it goes and whether it eats or stores the nut. It is possible that a squirrel will disappear behind vegetation or go onto a roof or otherwise leave your view; wait 5 min for the squirrel to reappear, continuing to look around at places where it might have gone (if the squirrel hasn't reappeared after 5 min, you may return to your team). If the squirrel reappears without the nut in less than 30 s, you can assume that the nut has been stored, but if the squirrel is gone for longer and reappears without the nut, you won't really know the nut's fate. For each nut you see eaten or stored, do your best to *estimate* the straight-line distance from the set of nuts it came from; if possible use metric units (centimeter and meter) in your estimates, but if not, you may use feet and inches and convert them to metric units later; do not attempt to actually measure these distances, as that would disrupt your experiment. Then return to the data recorder and report what happened to the nut and the distance at which the nut was eaten or stored. If the follower is unsure whether a nut was eaten or stored, the data recorder should record a "?" in both the "Eaten" and "Stored" columns. Occasionally, when there are two nuts in a shell, the nuts may have different fates, e.g., one may be eaten and the other may be stored; if this happens, then record the nut as 0.5 eaten and 0.5 stored (see the example in Table 11.1). If more than one squirrel visits your nuts look for markings that may allow you to recognize individuals (e.g., tail length or fullness, notches in the ears, or distinctive fur colors); you can let the data recorder know which nuts are taken by recognizable squirrels (you can give the squirrels descriptive names) in order to later check whether different individuals behave differently.

Students should take turns following squirrels, cycling nut by nut through everyone except the data recorder; if more than one squirrel is taking nuts, several students may be following different squirrels at the same time. About halfway through data collection (or more often, if you prefer) rotate the data-recording duty so that the data recorder has a chance to follow squirrels. Start a new data sheet each time you switch data recorders.

Over the course of the exercise, for two to five nuts from each set, the team should attempt to record how long it takes a squirrel to *eat* the nut. This will provide information on time (and correlated energy) costs for eating nuts from each set, which is evidence to support (or refute) hypotheses about the time and energy cost of shell removal. Sometimes nuts will be eaten quite close to the piles from which they were taken, allowing for easy recording by an observer with binoculars and a person keeping time—take advantage of this situation when it occurs. *For nuts without shells start timing when the squirrel first puts all or part of the nut in its mouth (provided that chewing motions follow soon after first contact with the mouth) and stop timing when chewing ceases and the nut has apparently been fully consumed. For nuts in a shell start timing* ONLY *when a squirrel first starts to remove the shell and stop timing when chewing stops after the nut(s) in the shell have been fully consumed.* If the squirrel takes a break of longer than 5 s while eating the nut do not include this time in your data. Report nut consumption times to the data recorder to be recorded directly onto the data sheet. If only one of your sets of nuts, or neither, has any nuts eaten, you may just leave that line(s) blank.

Stop collecting data if all the nuts in one of your sets have been taken before the end of your designated class time, as squirrels will no longer have a choice of nuts. On the other hand, if fewer than 10 nuts were taken during your scheduled class time, then have a member of your team return every hour up to sunset to record the number of nuts remaining in each of your sets. In the unlikely case that no additional nuts have been removed before sunset, please have a team member return for another count the next day. Your instructor will let you know whether squirrels are the most likely animals to be consuming the nuts you leave.

Finishing up

When you have finished observing squirrels for the day, use the measuring tape to measure the distance from the sets of nuts to where your team sat or stood to observe squirrels and count and record on your data sheet the number of nuts remaining in each set. Make sure that you have filled in all the other requested data and check that a student who didn't record data can correctly interpret everything recorded.

Results/discussion

The first step in analyzing your team's data is to fill in Table 11.2, entering the number of nuts in each category. Indicate whether the nuts that were taken followed the

Table 11.2 Data overview.

	Eaten	Stored	Fate unknown
Nut set 1			
Nut set 2			
Total in column			
Supports prediction? (Yes/No)			–

predictions you made in parts c and d of question 4 on your preactivity worksheet. For example, if you predicted that more of the nuts in set 1 than in set 2 would be eaten and you found that 10 nuts from set 1 were eaten but only 2 nuts from set 2 were eaten, those data would follow your predictions of which nuts would be eaten. Although nuts whose fate is unknown will not be included in statistical analyses, it is informative to know how many nuts were in the unknown category; if it is a small proportion of the nuts taken, you will have more confidence that the nuts with known fates are representative.

If more than a few nuts from each set were taken, a Fisher exact probability test will allow you to compare the relative proportions of nuts in the two sets that were eaten versus stored. If most of the nuts were eaten, or most of the nuts were stored, it may be more informative to conduct a chi-square test of goodness of fit to determine whether nuts from the two sets were equally common among the nuts that were eaten (or the nuts that were stored). In order to conduct each test of goodness of fit, you will need to have data for at least 10 nuts. For example, for nuts that were eaten, the null hypothesis would be that equal numbers of nuts from set 1 and set 2 were eaten. In the example described earlier, a total of 12 nuts were eaten; thus the null hypothesis would predict that six nuts from each set would be eaten. If you do not have enough data for any of these tests, but a total of 10 or more nuts were taken, you can use a chi-square goodness of fit test to determine whether squirrels had a significant preference for one set of nuts over the other, ignoring the fate of those nuts (the null hypothesis would be that squirrels had no preference for one set of nuts over the other).

If you were able to record nut-consumption times, then compare the ranges and averages for your two sets of nuts. Do differences (or lack of differences) help explain which nuts squirrels chose to eat? If both sets of nuts were eaten, a comparison of the eating distances from the nut piles may help explain the squirrels' choices (calculate the range and mean distance separately for each set of nuts). If both sets of nuts were stored, a comparison of how far they were taken for storage may show a difference that helps explain whether this behavior was adaptive (calculate the range and mean storage distance separately for each set of nuts). If multiple squirrels visited your nuts, you may have had the opportunity to observe whether squirrels behaved differently when potential competitors were nearby than when they were foraging alone. If you could recognize different individual squirrels, did they make different choices of which nuts to take or what to do with them?

Make a graph to illustrate one of the data sets that interests you—choose the format that you judge best to display the data you want to present. For example, a bar graph would make it easy to compare numbers of nuts eaten versus stored in the two sets of nuts you provided, while line graphs would be helpful for showing change over the course of your experiment.

Describe one or two experiments that would allow you to determine the potential effect of a single independent variable that was not controlled in your initial experiment, and that you believe may have affected your results. Question 3 on the preactivity worksheet includes a list of uncontrolled variables that could potentially have affected the results of your experiment, and others may have become apparent in the course of your experiment. For example, if in your experiment hazelnuts in shells were buried more often than peanuts without shells, you could test the effect of shell presence alone by comparing how often hazelnuts with and without shells were buried.

Conclusions

Decide which factors seemed to be important to squirrels in determining which nuts they preferred and in determining what they did with each kind of nut. Look back at your answers to questions 1 and 3 on the preactivity worksheet and discuss how each of the differences between your two sets of nuts may have affected the outcome of your experiment. Try to explain any behavior that you did not predict in advance.

Be sure to include explanations for the following choices made by the squrirrel(s) you observed:

- Which nuts the squirrels took first
- Which nuts squirrels ate
- Which nuts squirrels stored
- Why squirrels ate nuts from both sets (if they did that)
- Why squirrels stored nuts from both sets (if they did that)
- Why nuts in one set were stored farther away from the nut pile than the nuts in the other set (if this happened)
- Why nuts in one set were eaten farther away from the nut pile than the nuts in the other set (if this happened)
- Why squirrels took more nuts from one set than the other (if they did this)

If you observed any behavior that you did not predict, then propose an experiment or experiments that might clarify the causes of that behavior.

Option 2: Ants (family: Formicidae)

Materials needed (for each team of two to four students)

- Data sheets (one of each kind per student [forager count data sheet, forager behavior sheet, and observations of non-foraging ants sheet] plus at least two extra data sheets on nonforaging ants for each team).
- Supplementary ant keys (if provided) and nutrient content lists (one of each per student).
- Small quantities of common picnic foods. Food items should be selected from a list provided by your instructor.
- four paper or plastic cups (or 8-oz yogurt containers);
- one serrated knife (plastic picnic knives work fine);
- four paper towels;
- one small ruler, marked in millimeters;
- two clipboards;
- two pencils with erasers;
- a timing device (cell phone, stopwatch, or wristwatch with a second hand);
- one magnifying glass or hand lens (either 5× or 10× magnification works well);
- two small plastic (or glass) vials with screw tops, snap-on caps, or corks (plastic is preferred for the field);
- one small (~9–12 cm diameter) finger bowl and crushed ice to fill it;
- one blunt probe (or a pair of entomological forceps); a small soft paintbrush may also be useful;
- large plastic garbage bags for students who would like one to sit on;
- dissecting microscope.

Be sure to wear appropriate clothing and protection for fieldwork, including shoes or boots and sunscreen.

Step-by-step instructions

Leading up to the experiment day, be on the lookout for ant colonies in the area where you will be conducting your experiment; you may see colony entrances in sidewalk cracks or in lawns.

Designing your experiment

Your goal is to design and conduct an experiment on the food selection behavior of ants. You will investigate the choices ants make among four different foods you present around the entrances to their nests. Different components of foods may be needed for different purposes. When ants select a food item, there are many factors that must be balanced. Like people, ants use protein to build muscle and other structural body parts and also to construct enzymes to catalyze the chemical reactions in their cells, while fat and carbohydrates provide energy. Quickly growing and developing larvae of insects require more protein than adults (Dadd, 1985). Food preferences might change depending on how many larvae the colony is developing. Fat is the most space-efficient way to store energy, and therefore by weight is the richest energy source to consume; however, the energy in carbohydrates is more readily available. Water plays a key role in many metabolic processes, but excess water adds weight to food. Salt functions in helping maintain water balance, and vitamins serve as coenzymes.

Based on this information, what predictions would you make about food preferences of ants? For example, you may predict that foods with higher fat content will be preferred over foods with lower fat content because they contain more energy per unit volume.

Your instructor may supply a list of different picnic foods you may use in your experiment, with information on their percentage by weight of different nutrients, water content, and sodium (salt) content. Information on additional foods can be found in Gebhardt and Thomas (2002, pp. 1−103) and on package labels. Choose a set of food items that will allow you to best compare a single factor that you believe will be important to ants' choices. In addition to nutrient content, take differences in texture into account as you design your experiment. Note that if you are using different types of food, they are extremely likely to differ in more than one way, but try to control as many of the factors that you don't mean to test as possible. Predict the ants' order of preference for the food items. The preactivity worksheet will guide you through the process of designing your experiment. *It is essential to complete the worksheet before the activity; it will also be used to assign you to a team.*

Preparing for your experiment

Students will work in teams (two to four students if the class size permits) based on similarities in the experiments they propose in their preactivity worksheets. Within your team discuss and come to an agreement on details of your planned experiment. Be sure to consider how you have controlled for factors that you do not mean to test and how you will interpret results that could be explained by more than one factor.

Once you have designed the first experiment, you can plan to repeat it with or without modifications. The following are some examples of modification.

- Placement of food items can be rotated from the initial locations to test for unanticipated position effects, effects of lingering food or ant trail odors, and the effect of previous forager experience.
- More preferred foods may be placed farther away than less preferred foods to see whether preferences switch.
- Trails of "crumbs" can lead to a bigger piece of food to test whether ants discover it more quickly than a piece of food of the same type and size at the same distance without the trail of crumbs.

Once your plans have been approved by your instructor, prepare enough pieces of food for all the experiments that you will have time to conduct, plus a few extras for unanticipated disasters. You will want to be sure that there is enough of each food item to last 30 min in a hungry colony but not so much that ants can easily hide under it and be difficult to count and observe. Thus food items should be at least 3 mm^3 and no more than 1–2 cm^3. Plan the size and shape of your pieces of food and where you will place them (note that ants may have trouble finding foods placed more than 10 cm from the nest entrance, in the time allotted).

If ants with painful stings live in your area, ask your instructor for guidance in identifying these ants and be sure to avoid setting up your experiment anywhere near them.

Setting up your experiment in the field

Pack two clipboards for your research team's data sheets (forager count data sheets and forager behavior data sheets) as well as a ruler, a magnifying glass, four small vials, and a pen or pencil. Be sure that you have as many copies of each foraging data sheet as the experiments you intend to conduct. Also bring several data sheets for observations of nonforaging ants, just in case you need them (note that you will make observations of nonforaging ants only if the ants that you watch do not eat any of the food that you put around their nest and you are unable to collect foraging data). If nobody in your team has a cell phone or watch that can measure seconds, then borrow one or take a stopwatch if it is available.

Conducting your observations

Find your ant mounds—look in lawns and sidewalk cracks to find ant nest entrances. Avoid any areas where you see ants with painful stings and look for ant mounds with a few ants already wandering around them, which are at least five ant mound diameters from the closest other ant mound; this will reduce the confusion of competitive interactions between ants from different nests. Once you have found an ant mound with at least several ants outside the entrance, set up your experiment, record the start time and number of ants present at each food item at 0 min, and draw the locations and sizes of each food item on the forager count data sheet.

Use the ruler to measure the distance of each food item from the nest entrance; hold the ruler slightly above the ant mound and ground to avoid disturbing the ants while making these measurements. One team member can be responsible for timing the 3 min between count periods. All students should participate in counts so that counts for different food items occur as close together in time as possible.

Between count periods use the magnifying glass (held at least 2.5 cm above the ants) to observe the behavior of individual foragers and fill in the table on the forager behavior data sheets; work in pairs, with one person observing and the other writing down the observations, and trade roles partway through the experiment so that everyone has a chance to make observations. When using the magnifying glass be careful not to focus the sun's rays on your ant.

In the unlikely chance that you have waited for 20 min and no ants have visited any of your food items, you have two options. One option is to move to another ant mound, preferably of a different species, and attempt to do your experiment there. The other option is to stay where you are and fill out data sheets on the behavior of nonforaging ants (leave your food items in place until you have completed as many of these data sheets as you have time to complete).

Between experiments at the same ant nest be sure to clean up any leftover food from the previous experiment; if ants are on this food, gently knock them off (a finger flick will often suffice) or move the food with ants to 1–3 m away from the nest entrance (the ants should have no trouble finding their way home but won't disrupt your next experiment).

After your last experiment at an ant nest, trap one or two live workers in the vials (by holding the opening of the vial against the ground in front of a running ant, with the vial angled up at about 30 degrees and with light allowed to shine in through the vial walls and the end). Bring these ants back to identify under the dissecting microscope; you will release them back at their nest entrance when you are done.

Back in the classroom

Fill a small finger bowl with crushed ice and cover the ice with a moist (not wet) paper towel. Gently release the ant(s) onto the paper towel. Within a few seconds, the ants should stop moving. Place the finger bowl under the dissecting microscope; ants can be repositioned with gentle use of the blunt probe, entomological forceps, or a small soft paint brush, if necessary. Try to find all the anatomical features, shown in Fig. 11.1, that are present on your ants. Use the dichotomous key (Table 11.3) to try to identify your ants; in a dichotomous key, you proceed through a series of pairs of choices until you arrive at an identification. For example, if you have a little black ant, *Monomorium minimum*, the steps in using identification Table 11.3 would be 1a → 1b → 3a → 3b (at each step, if the characteristics listed in the "a" description do not match your specimen, then go on to the characteristics listed in the "b" alternative). The ants listed in Table 11.3 are common across much of the United States, but it is certainly possible that you may have a species that is not included, considering

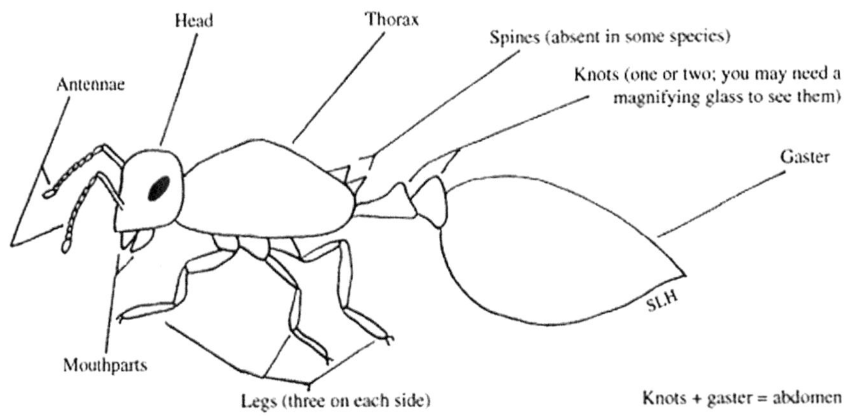

FIGURE 11.1

Generalized worker ant anatomy.

Table 11.3 Dichotomous key to workers of two common US ant species and two additional ant genera commonly found outdoors in the United States.

1) a. Single knot present 2
   b. Double knot present 3

2) a. Color black, or black with a red thorax; thorax rounded; 8-14 mm long.......... Carpenter Ant, *Camponotus* sp.
   b. Color light brown to dark brown; common at picnics!; about 4 mm long .... Cornfield Ant, *Lasius* sp.

3) a. Color dark brown; two very short spines present before knots; fine sculptured lines parallel to long body axis, on head and thorax; 3-4 mm long .... Pavement Ant, *Tetramorium caespitum*
   b. Color black; no spines present before knots; body shiny, without sculptured lines on head or thorax; about 2 mm long .... Little Black Ant, *Monomorium minimum*

that there are over 900 species in the United States, including over 85 that have been introduced from other countries. If you believe that your ant is not in the key you are using, then state which of the listed species (or genera) your ants most closely resemble and then describe how they differ from that species or genus. Your instructor may provide an additional or alternate key for you to use; Bolton (1994), Fisher and Cover (2007), and Ellison et al. (2012) cover far more species than the key provided here. Check your identification with your instructor, who may also want you to list the order of steps you followed in identifying your ants with the key.

Finishing up

Once you have identified your ants return them to the vials in which you captured them and tabulate your data. On the forager count data sheet, this will involve summing up the numbers of foragers counted at each food item, over all 11 counts (of course, many individual ants will have been counted multiple times, but this comparison should still give an accurate relative measure of food preference, with the realization that multiple (or prolonged) visits by the same individual are weighted as strongly as single visits by different individuals). Fill out the top of the forager behavior data sheet. *Be sure to return your ants to their colonies at the end of the class.*

Results/discussion

Make a graph to show how the number of foragers changed over time at different food items. Indicate the number of foragers on the vertical axis (from 0 at the origin, up to the maximum number of foragers observed) and the time of the count on the horizontal axis (0 min = time at which food was first placed, subsequent labels at 3-min intervals through 30 min). Graph four separate lines, each tracking the forager count at a different food item. All four lines should start at the origin, and they may well overlap for a part of their extent and/or cross one another; distinguish them by using different colors or patterns (e.g., lines that are continuous vs. dotted vs. dashed vs. alternating dashes and dots).

Conduct a chi-square one-sample test for goodness of fit to statistically compare the four food items in terms of the maximum number of foragers seen simultaneously feeding on that food item. The null hypothesis for this test is that each food item attracted the same maximum number of foragers (1/4 of the total obtained by adding together the maximum number of foragers for each of the four food items). The alternative hypothesis is that the maximum number of foragers was different for the different food items. If the maximum numbers of foragers seen at all four food items sum to <20 (making a chi-square test invalid) or if the maximum numbers of foragers are not very different for the different food items, conduct a chi-square one-sample test of goodness of fit to statistically compare each food item's summed ant counts from all 11 observation periods. The null hypothesis for this test is that each food item had the same total number of ants counted during the 11 observation periods (1/4 of the total of all the ants counted at all the food items). The alternative hypothesis is that the total number of ants counted during the 11 observation periods was different for the different food items.

Describe one or two follow-up experiments that could test for the potential effect of a single independent variable that was not controlled in your initial experiment, and that you believe may have affected your results. Your answer to question 3 on the preactivity worksheet contains a list of such uncontrolled variables that could potentially have affected the results of your experiment, and others may have become apparent in the course of your experiment.

Conclusions

There are a number of different measures that might be counted as indicators of food preference: the order in which different food items were discovered, the maximum number of foragers seen simultaneously at a food item, and/or a food item's sum of ant counts from each of the 11 observation periods.

- Can you think of other appropriate measures?
- Do all these measures give the same rankings for the four food items that you used? If not, attempt to explain why not and explain which ranking you believe best reflects the ants' food preferences.
- For your graph, attempt to explain any differences between the lines for the different food items; do higher ranked items have lines with different trajectories than those of lower ranked items?
- Which (if any) of your predicted rankings is the best match for your actual data?
- Which factor or factors do your data show are important in determining ant food preferences, and which can be ruled out?

- Are the differences in preference for your different food items statistically significant at the 0.05 alpha level?

Compare your data to those of other teams.

- Were the same factors preferred in food choices of all colonies of the same species of ants or were there differences between colonies?
- Were the same factors favored by different species of ants?
- Explain how the food preferences you observed may help fulfill the ants' nutritional needs at the time of year when you did the experiment.

If you made observations of nonforaging ants:

- What did these ants appear to be doing?
- Were their activities coordinated with those of other ants from the same colony?
- How did behaviors vary between individuals?
- Did you notice any morphological differences between ants doing different things, and if so, did these morphological differences appear to be adaptive for the behavior you observed?

Speculate about why some ants don't forage even when food is readily available near the colony entrance.

- Is the behavior that they are performing more important or are they morphologically or experientially unsuited for foraging?
- If no ants ate any of your food was the lack of foraging behavior restricted to the colony you observed or did it include all colonies of your ants' species or some or all colonies of all species?
- Did another team's ants eat the foods that were ignored at your colony? If so, were their ants of the same species as yours or a different species?

Questions for discussion

These apply to both the squirrel and ant versions of the exercise.

1. Do squirrels or ants need to be "aware" of the factors you have considered, in order to make the "decisions" that you would predict on the basis of quantitative calculations? Does use of the term "decisions" imply mental capacities that these animals may not have?
2. If animals are not individually identified and data from all individuals are summed together, individual animals will be represented in the data set in proportion to the number of visits to food (ants) or number of nuts they have taken (squirrels). If individual variations in behavior exist and all animals do not make the same number of visits/take the same number of food items while you are watching them, the variations shown by the individuals with the most visits/food items taken will be most strongly represented. How might this bias your data, and how does it limit the scope of your conclusions?

3. The nuts stored by squirrels and the seeds stored by ants may sprout and grow if the animals do not eat all of them. How might the seed-storing behavior that you observed be more beneficial to the parent plant than if the seeds were just left on the ground where they fell? Would you predict that these benefits could offset the cost to a plant of attracting seed-storing animals that may consume many of its seeds? Would variations you observed in seed-storing behavior, e.g., in whether a single seed was stored at each location or a group of seeds were stored together at the same location, and in distances that seeds were taken from the source, provide different amounts of benefit to the parent plant? See Steele and Smallwood (1994), Steele et al. (2014), and Lichti et al. (2020) for relevant information on this topic.

Acknowledgments

Peter Smallwood and Michael Steele generously provided logistical advice, hands-on training, and useful insights and ideas about studying the foraging behavior of squirrels. Michelle Scott provided SLH with instructions on handling and subduing ants for observation. Nathan Sanders, Bonnie Ploger, Ken Yasukawa, Thomas Mione, Peter Sherman, Thomas King, and Jeremiah Jarrett provided thoughtful reviews and advice on earlier versions of this exercise. Theresa Singer helped broaden the scope of the ant exercise. Valerie Kruse and Mary Waters wrote the main body of the key to ant species, with useful advice from Mary Salerno. We thank Susan Margulis and Heather Zimbler-DeLorenzo for inviting us to rework two exercises from the first published edition into this new version, and for their helpful editing. Students at Central Connecticut State University, Carleton College, and Moravian College provided data, good ideas, and help in testing different versions of these exercises.

References

Bolton, B. (1994). *Identification guide to the ant genera of the world.* Cambridge, MA: Harvard University Press.

Brown, J. H., Sibly, R. M., & Kodric-Brown, A. (2011). *Metabolic ecology: A scaling approach.* London: Wiley Blackwell.

Castillo-Guevara, C., Cuautle, M., Lara, C., & Juárez- Juárez, B. (2019). Effect of agricultural land-use change on ant dominance hierarchy and food preferences in a temperate oak forest. *PeerJ, 7*, e6255. https://doi.org/10.7717/peerj.6225

Cuthill, I. C., & Houston, A. (1997). Managing time and energy. *Chapter 5 in Behavioral ecology: An evolutionary approach* (4th ed.). Oxford: Blackwell Science Ltd.

Dadd, R. H. (1985). Nutrition: Organisms. In G. A. Kerkut, & L. I. Gilbert (Eds.), *Regulation: Digestion, nutrition, excretion: Vol. 4. Comprehensive insect physiology biochemistry and pharmacology* (pp. 313–390). Oxford: Pergamon Press.

Delgado, M. M., & Jacobs, L. F. (2017). Caching for where and what: Evidence for a mnemonic strategy in a scatter-hoarder. *Royal Society Open Science, 4*(9), 170958. https://doi.org/10.1098/rsos.170958

Duncan, F. D., & Lighton, J. R. B. (1994). Water relations in nocturnal and diurnal foragers of the desert honeypot ant *Myrmecocystus*: Implications for colony-level selection. *Journal of Experimental Zoology, 270*(4), 350–359.

Ellison, A. M., Gotelli, N. J., Farnsworth, E. J., & Alpert, G. D. (2012). *A field guide to the ants of New England*. New Haven, CT: Yale University Press.

Emlen, J. M. (1966). The role of time and energy in food preference. *The American Naturalist, 100*, 611–617.

Fischer, R. C., Ricter, A., Hadacek, F., & Mayer, V. (2008). Chemical differences between seeds and elaiosomes indicate an adaptation to nutritional needs of ants. *Oecolgia, 155*, 539–547.

Fisher, B. L., & Cover, S. P. (2007). *Ants of North America: A guide to genera*. Berkeley, CA: University of California Press.

Gebhardt, S. E., & Thomas, R. G. (2002). *Nutritive value of foods*. USDA-ARS Home and Garden Bulletin Number 72. Available online (last accessed 4 December 2020) at https://www.ars.usda.gov/arsuserfiles/80400525/data/hg72/hg72_2002.pdf.

Hansen, L. D., & Klotz, J. H. (2005). *Carpenter ants of the United States and Canada*. Ithaca, NY: Cornell University Press.

Hölldobler, B., & Wilson, E. O. (1990). *The ants*. Berlin: Springer.

Jacobs, L. F., & Liman, E. R. (1991). Grey squirrels remember the locations of buried nuts. *Animal Behaviour, 41*, 103–110.

Lichti, N. I., Dalgleish, H. J., & Steele, M. A. (2020). Interactions among shade, caching behavior, and predation risk may drive seed trait evolution in scatter-hoarded plants. *Diversity, 12*, 416. https://doi.org/10.3390/d12110416

Lima, S. L., Valone, T. J., & Caraco, T. (1985). Foraging-efficiency - predation-risk trade-off in the grey squirrel. *Animal Behaviour, 33*, 155–165.

MacArthur, R. H., & Pianka, E. R. (1966). On optimal use of a patchy environment. *The American Naturalist, 100*, 603–609.

Stahlschmidt, Z. R., & Johnson, D. (2018). Moving targets: Determinants of nutritional preference and habitat use in an urban ant community. *Urban Ecosystems, 21*, 1151–1158.

Steele, M. A., Contreras, T. A., Hadj-Chikh, L. Z., Agosta, S. J., Smallwood, P. D., & Tomlinson, C. N. (2014). Do scatter hoarders trade off increased predation risk for lower rates of cache pilferage? *Behavioral Ecology, 25*, 206–215.

Steele, M. A., Halkin, S. L., Smallwood, P. D., McKenna, T. J., Mitsopoulos, K., & Beam, M. (2008). Cache protection strategies of a scatter-hoarding rodent: Do tree squirrels engage in behavioural deception? *Animal Behaviour, 75*, 705–714.

Steele, M. A., & Koprowski, J. L. (2001). *North American tree squirrels*. Washington, DC: Smithsonian Books.

Steele, M. A., & Smallwood, P. D. (1994). What are squirrels hiding? *Natural History, 103*, 40–45, 90.

Stephens, D. W., & Krebs, J. R. (1986). *Foraging theory*. Princeton: Princeton University Press.

Tschinkel, W. R., & Kwapich, C. L. (2016). The Florida harvester ant, *Pogonomyrmex badius*, relies on germination to consume large seeds. *PloS One, 11*(1), e0166907. https://doi.org/10.1371/journal.pone.0166907

Whitney, E. N., & Rolfes, S. R. (1993). *Understanding nutrition* (6th ed.). Minneapolis, MN: West Publishing Company. Appendix H, pp. H-0–H-65.

Further reading

Brown, L., & Downhower, J. F. (1988). *Analyses in behavioral ecology: A manual for lab and field* (Two related labs: Lab 7, Food value and the foraging preferences of squirrels, pp. 28–30; Lab 5, Prey location in squirrels, pp. 22–23). Sunderland, MA: Sinauer Associates, Inc.

Delgado, M. M., Nicholas, M., Petrie, D. J., & Jacobs, L. F. (2014). Fox squirrels match food assessment and cache effort to value and scarcity. *PloS One, 9*(3), e92892, 8 pp.

Gurnell, J. (1987). *The natural history of squirrels*. New York, NY: Facts on File Publications.

Hopewell, L. J., Leaver, L. A., & Lea, S. E. G. (2008). Effects of competition and food availability on travel time in scatter-hoarding gray squirrels (*Sciurus carolinensis*). *Behavioral Ecology, 19*, 1143–1149.

Tamura, N., & Hayashi, F. (2008). Geographic variation in walnut seed size correlates with hoarding behavior of two rodent species. *Ecological Research, 23*, 607–614.

Thorington, R. W., Jr., & Ferrell, K. (2006). *Squirrels: The animal answer guide*. Baltimore, MD: Johns Hopkins University Press.

Vander Wall, S. B. (1990). *Food hoarding in animals*. Chicago, IL: University of Chicago Press.

Part II. Instructor notes

Squirrels

Classroom management/blocks of analysis

a. Brief instructions to students at least a week before the exercise is to be conducted, directing them to read the exercise description and any additional handouts and to complete the preactivity worksheet and submit it to the instructor in time for the instructor to sort students into teams based on the similarity of the experiments they suggested. (The instructor will also need to determine which nuts to acquire for each team, and to acquire those nuts before the day of the exercise.): 5–10 min. of time in class.

b. Whole-class discussion of introductory material and data collection methods (it will be useful to project and briefly discuss Table 11.1), assuming students have already read the activity and completed the preactivity worksheet: 15 min.

c. Within-team discussion to reach consensus on methods (assuming students are assigned to teams based on the similarity of the experiments they suggested on their preactivity worksheets): 15 min.

d. Nut preparation time within each team will vary depending on the types of nuts to be used and whether shells must be removed or damaged: 10–25 min.

e. Data collection: 1.5–2 h (will vary with the number of squirrels that participate in experiments). If after 45 min, one team has data for at least 15 nuts and another team has had no squirrels visiting their nuts, the successful team should remove their nuts right before the team with no data relocates their nuts to the tree that the successful team was using. Students should note that the starting conditions were different for the second team, e.g., squirrels may have been sated by eating the first team's nuts, and the second team may be using nuts with different properties than those the squirrels have become accustomed to finding in that location.

f. Data sheet checking: 5–15 min.

g. Compiling team data, team discussion of hypotheses to test, instructor moving between teams to provide statistical advice and possibly also graphing advice, students performing statistical tests: 1 h (up to 2 h if students are also making graphs and writing relatively short answers to questions).

h. Summary discussion: 30–45 min.

Steps b–f can occur in a standard $2\frac{2}{3}$ - to 3-h college lab period (with careful attention to timing). Steps g and h can occur in the next lab period, or g can occur during a lecture period if students can meet with their own teams, the instructor can move between teams, and the students are not making graphs during the same lecture period. Step h can occur in a subsequent lab or lecture period. If lab periods are shorter than $2\frac{2}{3}$ h, steps b–d can be done in advance; store nuts in sealed plastic bags in a refrigerator. Data collection can then take place during the next one or two lab periods.

Teaching the activity

Preclass preparation

On some campuses, Institutional Animal Care and Use Committee approval may be required to provide food to squirrels, so check with that committee to determine whether this is required on your campus. Instructors should ask students who are severely allergic to nuts to identify themselves to the instructor before the day of the exercise (or any activities involving the handling of nuts; the preactivity worksheet starts with a question on nut allergies). Provisions should be made for these students to be out of the classroom while nuts are being prepared by each team.

Recommendations regarding selection of species and/or setting for exercise

In the northeastern and central United States, eastern gray squirrels (*S. carolinensis*) readily forage on peanuts and hazelnuts provided by people during fall and winter. It can be more difficult to attract squirrels to nuts in spring and summer. If eastern gray squirrels do not live in your area (their range covers almost all the eastern half of the United States but barely extends into southern Canada [or Nebraska]), then experiment with your local species of *Sciurus* (at least one species of *Sciurus* lives in most parts of the United States where there are trees). Non-*Sciurus* squirrels may also be suitable subjects, but rather than burying nuts in the open, some of these squirrels store nuts in underground burrows, which prevents certainty about whether the nuts were stored or were eaten out of sight. If you plan to use non-*Sciurus* squirrels, you may want to look up relevant information about their foraging and caching behavior; some are "larder hoarders," which store large numbers of food items in just one or a few locations and, unlike scatter hoarders, actively defend those locations.

Good places to do this experiment are locations where you have regularly and recently observed foraging squirrels and, if possible, where the squirrels are relatively accustomed to the presence of people. A lawn with trees interspersed or on the edges should allow students to observe the squirrels eating nuts by the nut piles, carrying nuts away to eat in trees, burying nuts in the lawn, and possibly occasionally caching nuts in trees.

Squirrels tend to forage most actively in the hour or two before sunset, so it is convenient if you can schedule the class to overlap that time period. Squirrels often stay in their nests (dreys) on cold and windy days and rarely forage in rain or snow. As with all field exercises, it is a good idea to schedule several possible weeks during which this exercise could be done by the class and to conduct it starting in the first week with adequate weather conditions, with indoor or non-weather-dependent alternatives for weeks when the weather is unsuitable. This exercise works well if scheduled over 2 weeks of class time; all teams run their experiments in the first week. In the second week, any teams with fewer than 10 nuts taken by squirrels in the first week have a second chance to prepare additional nuts and collect more data—these teams can be sent to the most productive sites from the previous week (or other sites where squirrels are seen foraging in the second week; as long as the same experiment is conducted, data can be summed across both weeks). For teams with enough data at the start of the class, the instructor can work with them in turn on statistical analyses and graphing appropriate for their data, starting with the information they provide in Table 11.2. The instructor can work with teams collecting additional data when they return after their experiments.

It is important to do a pilot test to be sure that your squirrels will be cooperative at the time of year when you want to do the experiment. It is also important to accustom squirrels to come to the locations where students will be conducting their experiments. Under each tree the students will be using scatter 10 nuts of each species the students will be using (nuts should be provided in their shells). Do this for 1–2 weeks before the class exercise, daily if possible, at the time of day when your class will meet. It may be helpful to remain in the area and watch squirrels eating the nuts in order to accustom them to being observed. You may also toss (underhand) one or two nuts in shells close to foraging squirrels. Although they may initially run off a short distance, they will often return to retrieve the nut(s) and may then look around for more. In fall, you will often find squirrels foraging on acorns in and below oak trees; these squirrels will also eat peanuts and hazelnuts placed near the base of these oaks, although they may ignore acorns provided by experimenters, especially in years in which the local oaks have produced abundant acorns.

Unsalted peanuts and hazelnuts in the shell (these may be roasted or unroasted) are often available from a local supermarket. If not try a fruit market or store specializing in nuts. For peanuts, whose shells crack easily, make sure that most peanut shells are intact before purchasing them (they are generally sold in transparent plastic bags). One pound of peanuts with mostly intact shells will generally suffice for each set of two-nut peanuts. However, the cotyledons of peanuts tend to separate when they are removed from the shell, producing more individual food items; to reduce this, the red skin (integument) can be left on the nuts, but it would be good to have an extra 0.5 lb of peanuts/set if students won't be using nuts whose cotyledons have separated. Peanuts with a single nut in each shell are less common than peanuts with two nuts in each shell; if students want to use one-nut peanuts, more nuts may need to be purchased initially or one team may need to assemble nuts from the stock for the entire class. Three- and four-nut peanuts are even less common; if you would like to have these available, you may need to select them separately at a store that offers loose peanuts. Hazelnuts in shells (also sometimes called filberts) are most readily available in fall; if you would like to do this exercise at another time, you can store them for months in sealed plastic bags at normal household refrigerator temperatures. If students will be leaving hazelnuts in their intact shells, you need to purchase just a few more hazelnuts than the students plan to use in their experiments, as hazelnuts sold in shells normally have shells that are undamaged. However, for hazelnuts from which students will be removing the shells, you need to purchase about twice as many as the students will need, because some kernels will be broken in the process of removing the shells. It is best not to separately purchase nuts without shells because they may differ from the nuts you purchase in shells, in preparation method (roasting time, oil used, etc.), or other ways (size, genetic strain, nutrient composition due to growing conditions, etc.). When students are comparing a set of nuts in shells to another set of nuts of the same species whose shells have been removed, both sets of nuts should be as similar as possible before the shells are removed. For example, for peanuts, nuts left within their two-nut shells should be compared to nuts removed from two-nut shells, not nuts removed from shells that appear to contain a different number of individual nuts; this will help control for the size of the nuts inside the shells (however, if one-nut peanuts in shells are being used by one of your teams, you may not have enough of these to assemble a companion set of peanuts removed from one-nut peanut shells, adding another potential independent variable to the experiment). Some teams may need to repeat their experiments in a second week, if time allows, so you may want to purchase extra nuts initially or to wait to see what is needed. If any teams want to control for volume rather than the number of nuts with and without shells, you will need at least four times as many nuts for the set without shells (determine the equivalent volumes empirically in advance to make sure you will have enough nuts).

Nutcrackers are needed to open the shells of hazelnuts and other nuts with shells harder than peanut shells. No-jaw zinc-plated 4.5-inch nutcrackers are inexpensive, readily available, and work well; jawed nutcrackers are more difficult to manipulate and may not close tightly enough to crack hazelnuts. One to two nutcrackers are needed for each team that will be removing hard-shelled nuts from their shells.

Once nuts are set out, if ducks or gulls approach and appear to be interested in eating the nuts, one or two students posted on the periphery of the study area can "shoo" them away. On the other hand, if you don't have squirrels available, these birds could be the experimental subjects (see later discussion of extensions of this exercise).

If 10 nuts have not yet been taken by squirrels by sunset on the day of your class, students can return the next day to count the remaining nuts, ideally around mid-morning. If you are not familiar with the local mammals consult an expert to determine whether squirrels are the most likely animals to have eaten the nuts (other possibilities include other rodents, raccoons, opossums, and dogs).

In-class preparation

Analytical approach

See instructions for students on the use of Fisher exact probability tests and chi-square tests of goodness of fit. Suggestions of statistical tests are also provided with ideas for exercise extensions. For comparisons of sets of times to eat nuts or sets of distances at which nuts were eaten, or were buried, the appropriate nonparametric statistical test is a Mann–Whitney U test. The VassarStats website (VassarStats.net) will easily calculate many commonly used statistical tests and provides some useful background information on tests. Also see the appendix in this book.

Information on distances at which nuts from each set are eaten and buried may help explain some unpredicted choices. For example, when squirrels are eating nuts at locations relatively far from the source or are climbing trees to eat nuts, they may choose two-nut peanuts in shells over single peanuts without shells. Although they will need to spend extra time and energy removing the shells before they eat the nuts, they will save on the time and energy that would be required to make twice as many trips with single nuts without shells (eastern gray squirrels rarely carry multiple food items together in their mouths). In another example, nuts without shells may be buried for short-term storage close to the place where they have been provided, while nuts in shells may be buried at greater distances for longer-term storage with a reduced probability of pilferage.

Areas of potential confusion or difficulty for students

Students sometimes do not understand that all the nuts within each set should be as similar as possible; you may need to clarify this. Students may find it challenging to think about other variables correlated with the independent variable they are intending to test. Students who do not completely understand the experiment may combine data from both sets of nuts when reporting times to eat nuts, distances from the sets of nuts at which nuts were eaten, or distances from the sets of nuts at which nuts were buried. The meanings of the terms "shelled" and "unshelled" are ambiguous; students should be directed to avoid these terms or clearly define them; substitutes include "nuts with shells" and "nuts without shells" or "deshelled nuts."

Recommendations for extensions or continuations for more advanced classes

Examples of experiments to isolate correlated independent variables

Color: Food coloring could be used to make the less preferred nuts the same color as the more preferred nuts (to control for the effects of handling and food coloring ingredients, the same coloring and process should be used on both sets of nuts; however, adjustments may be needed if both sets of nuts do not turn the same color with the same treatment.).

Volume of the entire set of nuts: The most straightforward way to test the effect of volume alone would be to present a larger and a smaller pile of nuts with the same characteristics as one of the sets of nuts tested, e.g., to follow up on an initial comparison of nuts with versus without shells, one pile of 30 nuts without shells could be presented 20 cm from another pile of 60 nuts without shells or one pile of 30 nuts with shells could be presented 20 cm from another pile of 45–60 nuts with shells. To test for interactions between volume and other characteristics, students could reverse the original volume differences between the sets of nuts, e.g., for a comparison of nuts with versus without shells, presenting fewer nuts in shells, versus a larger volume of nuts without shells. As noted in the student instructions, this may involve a lot of work in removing nuts from shells to reach the desired volume.

Size of individual nuts: This could be compared by taking advantage of the natural variation in the sizes of peanuts, which vary a fair amount both in shell size of two-nut peanuts and in the sizes of individual nuts removed from shells. To compare larger and smaller nuts of different species, larger nuts without shells can be cut into smaller pieces, to match the size of smaller nuts without shells; the effects of exposing cut surfaces can be controlled by shaving the surfaces of the smaller nuts, as well.

Additional options We have had the most success in getting squirrels to take hazelnuts and peanuts. Squirrels tend to be less interested in almonds than in peanuts or hazelnuts, although almond shells are usefully intermediate between peanuts and hazelnuts in hardness and ease of damage, and it is easy to remove intact nuts from almond shells. We haven't used English walnuts or pecans in these experiments, but they also tend to be readily available in supermarkets; it would be quite difficult to remove the intact nuts from their shells. The approximate caloric contents of these species are reported as follows:

Almonds (assumed to have one nut inside each shell): approx. 7.7 kcal/nut
English walnuts: approx. 26.4 kcal/nut
Pecans: approx. 19.6 kcal/nut (9.8 kcal/half pecan)

Acorns tend to be less attractive to squirrels than these nut species, when both are present, especially in the years when local oaks are producing large numbers of acorns; while it is interesting to compare what squirrels do with acorns of different species (e.g., as reported in Steele & Smallwood, 1994), such experiments will likely take longer. Native species of hickory nuts or American black walnuts could also provide interesting data, and these could be presented with or without the thick fleshy outer shell that surrounds the hard inner shell. It would be quite challenging to extract intact hickory nuts or American black walnuts from within the inner shell.

If squirrels consistently approach the sets of nuts from the same direction, they might tend to take nuts from the set they encounter first. To control for this possibility, nuts can be presented in a ring, alternating between the two nut types/preparations (Philip K. Stoddard, personal communication). This presentation will make it more challenging to determine which nuts have been taken, especially if multiple squirrels are visiting, because they may well displace some of the nuts they don't take.

Students could test whether squirrels are more likely to climb trees to eat nuts (as opposed to eating the nuts while standing on the ground) when both sets of nuts are placed within several centimeters of the trunk of a tree compared to when both sets of nuts are several meters from the nearest tree and additional time and energy would be required for the squirrels to reach the tree trunk.

Students could provide two sets of peanuts without shells, one pile with the red integument layer left on and the other pile with that layer removed, to determine which the squirrels prefer.

Students might compare the amount of time required to eat peanuts with and without the integument layer. If the cotyledons of nuts without the integument tend to fall apart while the squirrel is eating them, in a way that requires the squirrel to take extra time in retrieval of separated cotyledons, then nuts without integument may take longer to eat. For any experiments comparing nuts with and without the integument layer, it will also be relevant to note whether the integument layer is consumed or discarded.

To determine whether the scent/taste of food is an important cue about whether shells are intact, students could apply peanut oil or a small amount of peanut butter to the outside of peanuts in intact shells in one pile and compare the fate of these peanuts to the fate of peanuts in another pile with intact shells and no peanut oil or peanut butter on the outside.

Squirrels go through an interesting, and relatively stereotyped, sequence of behavior when they bury an acorn or nut; students may be asked to describe this sequence and to compare observations of different burying events and/or different squirrels to try to determine the degree of stereotypy involved.

If squirrels can be individually identified, the behavior of different individuals may be compared. Individuals may be identifiable by natural variations in pelage or irregularities in the outlines of the ears caused by past injuries. Bonnie Ploger marked individual squirrels by dripping black permanent hair dye onto them as they fed at a station she had set up below an openable window. If you would like to mark individual squirrels, first check with your state Department of Natural Resources (or an analogous agency) to determine which permits may be required. Either a chi-square test of independence or a Fisher exact probability test is appropriate for comparing the choices of different individuals feeding at the same sets of nuts. With individually identified squirrels, the Wilcoxon matched-pairs signed-ranks test can be used to make a stronger comparison of the ratio of eaten to stored nuts (or any other binary set of nut fates) and for nuts with versus without shells (or any other binary set of nut types).

Students may observe squirrels performing one or more of the types of pilferage-averting behavior described in Steele et al. (2008) or may notice squirrels caching nuts of the same type in close proximity (Delgado & Jacobs, 2017) and could attempt to replicate the experiments described in these articles. However, students focused on this behavior would be apt to miss some of the information they need to report for the current exercise, so these activities should not be combined.

It is also possible to do this exercise by presenting peanuts to gulls (*Larus* spp.) or mallard ducks (*Anas platyrhynchos*) that are accustomed to being fed by people. Of course, neither of these species will bury or otherwise store nuts, but both may be more likely to eat more nuts without shells than with shells. However, both nuts with and without shells are taken and eaten, perhaps because birds feeding in a flock with a lot of competition may simply choose the closest food items. The birds must be chased away as soon as one pile of nuts is about to disappear (with no nuts in one pile, the birds would not have a choice to make). The remaining nuts must be counted to determine the number of nuts of each type that were taken (birds tend to have much more of a "feeding frenzy" than squirrels, especially if they are present in large flocks, and counting each nut as it is taken is virtually impossible).

Likely results and samples of results

The results of this experiment are likely to indicate that energy, time, predation risk, shell presence, and competition (if multiple squirrels are foraging together) all factor into squirrels' decisions of whether to take nuts with or without shells and what to do with them. Probably the simplest test is to compare nuts of the same type, with and without shells. If students choose to make this comparison, they will probably predict that squirrels will immediately eat the nuts without shells and will store the nuts in intact shells. This may happen, and squirrels may eat nuts without shells close to the location where they are provided, saving travel time and energy. However, if multiple squirrels are visiting the nuts, both sets of nuts may be stored: this behavior is probably adaptive because storing a nut often requires less time than consuming it, and the nut-storing squirrels are moving nuts out of easy access and the view of competitors (squirrels may chase one another away from the nut piles). Even in the absence of competitors, squirrels that have eaten their fill may store some nuts without shells for convenient consumption in the near future. Presumably the nuts without shells will be retrieved and eaten by the individuals that buried them before they retrieve the nuts with shells, but it is very unlikely that the exercise session will last long enough for students to see any nuts being retrieved and it is also unlikely that students will be able to identify most individual squirrels or to reliably keep track of the locations where those individuals buried nuts.

When shells contain more than one nut, and a set of nuts in shells is paired with a set of nuts without shells, squirrels may also take nuts in shells to be eaten at a distance from the source. Squirrels that carry only one object in their mouths at a time can carry multiple nuts inside a shell but only one nut outside of a shell. Carrying two or more nuts together in a shell saves squirrels the time and energy of an extra trip to the nut piles. Squirrels may move away either to get away from competitors or to go to places where the risk of predation appears to be lower than at the nut piles: climbing trees is probably done to reduce the risk from ground predators (and squirrels eating in trees generally are in areas of the tree with enough branches and foliage [if present] to protect them from aerial predators as well). Lima et al. (1985) found that squirrels go to more protected places to eat larger food items that take longer to consume, presumably because eating these in the open would expose them to predators for a longer period. So once a hungry squirrel has chosen a set of nuts in a shell, it may be more likely to go to a protected area to eat them than if it had chosen a single nut with no shell.

Eating nuts permanently removes them from access to other squirrels, and squirrels usually eat some nuts before they bury any. Eating nuts in shells is also a way to sample their quality; generally squirrels eat one or more nuts in shells before burying any, or after burying just one or two, before burying more. Nuts buried are usually those with shells; the shells are likely to help protect nuts from seed predators and decomposers while they are buried, leaving the squirrel with more food when it retrieves them. However, squirrels sometimes seem to ignore even fairly large holes or cracks in peanut shells, in which case the fate of such nuts does not differ from the fate of nuts with intact shells; peanut shells may be so thin that holes or cracks in the shell do not significantly reduce the protection that the shells provide from decay or consumption by insects, but it seems likely that holes or cracks in the shell would increase the olfactory conspicuousness of the nuts to potential pilferers. Sometimes squirrels appear to prefer nuts in intact shells for both storage and immediate consumption; my students have suggested that squirrels may have learned that nuts in shells are less likely to have been partially eaten or degraded by decay. It's fine for the instructor to try to lead students to potential explanations. This is another opportunity to ask students to propose experiments that might be able to discriminate among possible explanations.

Sample of results Table 11.4 shows data from an experiment conducted in Connecticut on February 17 from about 3–4 p.m. Thirty hazelnuts were initially provided in each set, and nuts were taken by two different eastern gray squirrels.

Table 11.4 Sample of results; see text for further information.

	Eaten	Buried	Unknown
Hazelnuts with shells	2	12	0
Hazelnuts without shells	15	1	4

Null hypothesis: The same proportions of nuts were eaten and stored, regardless of whether the nuts had shells. Two-tailed Fisher exact probability, 1 degree of freedom (ignoring nuts with unknown fates) = 1.7×10^{-5}

Had sample sizes been inadequate to compare proportions of each set of nuts with each fate, but at least 10 nuts had been eaten, or at least 10 nuts had been stored, chi-square tests of goodness of fit could be done on either eaten or stored nuts to test the null hypothesis that of the nuts eaten/stored, half were from one set and half were from the other set. Failing that, if at least 10 nuts were taken, it would be possible to test whether there was a preference for one set over the other (regardless of the fate of those nuts).

For the same experiment, the following consumption times were recorded:

Nuts with shells: 188 s, 230 s
Nuts without shells: 124 s, 90 s, 30 s, 35 s

While the time-to-eat sample sizes are usually too small for statistical testing, they often, as here, clearly demonstrate that it takes squirrels more time (and presumably more energy) to eat nuts with shells.

Ants (family: Formicidae)

Classroom management/blocks of analysis

a. Brief instructions to students at least a week before the exercise is to be conducted, directing them to read the exercise description and any additional handouts, complete the preactivity worksheet, and submit the worksheet to the instructor in time for the instructor to sort students into teams based on the similarity of the experiments they suggested. (The instructor will also need to determine which foods to acquire for each team, and to acquire those foods before the day of the exercise.): 5–10 min. of time in class.

b. Whole-class discussion of introductory material and data collection methods (assuming students have already read the activity and completed the preactivity worksheet): 15 min.

c. Within-team discussion to reach consensus on methods (assuming students are sorted into teams based on the similarity of the experiments they suggested on their preactivity worksheets): 15 min.

d. Food preparation within each team: 10–25 min (will vary depending on the amount of preparation needed).

e. Data collection: 1.5–2 h (will vary with how long it takes for the team to find an ant colony to conduct experiments). If after 30 min, a team does not have any ants approaching their food, they should start on the nonforaging data sheet.

f. Data sheet checking: 5–15 min.

g. Compiling team data, team discussion of hypotheses to test, instructor moving between teams to provide statistical advice and possibly also graphing advice, and students performing statistical tests: 1 h (up to 2 h if students are also making graphs and writing relatively short answers to questions).

h. Summary discussion: 30–45 min.

Steps b–f can occur in a standard $2^2/_3$ - to 3-h college lab period (with careful attention to timing). Steps g and h can occur in the next lab period, or g can occur during a lecture period if students can meet with their own teams, the instructor can move between teams, and students are not making graphs during the same lecture period. Step h can occur in a subsequent lecture or lab period. If lab periods are shorter than $2^2/_3$ hours, steps b–d can be done in advance; store food in cups in sealed plastic bags in a refrigerator. Data collection can then take place during the next one or two lab periods.

Teaching the activity

Preclass preparation

Institutional Animal Care and Use Committee approval is not required to conduct experiments on ants. If compositional data on foods is provided to students in an electronic spreadsheet, this will facilitate sorting of foods by different aspects of their nutrient content. The instructor will decide which food options will be available to students and will need to edit the list we have provided depending on what you would like to include. Instructors can use the students' completed preactivity worksheets to determine which foods the students have chosen to use. These foods will need to be provided on the day(s) of the exercise, along with any tools needed to prepare the food (e.g., a can opener for a can of tuna, and food transport and storage containers [lidded plastic containers or sealable plastic bags work well]). Our list of foods includes many we have previously offered to ants and a few we have not tested. Sesame seeds were included as the items most suitable for longer-term storage, but we have not previously offered these to ants; if they do not take them, other small seeds could be substituted.

Prepare in advance by finding nearby places with ant mounds *and ants actively moving in and out of them* at the season and time of day when you plan to do this exercise. It is best to do this exercise when the outside temperature is at least 70°F (21°C), the sun is shining (especially if the temperature is relatively low), and the wind is relatively calm. If it is very hot and dry, ant colonies in the shade may be active, whereas those in the sun are not.

If you live in the southern or southwestern United States, learn to identify fire ants (*Solenopsis* and *Wasmannia* spp., including *Solenopsis invicta* and *Solenopsis richteri*), harvester ants (*Pogonomyrmex* species), and any other local ants that are likely to inflict painful stings. Do not conduct these experiments anywhere near their nests. If you do not know a local entomologist who can help you to identify these species, the Agricultural Extension department of your state university should be able to help or refer you to someone who can. Students with extreme allergies to bee or wasp stings should avoid handling ants and should bring their epinephrine kits to this class (and to all field exercises). As with all field exercises, it is a good idea to schedule several possible weeks during which this exercise could be done by the class and to conduct it starting in the first week with adequate weather conditions, with indoor or non-weather-dependent alternatives for weeks when the weather is unsuitable.

The key provided in the exercise covers two ant species and two additional ant genera that are common across much of the United States but may not include species common in your area. Ellison et al. (2012) provide comprehensive keys to identify ants found in the New England area, that can be used for many of the species in other locations. Keys to ant subfamilies and genera provided by Bolton (1994) require familiarity with characteristics of the ants illustrated in drawings in the glossary and in the remarkable electron microscope photography that makes up a large part of his book. His keys to subfamilies are only a couple of pages long. As the two subfamilies Formicinae and Myrmicinae contain most of the common ant species in many regions of the United States, identification just to the subfamily may be inadequate for the purposes of this activity. Local entomologists, or the Agricultural Extension department of your state university, may be able to provide you with a key to ants common in your locality.

In-class preparation

Analytical approach

See instructions for students on the use of the chi-square tests of goodness of fit. The VassarStats website (VassarStats.net) will easily calculate many commonly used statistical tests and provides some useful background information on tests. Also see the appendix in this book.

Areas of potential confusion or difficulty for students

Instructors should remind students to be careful not to harm any of the ants in the course of the experiments. This will avoid cruelty to the ants, as well as not injuring the ants, which would release alarm pheromones that will alter the behavior of nestmates and disrupt feeding experiments. Ants in the activity should also be handled with care to ensure they do not curl up, making them difficult to identify. At the end of the activity, ants should be returned near the entrance of the colony from which they came; they cannot be deposited at another colony because they will be attacked as intruders.

It is known that different species of ants prefer different food types: some species show the highest preference for sugary foods, while others prefer oily foods or foods high in protein. Food preferences may also vary with the time of year and the colony's needs. We have found that ants may prefer foods with high water content on hot, dry days. If students use multiple food types in their experiments, the foods are certain to differ in more than one way. Check the students' plans to make sure that they have minimized variation in factors (independent variables) other than the one that they are attempting to test.

Before students start their experiments, it is helpful for the instructor to demonstrate how data are to be entered on data sheets (drawing a sample diagram of food locations around an ant mound, filling out one or two lines of data tables). You should also note that ants may go underneath the pieces of food and be difficult to see. Although students should try to place food so that ants are unlikely to crawl underneath, the class should discuss what students will do if this happens. Gently moving pieces of food to count the ants underneath is probably the best thing to do, although the movement may disturb some feeding ants. If ants have crawled under one or more food items, students should try to move each of the four food items in the same way to try to avoid creating differences due to disturbance, but depending upon the food items used and the dexterity of the students, differences may be unavoidable (and if so, these differences should be acknowledged and described on the back of the forager count data sheet). No attempt should be made to move wet foods.

Students may need guidance on using a dichotomous key for identification. If so, it is best to go over how such keys work just before the students are about to use them; this is also a good time to review the salient features of ant anatomy. It will be useful to project the ant key and illustration of an ant to communicate with multiple teams simultaneously.

Recommendations for extension or continuations for more advanced classes

Students could investigate the importance of various nutritional components of food by comparing foods with different proportions of protein, fat, carbohydrates, water, or salt. For example, beef and turkey hot dogs may have similar proportions of protein, carbohydrates, and salt, but about $^1/_3$ of the fat in beef hot dogs may be replaced by water in turkey hot dogs. Their tastes, smells, visual appearance, and textures are similar, so those independent variables would also be controlled in this comparison. American and cheddar cheese vary mainly in their salt content (check the packages for the brands you buy to assure that this is true in your particular case). The nutrient content of some corn chips and milk chocolate differs mainly with respect to salt and the nature of the carbohydrates they contain (more sugar in chocolate, more complex carbohydrates in corn chips); the appearance, taste, and texture differences are also large. Carrots and lettuce both have very little protein or fat, and low salt content, but carrots have far more carbohydrates and less water than lettuce. The textures of some foods might be made more similar by mashing or grinding them into pastes, but this tends to be messy and disturbingly unappetizing; I don't have my students do this. Water or salt content could be manipulated by adding measured amounts of water or salt to a paste.

To investigate how travel and handling time affect foraging decisions, students could compare identical food items placed at different distances from the nest entrance, or preferences for food broken into little pieces or flakes, versus a single big piece of the same food. Combinations of foods (e.g., a piece of peanut butter and jelly sandwich) could be compared to each of the constituents by itself (peanut butter, jelly, bread alone).

Students could investigate whether different species of ants prefer different foods by repeating their experiment at colonies of different species, or for a less structured survey of preferences of local ant species, students could set out a wide variety of foods in an area likely to be inhabited by multiple species. Foods known to attract a variety of ant species include peanut butter, jelly, honey, canned tuna, honeyed tuna (canned tuna lightly covered in honey), and Pecan Sandies cookies (crumbled or not crumbled). The baits should be placed in a "neutral" area not immediately adjacent to a nest entrance, as ants may defend an area around their colony from other ants. It may take several hours or longer for different species of ants to show up at the different foods.

The same experiment could be conducted multiple times throughout the semester on the same colonies to compare the possible effect of season on food preferences (note that ants are only likely to be active at temperatures of 70°F [21°C] or above). The same experiment could also be repeated to compare food preferences at different times of the day.

Sample of results

Table 11.5 shows forager count data collected during an actual run of this exercise. The order of food discovery by the ants was apple → jelly → bread → hot dog. For a

Table 11.5 Ant forager count data.

# Foragers at (min)	Apple	Jelly	Bread	Hot dog
0	2	0	1	0
3	1	2	1	0
6	0	1	1	0
9	1	2	1	0
12	2	5	1	0
15	3	9	0	0
18	2	10	0	0
21	1	11	0	0
24	5	10	0	0
27	7	16	0	0
30	7	17	0	0
Total	31	83	5	0

chi-square one-sample test of goodness of fit comparing the four food items in terms of the maximum number of foragers seen simultaneously feeding on that food item, the expected value for each food item is $(7 + 17 + 1 + 0)/4 = 6.25$. Note that the maxima for different food items occurred during different observation periods. The results of this test reject the null hypothesis that the maximum number of foragers was the same for each food item (chi-square one-sample test of goodness of fit, $X^2 = 29.24$, 3 d.f., $P < .001$). Possible factors contributing to the preference for jelly could be its high sugar content or a texture that was easy to ingest. The lack of foragers at the hot dog might indicate that acquisition of foods with high fat or protein content was not a priority for this ant species at this time, that they were repelled by the relatively high salt content of hot dogs, or simply that the hot dog did not happen to be found until late in the experiment.

Answers to discussion questions (these apply to both the squirrel and the ant exercises)

1. Do ants or squirrels need to be "aware" of the factors you have considered, in order to make the "decisions" that you would predict on the basis of quantitative calculations? Does use of the term "decisions" imply mental capacities that these animals may not have?

Although animals behave as if they have some awareness of time and energy costs, predation risk, and suitability of items for underground storage, this behavior may be based on simple cues and rules that lead to adaptive behavior. They surely are not performing the same mathematical calculations of energy trade-offs that people can perform, even when they behave in accord with predictions based on these calculations. The precision of their match to the behavior predicted by models is informative for identifying which factors animals are taking into account and how they are weighting them, and for identifying factors included in models that the animals must be able to somehow assess. The term "decisions" may imply mental capacities that ants and squirrels may not have (by the dictionary definition of a decision as "making up one's mind" [Webster's New World Dictionary, 1964]); "choice" might be a more neutral term with regard to mental capacities, but it carries fewer connotations of consequences varying with choice. "Strategies," another term commonly used in behavioral ecology optimality modeling, is even more loaded than "decisions," with implications of humanlike mental operations that animals may not share.

2. If animals are not individually identified, and data from all individuals are summed together, individual animals will be represented in the data set in proportion to the number of visits to food (ants) or number of nuts they have taken (squirrels). If individual variations in behavior exist, and all animals do not make the same number of visits/take the same number of food items while you are watching them, the variations shown by the individuals with the most visits/food items taken will be most strongly represented. How might this bias your data, and how does it limit the scope of your conclusions?

The data collected will represent the animals that contributed, in proportion to the number of visits to food or food items taken by each individual. We cannot know whether their choices are representative of the behavior of their species in general, and in ants, choices may depend upon the current food stores in the colony. For data from multiple animals, rigorous statistical testing is only possible if animals are individually identifiable and each animal contributes only a single data point to each data set, or else data are sorted by individual as for the Wilcoxon matched-pairs signed-rank test. Chi-square tests, Fisher exact probability tests, and Mann–Whitney U tests assume that the data samples are independent, but this would be the case only if each visit/food item taken involved a different individual (or if all visits made or food items taken were by the same individual).

3. The nuts stored by squirrels and the seeds stored by ants may sprout and grow if the animals do not eat all of them. How might the seed-storing behavior that you observed be more beneficial to the parent plant than if the seeds were just left on the ground where they fell? Would you predict that these benefits could offset the cost to a plant of attracting seed-storing animals that may consume many of its seeds? Would variations you observed in seed-storing behavior, e.g., in whether a single seed was stored at each location or a group of seeds were stored together at the same location, and in distances that seeds were taken from the source, provide different amounts of benefit to the parent plant? See Steele and Smallwood (1994), Steele et al. (2014), and Lichti et al. (2020) for relevant information on this topic.

Germinating seeds that fall beneath the parent plant will compete with the parent and with one another for light, water, soil nutrients, and space. Those transported to greater distances may escape from this competition, and wider dispersal of seeds increases the probability that some will escape localized disasters (e.g., fire). Lichti et al. (2020) *suggest that coevolution between scatter-hoarding birds and rodents and the plants that produce the seeds they bury may have selected for seed characteristics (including high energy content) that induce scatter hoarders to expend extra time and energy to move preferred seeds to open locations where not only is seed pilferage less likely but also seeds that grow will escape competition from nearby plants. Animals that store seeds in groups provide less benefit to the plants because seedlings that sprout in close proximity will compete with one another.*

Part III. Supplementary material

Supplementary worksheets and information related to this chapter can be found online at https://doi.org/10.1016/B978-0-12-821410-7.00027-3.

CHAPTER

Predators strike and prey counterstrike

12

Eduardo Bessa

Graduate Program in Ecology, Life Sciences Department, Campus of Planaltina, University of Brasília, Brasília, Brazil

Chapter outline

Exploring Animal Behavior in Laboratory and Field. https://doi.org/10.1016/B978-0-12-821410-7.00015-7

Part I. Student instructions

Learning goals, objectives, and key concepts

- To learn how predators deal with different strategies used by prey to avoid being eaten.
- To simulate situations in which camouflage, escape, and resistance work against predation.
- To understand how predators and prey coevolve.

Background

It is a common misconception that the world is not fair to prey species. In general, we are used to wildlife videos of predators' strikes ending in a blood bath, but success in capturing their prey is actually the exception, not the rule. This is because prey have evolved a number of very effective strategies to avoid predation (Abrams, 1990). These strategies will be the topic of the activity in this section.

One can divide antipredator strategies used by prey into three broad categories (Breed & Moore, 2015). First, prey can avoid being detected by predators by camouflage, especially against visually oriented predators (Duarte et al., 2019). Second, once the prey is detected, it can avoid being caught. One way of doing this is by escaping your predator faster than it can chase you (Moore & Biewener, 2015). Finally, even if caught, the prey can avoid being killed. For example, prey may be covered in hard armor or thorns (Sugiura, 2016) or can retaliate and fight back (Carlson et al., 2018).

On the other hand, predators have been under selective pressure to deal with such strategies for many generations (Brodie & Brodie, 1990; Mougi & Iwasa, 2011). Detection can be enhanced by predators with keener senses of sight, hearing, and smell (Ben-Tov et al., 2018). Capture becomes more effective with more agile predators directing their efforts toward less agile prey (FitzGibbon & Fanshawe, 1988). Predators may also increase their probability of success via selection for stronger claws, teeth, or jaws; skilled hands or beaks; and even tool use (Ottoni, 2015).

This cyclic process of predators imposing a selective pressure on prey by means of predation and prey imposing a selective pressure on predators through starvation is a classic example of coevolution (van Velzen & Gaedke, 2017). Every adaptation of the prey to avoid predation will select for variants among the predators who can deal with that prey more efficiently. The reverse is also true: more effective predators select for masters of survival among the prey. This process may continue and escalate throughout generations. Many other evolutionary forces or ecological interactions may interrupt or disturb this escalation; alternatively, one of the two populations, predator or prey, may go extinct due to starvation or overpredation (respectively), ending this coevolutionary process. In general, scientists believe that prey are a little bit ahead in this dispute (Dawkins, 1976), as failure for the predator means the loss of one feeding opportunity, while failure for the prey means immediate death.

Purpose

In this activity, you will play the role of a predator. The prey will be represented by tokens such as beans or candies. You will experience dealing with prey that are camouflaged, that escape, and that require handling in order to subdue. You will evaluate predation success in each of these conditions.

Methods

No species will be needed for this exercise, as it is a simulation. Nevertheless consider which species each part of the simulation refers to, as this is fundamental to make clear that this is not just a game.

Materials

Each group of three students will have three feeding stations with the following materials:

- one small tray,
- one tall drinking glass,
- one pair of chopsticks,
- one stopwatch or timer app,
- one worksheet,
- a mixture of tokens (such as candy or beans) (your instructor will have this ready for you) mixed with buttons.

Procedure

First, the class will be divided into three groups, with as close to the same number of students per group as possible. Your goal is to find the tokens amid the buttons before "eating them". Each of you will be the "predator" for one of the three trials. If you are not the predator, your job is to count how many tokens were retrieved. The instructor will give a "go" signal, and for the first strategy, you will "hunt" the tokens on the tray for 1 min. When a token is found separate it on another plate. You should log in the worksheet how many tokens were found by all the groups.

For the second strategy, tokens have been placed in a tall drinking glass. You will pick up the tokens using only the chopsticks and without touching the glass. Again, the instructor will give 1 minute to pick as many tokens as you can, while the other members of the group record the number of tokens "eaten" in the worksheet.

For the final simulation, the instructor will place tokens wrapped with Scotch tape on your tray. Once again, the "hunter" will have 1 min to manipulate and retrieve the tokens, while the other members count and log how many were found in the worksheet.

After you have completed all three of the feeding trials write on the board the number of tokens "captured" by your group in each of these simulations. Use these data to plot a bar chart with mean and standard deviation bars for the number of tokens found by the "hunting" students for each strategy.

Results/discussion

You should have been successful in capturing at least some tokens, but hardly all of them. It is also likely (and extremely helpful in discussion) that everyone varied in "hunting" efficiency.

In the first strategy, you had to find the tokens amid the buttons. This simulates how camouflage works, with prey trying to conceal themselves as inedible items in the landscape. It could represent a background-matching species, such as *Uroplatus phantasticus* geckos whose colors are similar to the leaves they live on or the peppered moth (*Biston betularia*), a famous example of how evolution works on color phenotypes. This could also represent mimicry, such as in case of the *Papilio xuthus* caterpillar, which resembles bird droppings, something not many predators would be interested in. A symbol for sharp vision, the eagle, is an example of a predator who deals well with finding its prey.

The second strategy poses a challenge in capturing prey that are easily spotted in the glass. This simulates escaping prey. It could simply be a very slippery item, such as a catfish or a hagfish. It could also represent good runners, such as roadrunners (*Geococcyx californianus*) or Thomson's gazelles (*Eudorcas thomsonii*). Schooling fish, such as sardines, or flocking birds, such as starlings, could also be represented by this strategy. Cheetahs and swordfish are two good examples of animals good at capturing fast escaping prey.

In the third strategy, you had to manipulate wrapped tokens. Most obviously, this represents armored animals, such as a turtle, an armadillo, a pill bug, or a snail. This could also represent species covered in thorns or spines, such as the porcupine (*Coendou prehensilis*) or the Australian echidnas (*Tachyglossus aculeatus*). Strong predators such as jaguars, who can chew a turtle shell, or skillful ones such as chimpanzees, who instead of breaking a termite nest "fish" for them using twigs, are examples of predators who deal with armored prey. Alternatively, this could represent not a mechanical difficulty in ingesting but a digestive challenge, such as the presence of toxins or venoms including those found in vipers or in pufferfish. The mussurana (*Clelia clelia*), a black snake, is immune to rattlesnakes' venom, fighting viciously against the viper and sometimes eating it.

Questions

1. What aspect of predator eviction does each simulation represent?
2. Name prey species that rely on each of these strategies and predators that can cope with them.
3. Some tokens were not captured during the simulations. If this was not a simulation, this means some individuals avoided predation. In a real situation, is it the case that which individuals survive and which get eaten is exclusively by chance? Why?
4. What about the success of the "hunters"? Did they succeed merely out of luck? Was there a more effective strategy in the different simulations? How could this more effective strategy spread throughout the predator population? Can such a strategy be hereditarily inherited? Can it be learnt?
5. In your opinion, who should be more effective: predators or prey? Are they equally effective? Why is that so?
6. We humans are very visually oriented animals. How could the prey of a chemically driven predator and the prey of an auditory predator avoid detection? Give some examples of such strategies.
7. Can you imagine a different kind of predator-avoidance strategy from the three mentioned here? Is there an example in nature of a species who uses this strategy? What about a predator who deals with it?

References

Abrams, P. A. (1990). The evolution of anti-predator traits in prey in response to evolutionary change in predators. *Oikos, 59*, 147–156.

Ben-Tov, M., Ben-Shahar, O., & Segev, R. (2018). What a predator can teach us about visual processing: A lesson from the archerfish. *Current Opinion in Neurobiology, 52*, 80–87.

Breed, M. D., & Moore, J. (2015). *Animal behavior.* Academic Press.

Brodie, E. D., III, & Brodie, E. D., Jr. (1990). Tetrodotoxin resistance in garter snakes: An evolutionary response of predators to dangerous prey. *Evolution, 44*(3), 651–659.

Carlson, N. V., Healy, S. D., & Templeton, C. N. (2018). Mobbing. *Current Biology, 28*(18), R1081–R1082.

Dawkins, R. (1976). *The selfish gene.* Oxford University Press.

Duarte, M., Gawryszewski, F. M., Ramineli, S., & Bessa, E. (2019). Disruptive coloration and habitat use by seahorses. *Neotropical Ichthyology, 17*(4), e190064.

FitzGibbon, C. D., & Fanshawe, J. H. (1988). Stotting in Thomson's gazelles: An honest signal of condition. *Behavioral Ecology and Sociobiology, 23*(2), 69–74.
Moore, T. Y., & Biewener, A. A. (2015). Outrun or outmaneuver: Predator–prey interactions as a model system for integrating biomechanical studies in a broader ecological and evolutionary context. *Integrative and Comparative Biology, 55*(6), 1188–1197.
Mougi, A., & Iwasa, Y. (2011). Unique coevolutionary dynamics in a predator–prey system. *Journal of Theoretical Biology, 277*(1), 83–89.
Ottoni, E. B. (2015). Tool use traditions in nonhuman primates: The case of tufted capuchin monkeys. *Human Ethology Bulletin, 30*(1), 22–40.
Sugiura, S. (2016). Bagworm bags as portable armour against invertebrate predators. *PeerJ, 4*, e1686.
van Velzen, E., & Gaedke, U. (2017). Disentangling eco-evolutionary dynamics of predator-prey coevolution: The case of antiphase cycles. *Scientific Reports, 7*(1), 1–11.

Part II. Instructor notes

Classroom management/blocks of analysis

This activity can readily be completed in a standard class period. If done in a 2- to 3-h laboratory period, additional time for data analysis and discussion may be included.

Teaching the activity

Preclass preparation

All materials should be compiled by the instructor in advance of the lab. We recommend that the tokens used in the activity NOT be ingested due to food intolerance, diet restrictions, or health concerns. If candy is used, then eating candies NOT used in the activity can still be done and should take place after all statistics are done. Make sure you communicate to the students what the candies are, especially if there are ingredients in them that could cause allergies or offend students with cultural diet restrictions. If you want to avoid this problem completely, the candies can be replaced by tokens such as poker chips, dried beans, or something similar that the teacher could exchange for any other reward after the activity is done if they so choose. For the final simulation, the instructor can wrap two or three tokens with a piece of napkin and then wrap this with tape to avoid having the tokens stick to the tape. Do not overwrap each token to avoid frustrating the students in this strategy.

Materials

Each group of three students will have three feeding stations with the following materials:

- one small tray,
- one tall drinking glass,
- one pair of chopsticks,
- one stopwatch or timer app,
- one worksheet,
- a mixture of tokens, such as candy or beans, and buttons. The total number of buttons needed for a class of 30 (for example) would be 300–500; the total number of tokens needed is about 200–300.

- For scenario 1, each tray should have 10 tokens mixed with 30–50 buttons (whatever number you choose make sure each station has the same number).
- For scenario 2, each glass should have 10 tokens added.
- For scenario 3, wrap two tokens together in tape and add five such wrapped items to each tray when the students are ready for this scenario.

Data analysis

This is a good opportunity for students to work on data presentation. The teacher could ask them to propose how to clearly present it. Make sure the data compiled on the board are available to all students. You can photograph the board, enter it into a spreadsheet, or have students work on the analyses in the lab. We suggest bar graphs with standard deviation bars, but you can let students' creativity work in this moment and then discuss what representation worked best. If you have a teacher assistant or enough time, then build your own bar graph with standard deviation and include it along with the students' versions to discuss how clearly each conveys the necessary information. Otherwise suggest the bar chart to the groups. Even with all the groups working with bar charts, you could still discuss how grid lines, color schemes, and 3D effects help or hinder comprehension. This exercise is also a good opportunity to apply central tendency and dispersal measures; we recommend calculating the mean and standard deviation for predator efficiency (number of "prey" captured). The teacher could discuss why not to use the mean value alone and what each measure represents.

Areas of potential confusion or difficulty for students

The main difficulty your students could face is in transposing the simulations to real-world animal behaviors. Specifically discussing what the simulations represented and asking the students to indicate species which use such predator avoidance techniques is the best way to avoid such confusion. Another problem I sometimes face is that many students do not understand or accept that an animal having a cryptic coloration, thorns, or a shell is a subject for the animal behavior class. This kind of comment is particularly useful to show how wide the scope of animal behavior is.

The instructor can discuss what the best defense strategy appeared to be (i.e., in which strategy were the fewest candies captured). It is also interesting to discover which student is the best finder, the best catcher, and the best manipulator. Then the instructor can discuss what the different strategies represent in the natural world. Ask the students for an example of an animal that is good in matching the background like the candies or beans do with the buttons. Then ask for an example of an animal that has good sight or another sensory mechanism or behavioral adaptation to circumvent this background matching. Continue exploring examples of effective escapers and chasers and finally examples of shelled versus skillful or strong animals.

Recommendations for extensions or continuations for more advanced classes

- There are several natural history films that could be used after this class to help clarify how the simulations experienced by students find correlates in the natural world. My recommendation, if you can tolerate the sensationalistic tone it sometimes assumes, is Bob Brisbane's 72 Dangerous Animals series.
- Behaviors related to feeding will be discussed again in Chapter 16. Make sure you discuss how camouflage may prevent a predator from perceiving a place as a good feeding site. If you have not done so, then consider using the activity in Chapter 11 to discuss how antipredator behavior affects the predators' foraging decisions.
- For a class in evolution, you could use this as an opportunity to discuss runaway selection. Runaway selection suggests that every mutation that allows prey to run faster would pose a selective pressure on predators to run faster too. It is easy to assume that, given enough generations, this could result in ever faster prey and predators running around at ever-increasing speeds. Why does this not happen?
- You could expand the discussion of coevolution. For example, during his Beagle voyage, Charles Darwin found an orchid with a very long pistil. Allegedly, he predicted the existence of a pollinator with a similarly long proboscis. Could Darwin have made a similar prediction about a very strong predator after finding the thick-shelled giant tortoises of the Galapagos Archipelago? Why?
- Finally, if you feel that your class is interested in applications of behavioral research, then discuss with them how different antipredator strategies could be applied to practical problems. How could these strategies help fight obesity? What could be the military applications of antipredator strategies?

Answer key

1. Simulation 1 represents camouflage; simulation 2, escaping; and simulation 3, a shelled or thorny body.
2. Just a few examples. Simulation 1: prey, rabbit; predator, lynx. Simulation 2: prey, gazelle; predator, cheetah. Simulation 3: prey, armored catfish; predator, giant otter.
3. Not by chance alone. It could reflect that some individuals are better disguised, faster escapers, or harder shelled. This is an adaptive advantage.
4. Not merely out of luck. Perhaps a student can invent a more effective strategy, such as sliding the candies against the glass walls instead of holding it with the chopsticks in simulation 2. Sometimes such more advantageous strategies can be not only genetically inherited but also culturally transmitted.
5. Preys tend to be more effective because one failure in predation means a lost feeding chance, whereas one failure in predator evasion means death. So it has a higher cost.

6. By disguising its smell, some animals (even domestic dogs) rub their bodies on feces, for example. Preys of auditory predators become quieter, having soft pads under their feet, for example.
7. My students usually refer to poison and venom and predators immune to them, and you can discuss how it differs from simulation 3. Some also refer to Batesian mimicry and predators capable of recognizing the mimic, and you can discuss how it differs from simulation 1.

Part III. Supplementary material

Supplementary data related to this chapter can be found online at https://doi.org/10.1016/B978-0-12-821410-7.00015-7.

CHAPTER

13

The circle game: intergenerational transmission and modification of solutions to a universal need

Andrew Goldklank Fulmer

Department of Psychology, Lehman College, City University of New York, Bronx, NY, United States

Chapter outline

Exploring Animal Behavior in Laboratory and Field. https://doi.org/10.1016/B978-0-12-821410-7.00006-6

Part I. Student instructions

Learning goals, objectives, and key concepts

- Understand that modification in life strategies requires interaction among generations.
- Describe the evolutionary pressures which cause populations to evolve stability and/or flexibility in reproductive tactics.
- Summarize how intergenerational transmission of information often relies heavily on individual variation in behavioral type/personality/ontogenetic experience.

Background

Dichotomizing "nature" and "nurture" (typically referring to genes and the environment, but rarely taking into account the feedback cycle of genes, modification of gene expression, and environmental experience) is one of the most pernicious oversimplifications in biology education. While the colloquial view is that these influencing factors in development are in some way discrete, they have long been established as interdependent (e.g., Sasaki & Kim, 2017).

Experience can influence gene expression, for example, via epigenetic modification, which may be seen as a "dimmer switch" control on genes, where environmental influences increase or decrease the way in which a genotype is expressed in a phenotype relative to its existing capacity (Goldberg et al., 2007). Experiences begin long before an organism leaves the womb or egg, particularly illustrated by the presence of the egg inside a grandmother (in humans, oocytes can develop and be influenced by chemical factors within the first 25 weeks of embryonic development) (Jamnongjit & Hammes, 2005). After birth, experiential elements such as exercise, nutrition, and trauma may regulate the traits passed on to the offspring (Jawaid et al., 2018).

Life history traits refer to a spectrum of reproductive styles among *and within* species. Put simply, there are two ends of this spectrum: the K selected species and the *r* selected species. It bears emphasis that in many species, some individuals will follow a more K selected strategy and some individuals of that same species will follow a more *r* selected strategy. K selection refers to the reproductive tactic of producing relatively few young animals, given the lifespan of a particular animal, and providing a relatively long period of care for them. *r* selection refers to the reproductive tactic of producing a relatively large number of young animals, given the lifespan of a particular animal, and providing a relatively short period of care for them.

These life history styles are highly context dependent, typically related to mortality risk and resource availability, with individuals exhibiting the K strategy existing in resource-rich environments and having low mortality and *r* strategists existing in resource-poor environments and having relatively high mortality. When resources are abundant, all eggs can be put in one basket, whereas when they are not, it is optimal to invest relatively little in each offspring but have as many as possible to maximize the potential for some to survive (e.g., Mace, 2000; Dillon & Conway, 2018). This cycle, in turn, can produce both short-term ontogenetic shifts in individuals and long-term evolutionary shifts in populations. In the family tree of life, it is the mechanism for the life cycle of the whole family (McGoldrick et al., 2016).

This is an adapted version of a game initially proposed by Fulmer (1983), which was intended to broaden the taxonomic scope and to explore the natural history of family development.

Purpose

We are going to play House. This will involve small groups forming "families" including parents and children, who will enact the family life cycle: the process by which families begin (mate selection), reproduce (have offspring), and influence those offspring (with an emphasis on the idea that it is not possible to meaningfully distinguish between nature effects and nurture effects; all behaviors are a combination of both nature and nurture) and how those offspring take that influence into their own search for mates. The fundamental aspect is for each individual, and each generation, to answer the question: what is the safest kind of home? Does this vary by the nature and structure of a family group?

Methods

Step-by-step instructions

Begin by writing down a number (any number will do).

Your instructor will sort you into groups. Within each group share your numbers. Those with the two highest numbers are assigned as "parents" to the other members of the group (we will refer the groups as "parents" and "offspring" from here on).

Parents: Parents are asked to provide their personal answers to the question "What is the safest kind of home?" Parents must then agree, together, on the answer to this question. Some discussion of which elements make it safe, and how safety is defined, would be appropriate here. Parents write their agreed-upon answer in the form of one sentence. Both parents must sign the final sentence.

Offspring: The offspring are presented with the sentence and asked to discuss whether or not they agree. They are invited to make some modifications, in discussion with the parents, at this time, but all members of the family must sign each child's final version of the sentence. Carrying their own versions of the statement, the offspring are instructed to "leave home" and pair off.

Depending on how much time is available, your instructor may ask you to "speed date" until you select a partner. These new couples must agree with one another on a compromise between their two statements (it must at least make sense; it cannot, for example, be the two sentences mashed together). In turn, they sign this new combined statement. The new couples visit both sets of parents and present their combined statements, discuss the concepts, and receive or fail to receive approval.

Your instructor will tell you how many times this cycle will be iterated, with parents becoming children, etc.

Results/discussion

Following the end of the cycle prepare a brief summary of your experience, particularly focusing on how (or if) the statements evolved across the generations. Did you see any shift in the type or location of the ideal "safe" home? In your discussion with parents, mates, and/or offspring, did you get new ideas about what constitutes safety? About what constitutes a home? Did a parent's, a mate's, or an offspring's experience differ from your own in an interesting way (hint: this might suggest that under different selective conditions, different definitions of safety develop and/or evolve)?

Discussion questions

1. First and foremost, why did you choose this definition of safety? Remember that physically safe and emotionally safe are concepts with blurry lines.
2. What sorts of nonhumans might have made the choices you did? Did you base your model on a nonhuman (e.g., if you chose to overwinter through a sedentary behavior, you might have wanted a secure cottage or access to thin ice for fishing).
3. What changes to your environment could you imagine affecting your choices? What might change to make you choose something else?
4. Did you see any variations which might reflect on different life history strategies (keep in mind factors such as expectation of safety for offspring and resource control)?
5. How many generations, following you, do you expect to use your definitions of safety? Of a home?
6. Do you want to live in the same type of home in which you want your offspring to live? Do they want to live in the same type of home in which you want to live?
7. Did the experience of "leaving the nest" influence the way in which you chose to redesign your own nest with your new partner?
8. What is a home?

Acknowledgement

I wish to thank my father, Dr. Richard Huntington Fulmer, who inspired this chapter with his deep pedagogy on the family life cycle both professionally and personally.

References

Dillon, K.,G., & Conway, J. C. (2018). Nest predation risk explains variation in avian clutch size. *Behavioral Ecology, 29*, 301–311.

Fulmer, R. H. (1983). Teaching the family life cycle: A guide for a workshop using simulated families. *American Journal of Family Therapy, 11*, 55–63.

Goldberg, A. D., Allis, C. D., & Bernstein, E. (2007). Epigenetics: A landscape takes shape. *Cell, 128*(4), 635–638.

Jamnongjit, M., & Hammes, S. R. (2005). Oocyte maturation: The coming of age of a germ cell. *Seminars in Reproductive Medicine, 23*, 234.

Jawaid, A., Roszkowski, M., & Mansuy, I. M. (2018). Transgenerational epigenetics of traumatic stress. *Progress in molecular biology and translational science, 158*, 273–298.

Mace, R. (2000). Evolutionary ecology of human life history. *Animal Behaviour, 59*, 1–10.

McGoldrick, M., et al. (2016). The life cycle in its changing context: Individual, family, and social perspectives. In N. A. McGoldrick Garcia Preto, & B. Carter (Eds.), *The expanding family life cycle* (pp. 1–40). Pearson Education.

Sasaki, J. Y., & Kim, H. S. (2017). Nature, nurture, and their interplay: A review of cultural neuroscience. *Journal of Cross-Cultural Psychology, 48*, 4–22.

Part II. Instructor notes

Classroom management/blocks of analysis

This exercise can take place in virtually any setting and can be modified for a range of class sizes. It might be difficult to accomplish with fewer than six participants, but in theory, participants could generate fictional characters to manipulate in a story format. Typically students use paper and writing utensils, but this can also be accomplished with phones, laptops, etc. It is useful if the area in which the exercise takes place is not too cramped or cluttered.

Students will enact the exercise themselves. This exercise can be modified for the outdoors as well, time permitting. This activity can be completed within approximately 45 min, but I have also stretched it out over multiple classes with multiple iterations, to good effect (see earlier discussion for possible iterations that can prolong and enhance the experience).

Teaching the activity

Preclass preparation

A review of K and *r* selected life history strategies is a good lead-up to this game. Habitat selection is also, of course, relevant (particularly in encouraging students to think about the many different kinds of home found even within our own species). You may also opt to have students respond to the discussion questions and submit their answers.

In-class preparation

Organize students into class-size appropriate groups, a minimum of three students per group (preferably four or more). Mobility is useful. The faster students can group (and move among groups) the better. This could be done well outdoors with students sitting in groups. The alternative version I would recommend, in the event that there are relatively few students, would be to have students each create "families" with multiple characters (each character represented with a different definition of safety in a home) and make exchanges/form compromises among each other's characters. The game can otherwise proceed as described earlier. I have also had situations in which groups of more than two form; typically, this situation is approached as a single parent and multiple children.

Areas of potential confusion or difficulty for students

It is important to remind students to consider the concepts of "safety" and "home" somewhat broadly. The question works best if it is answered in a very open-ended fashion, since if it turns into excessively specific requirements (i.e., particular structures such as walls or access to light), it can become too easy for sentences to be combined as lists, rather than integrated in a meaningful way. Students find the discussion and compromise more difficult than they seem to expect, in many cases, and may be tempted to mash up concepts. Steering them away from this style of solution is essential to the integrity of the exercise. Help them to conceptualize their core goals in a home, rather than the superficial trappings.

Recommendations for extensions or continuations for more advanced classes

Students can be asked to lean further into the "characters" they produce. I find that they become emotionally invested (often to their own surprise) in their "families." Given enough time, or students of the appropriate level of advancement, this could provide a good basis for adding more decisions to make, for example, extending the mate selection process or permitting more than one question to be answered per individual.

To form new iterations of the cycle without being repetitive and to educate students about the evolution of life history formats, I recommend forming different kinds of families seeking ideal homes.

Some (flexible) examples are discussed in the following.

First, in what biome is the family located? This can change with iteration, from taiga, to tundra, to desert, to rainforest, etc. Within those iterated conditions can be nested social environments, as discussed in the following.

- The size of families can be altered (larger or smaller groups may find safety in different contexts—a good chance to introduce selfish herd hypothesis for advanced students and to contrast the needs of, for example, a wolf pack vs. a more solitary fox).
- Family structure can also be adjusted, i.e., cooperatively breeding groups with "helpers at the nest" versus a human nuclear family or a commune (which might be akin to a shoal of fish).
- The goals of the family might change: do they need to migrate? Do they over-winter while active or go into a sedentary state?
- McGoldrick et al. (2016) discuss the family life cycle, which is core to the basic exercise. Varying the specific ages and health conditions of family members can provide further incentive to get creative with, and show a diversity of, ideal home choice.

Samples of results

Please find below the positive and negative examples of potential results (assuming only one class-period's worth of iterations).

Positive (integrates ideas): One member of a couple has "left the nest" and approaches the partner with the idea that the ideal home is distant from other people and allows for reliable access to resources (a compromise reached by their natal family). Their intended partner arrives with the idea that a good home allows for strong community support and tradition. Their compromise might resemble living on a farm near a small town.

Negative (lists ideas): One member of a couple has "left the nest" and approaches the partner with the idea that the ideal home is a castle in the city (a list reached by their natal family). Their intended partner arrives with the idea that the ideal home is on an island. Their list might resemble a castle in a city on an island.

One member of a couple has "left the nest" and approaches the partner with the idea that the ideal home is a wealthy neighborhood with a good school district. The intended partner approaches with the idea that the ideal home is a rustic cottage with wildflowers. Their compromise is that they have a rustic cottage with wildflowers in a wealthy neighborhood with a good school district.

If you opt to use the discussion questions as an assignment for students to complete, some example responses/directions of thought are offered here.

Discussion questions

1. First and foremost, why did you choose this definition of safety?
   A student might say that emotional safety could be generated in many ways, by many different arrangements of family, but that this cannot be achieved without physical safety. Alternatively, the idea that emotional safety is a prerequisite for self-care and risk-avoiding behavior in offspring might be offered.
2. What sorts of nonhumans might have made the choices you did?
   A student might suggest that relative isolation and self-sufficiency constitute safety in a home. This student might discuss animals that have adapted to harsh or unpredictable environments with relatively small social networks. The student might discuss corresponding adaptations (e.g., hibernation, food caching, long-distance communication, large territory size per individual). Alternatively, a student advocating for city apartments as the safest form of home might discuss colonial animals and how their social spacing might be reflected in real estate trends, demographics, and pricing. Another student might discuss fission–fusion groups and the advantages of behavioral flexibility.
3. What changes to your environment could you imagine affecting your choices? What might change to make you choose something else?
   Student responses might include changes from change in food availability leading to a change in foraging strategy or environment, changes in weather or water level influencing decisions about shelter or transportation, or social factors, such as the availability of educational opportunities, cultural resources, or shared values/goals with coinhabitants.

4. Did you see any variations which might reflect on different life history strategies (keep in mind factors such as expectation of safety for offspring and resource control)?
Students are given the opportunity to evaluate the expectations others have of creating safety. A student response could include speculation on how others expect to be prepared for conflict or dangerous environments (as reflected in homes designed around fortification or self-sufficiency) or if other students are expecting to have the relative luxury of devoting resources to advancing resources across generations.
5. How many generations, following you, do you expect to use your definitions of safety? Of a home?
Depending on the level of the student, either simple numbers (perhaps explained by their own experience during the activity) are to be expected or this question can be used to explore larger questions involving life history strategies and changing environments. For example, students might discuss the adaptive value of facultative behaviors under unpredictable resource conditions.
6. Do you want to live in the same type of home in which you want your offspring to live? Do they want to live in the same type of home in which you want to live?
These responses are a chance for students to discuss intergenerational inheritance. Students can reflect on their own experiences in each role (if the game has been iterated more than once) and/or discuss amongst the class. A student leaning more on developmental safety/resource stability might be choosing a home with a long view, something to protect and ensure resources for future generations.
7. Did the experience of "leaving the nest" influence the way in which you chose to redesign your own nest with your new partner?
Students may cite their perception of their parents' response to their departure as well as to the edits/contributions they, as offspring, used in modifying the family statement. Some may comment on how they wanted to create a different deliberative process when starting their own families and choosing their new definition of homes, some may comment on how they adapted ideas presented by their own parents, and some others may comment on how they may have felt relief at having increased authority in defining the home and its purpose.
8. What is a home?
These answers can range broadly in length and focus. The instructor may wish to add guidelines as appropriate to the level and length of the class. Some students may opt for an aphorism such as "where the heart is," while others may explain in detail the choices they have made and encountered during the activity. For example, one student may have a more migratory definition, centered around relationships and resource security across seasons, whereas another might focus on physical structures/objects necessary for prolonged dwelling.

CHAPTER

14 Demonstrating strategies for solving the prisoner's dilemma

Heather Zimbler-DeLorenzo[1], Kathleen Morgan[2]

[1]*Department of Life and Earth Sciences, Perimeter College at Georgia State University, Decatur, GA, United States;* [2]*Department of Psychology, Wheaton College, Norton, MA, United States*

Chapter outline

Exploring Animal Behavior in Laboratory and Field. https://doi.org/10.1016/B978-0-12-821410-7.00024-8

Part I. Student instructions

Learning goals, objectives, and key concepts

- To be able to describe how game theory could be used to predict behavior.
- To use mathematics to compare behavioral strategies.
- To create visual representations of how behavioral strategies may change over time.

Background

Animals organize their behavior with respect to conspecifics in a variety of ways. In some species, individuals actively avoid one another, coming together only to mate. In others, individuals gather in groups. For any individual, the advantage of living in a group may be a little more than reducing the likelihood of being eaten by a predator (the "dilution effect" hypothesis). However, in some species, group living may provide more benefits than simply improvement of odds in escaping predators. Group members may actively help one another to find and catch food, to harvest food, to build and maintain shelters, and/or to care for young.

Why animals that are unrelated to one another would cooperate has been a puzzle for evolutionary biologists. Behaving **altruistically**—doing something that costs you while benefiting another—simply does not "make sense" in the evolutionary scheme of things. A truly altruistic individual that repeatedly helps unrelated others is jeopardizing its survival and may not survive to pass on any of those "altruistic genes" to the next generation.

Nonetheless, some animals do behave altruistically. For example, female vampire bats roost during the day in communal groups of related and unrelated individuals. An individual bat might not always be successful at night in finding a host for the blood meal it needs, and a vampire bat cannot survive a missed feeding for very long. However, bats that roost together will share food with a roost-mate that has been less successful on a given night by regurgitating blood for her (Wilkinson, 1984). In this situation, a bat that fed successfully is sacrificing some of her own gain to benefit another, perhaps an unrelated individual.

By definition, behaving altruistically means that the behavior entails a cost of some kind. What is to keep other individuals in a group from "mooching" off the kindness of an altruist? You may have had the experience of working on a group project for a class, in which there was a member of your group who did nothing and yet received the same credit as everyone else in the group. Imagine, for example, that you are assigned a project to work on as a pair. Your own grade, then, is dependent on the joint grade earned for the project that you do with your partner. If you and your partner both work hard, you should do well. However, if your partner does none of the work, you have a dilemma. In order to get a good grade, you would have to do *all* the work yourself, thus giving your partner a "free ride" at your expense. But if you retaliate against your partner's laziness by *also* not doing the work, *both* of you will get a poor grade (This "damned if you do, damned if you don't" situation is called the "cruel bind" by behavioral ecologists.). How could you maximize the advantages of group living (or in this case, group working) without incurring the cost of being taken advantage of? What might you do if you were assigned to work with that partner again?

Trivers (1971) suggested that animals would behave altruistically in situations in which it was highly likely that the favor would be returned at some time in the future—sort of an "I'll scratch your back now if you scratch mine tomorrow" scenario. However, such a strategy would be highly susceptible to sabotage by "cheaters"—individuals who gladly accepted aid but failed to return it in the future. Trivers recognized this problem with his idea and considered that **reciprocal altruism** (behaving altruistically and having that behavior reciprocated at some time in the future) would evolve only under conditions in which individuals recognized one another and in which the likelihood of two individuals meeting repeatedly was high.

Still another view of why and under what circumstances animals would cooperate with one another was developed by Axelrod and Hamilton (1981). Their approach was to apply the classic social science model of the "prisoner's dilemma" (PD) to the problem of reciprocal altruism. The PD posits that two suspects are captured by the police and held in prison. Neither one may communicate with the other. Each is told that the case against them is weak and that, because of this, each will be held in jail for only a little while. However, each is also told that if they would be so kind as to inform on the other (i.e., to defect), the stool pigeon would be released immediately, whereas the other would go to jail for a long, long time. Finally, each is told that if both prisoners inform against each other, they will both remain in jail for an intermediate period.

Fig. 14.1 shows the relative "payoff" to a given prisoner, in this case, prisoner A, in a single instance of PD. In this situation, prisoners have only one chance to make a decision. Under such circumstances, the **evolutionarily stable strategy** (ESS, a behavioral strategy that cannot be bettered once it is widely adopted) is for both prisoners to defect (to inform against one another). Imagine that you are one of the prisoners. Remember what the police have said about the consequences of your decision. If you cooperate with your fellow prisoner and do not inform, and they are equally cooperative, both of you will be in jail a little while. This is R, the payoff for mutual cooperation. Because the jail term here is not very long, the payoff has an intermediate value. However, if you cooperate and your colleague defects, you will be in jail for a much longer time, while they go free. This situation represents S, the sucker's payoff, with you as the sucker! In this situation, you gain nothing and in fact "lose" the game because you will receive the maximum penalty.

Prisoner A Strategy	Prisoner B Strategy: C Cooperation	Prisoner B Strategy: D Defection
C Cooperation	$R = 3$ Reward for mutual cooperation	$S = 0$ Sucker's payoff
D Defection	$T = 5$ Temptation to defect	$P = 1$ Punishment for mutual defection

FIGURE 14.1

Payoffs to prisoner A in the prisoner's dilemma, for each possible strategy employed by prisoner B. Payoffs are represented with arbitrary numerical values for illustrative purposes (Axlerod and Hamilton 1981).

Suppose that you decide to inform on your colleague, however, and defect. If they do the same, you will both stay in jail for a while, but not for as long as you would have if you remained loyal while your comrade ratted on you. This is *P*, the "payoff" or punishment for mutual defection. However, if you defect and your colleague does not, you gain the most because under these conditions you will go free while they stay in jail. This situation represents *T*, the payoff that results in temptation to defect. In cases in which the prisoners may make only one decision (in other words, in cases in which the "game" of the PD played only once), defecting is the best strategy. No matter what the other prisoner does, it always pays in the short run to defect, because *T* (the reward for defection when the other prisoner cooperates) is greater than *R* (the reward for mutual cooperation). This, evolutionarily speaking, cooperation is a suboptimal strategy when individuals meet only once in a situation in which cooperation is possible.

But what about under those conditions specified by Trivers (1971) as the exceptions? When individuals are able to interact repeatedly, as in a social species, then a different ESS might occur. Imagine two individuals who exchange goods as part of a trade system (Axelrod and Hamilton, 1981). The two never meet but, rather, leave the goods at a predetermined place in the forest. When each comes to pick up what the other had left, should one simply take what is left and leave nothing in return? Does it profit the individual to reciprocate if there is the possibility of another exchange in the future? If one individual should opt on occasion to cheat and leave nothing for the other, what should the offended party do when he or she returns next and finds an empty larder? Such a situation is essentially an **iterated prisoner's dilemma** (IPD), a repeated playing of the same decision game over and over again.

Using a series of tournaments in which a computer played a string of IPDs, Axelrod and Hamilton (1981) concluded that the strategy that would result in a cooperator gaining the most while cutting losses was what they called **tit for tat**. In this strategy, a player's first move should be to cooperate, but the player on his or her next move should do whatever the other player did. In other words, if after you cooperated the other play responded in kind, your next move should also be cooperative. If, on the other hand, you cooperated but the other player defected and cheated on you, then the way to cut your losses is to also defect on your turn.

Purpose

The purpose of this investigation is to reveal some possible strategies that humans might use when presented with an opportunity either to cooperate or to behave selfishly and to explore which strategies appear to be more effective. Your mission is to ask pairs of humans to play a game. The purpose of the game is to move a penny from the center spot of a game board to a goalspot (different for each player). The game is similar to checkers, except that there is only one playing piece and the number of moves that each player can use is limited. Be sure you read through all the instructions before you begin collecting your data.

Methods

There are two types of games that will be run for this activity: IPD, which requires two human game players per observer, and PD, which needs 20 volunteers (10 pairs of game players per observer). Each observer will need a game board, a penny or similar coin, and a writing utensil.

Before you find your subjects and begin take a few moments and make some predictions about what you think will happen.

- PD: What do you think people will do when confronted with this game for the first time? Do you think they will play in such a way that one person wins while the other loses? Will the players compete or cooperate? How about after they have played a few rounds?
- IPD: What do you think people will do when you confront them with the game? Do you think they will play in such a way that one person wins while the other loses? Will the players compete or cooperate?
- For both: What kinds of factors do you hypothesize might influence the decisions that your subjects make? Do you think that your subjects will figure out how to maximize their winnings? Why or why not?

Respond to these questions *first* and then collect all your data.

Complete the top part of the data sheet for the game, taking care to record your name, the date, and other relevant information. Sit so that the game board can be placed between the game players, with one player seated before goalspot A and one player seated before goalspot B. Place the penny in the center spot.

Procedure 1 (iterated prisoner's dilemma)

The observer should read the following instructions to the game players.

> *The purpose of this study is to see how people solve the problem of how to play this game so they maximize their winnings and minimize their losses. The game is fairly simple, but winning it can be difficult. I'll ask you to play the game 10 times, taking turns at going first. In each game, each player has only 10 moves to try to move the penny into his or her own "goalspot," for a total of 20 moves per game. The penny can be moved only one circle at a time and only by following the lines connecting two circles. If after 20 moves neither of you has moved the penny into your own goalspot, the game ends in a draw and neither player wins. I need to count the number of moves you make in this game, so please don't move too quickly! Also, the game must be played in total silence. Please do not talk to each other or otherwise signal one another before or during the game.*

The observer should ask the players if they have any questions. Do not continue until everyone understands how to play the game. Toss the penny, and have a player call heads or tails. The player who wins the toss should begin the first game. Fig. 14.2 shows how to use a game data sheet to keep track of the moves made during the game. As each player makes a move, the observer should write down the number of that move in the circle that represents it on the data sheet for that game. When one player wins, or when the 20 moves allowed for the game have been made, end that game. Record the outcome of the game in the space provided on the data sheet for game 1.

Observers should remind the players that the object of the game is "to see how many people solve the problem of how to play this game so they maximize their winnings and minimize their losses." Again, remind them not to speak or discuss the game with you or with each other but, rather, simply to play the game again. This time the person who lost the toss for the first game should go first. Just as for game 1, as each player makes a move, the observer should write down the number of that move in the circle that represents it on the data sheet for that game. When one player wins, or when the 20 moves allowed for that game have been made, end that game. Record the outcome of the game in the space provided on the data sheet, and then begin the next game by replacing the penny in the center spot. The observer should allow the players to take turns starting each new game and should remind them not to communicate with one another until the 10 games are over. If the players are volunteers, the observer should be sure to thank them for helping with this assignment after finishing!

For each of the 10 games, the observer should count the number of moves made and should end the game if it is not won within 20 moves. Record the outcome of each game. At the end of the 10 games, the observer should ask the players to describe any ways in which someone could win the game. The observer should record any answers on the data sheets.

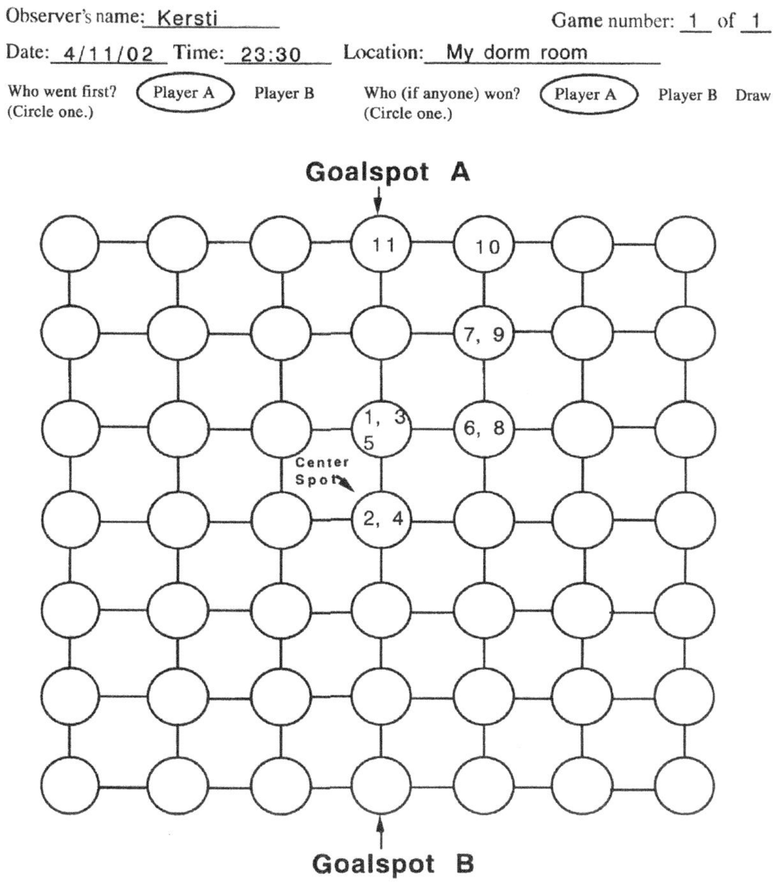

FIGURE 14.2

Example of a completed game data sheet. Numbers indicate each sequential move in the game. In this game, Player A (move 1) begins by moving the game piece directly toward their gamespot. Player B (move 2) retaliates by moving it back to the center. A few exchanges of competition take place (moves 3-5), and then Player B tries to avoid directly confronting Player A by moving to the side in move 6 (showing the "staircase" pattern associated with conflict avoidance). Player A responds by moving the game piece once more in the direction of their goalspot (move 7). Player B attempts to move it back (move 8), but Player A continues to be competitive (move 9). Player B finally submits to Player A in move 10 so that Player A wins the game in their next move (move 11). This is an example of a game showing mixed strategies; it includes evidence of conflict avoidance and submission.

Procedure 2 (prisoner's dilemma)

The observer should read the following instructions to the game players.

> *The purpose of this study is to see how people solve the problem of how to play this game so they maximize their winnings and minimize their losses. The game is fairly simple, but winning it can be difficult. I'll ask you to play the game only once. In the game, each player has only 10 moves to try to move the penny into his or her own "goalspot," for a total of 20 moves per game. The penny can be moved only one circle at a time and only by following the lines connecting two circles. If after 20 moves neither of you has moved the penny into your own goalspot, the game ends in a draw and neither player wins. I need to count the number of moves you make in this game, so please don't move too quickly! Also, the game must be played in total silence. Please do not talk to each other or otherwise signal one another before or during the game.*

The observer should ask the players if they have any questions. Do not continue until everyone understands how to play the game. Toss the penny, and have a player call heads or tails to determine which player will start the game. Fig. 14.2 shows how to use a game data sheet to keep track of the moves made during the game. As each player makes a move write down the number of that move in the circle that represents it on the data sheet for that game. When one player wins, or when the 20 moves allowed for that game have been made, end that game. Record the outcome of the game in the space provided on the data sheet for game 1. If the players are volunteers, the observer should thank them for helping with this assignment and then go look for another two volunteers. The observer should repeat this procedure until he or she has tested 10 pairs, each pair playing the game only once. The observer must be sure to treat all pairs identically, including testing each in the same room and reading each the same set of instructions.

For each of the 10 games, the observer should count the number of moves made and should end the game if it is not won within 20 moves. Record the outcome of each game. At the end of each game, the observer should ask the players to describe any way in which someone could win the game. The observer should record any answers on the data sheets.

Results and discussion

DO NOT READ THIS SECTION until after you have completed all the data collection.

In solving the problem of how each player can win the game the most times while losing the least, people may have adopted one of several strategies. You will need to be familiar with each of these so you can determine which strategies were used.

- Cooperation
  In the original PD, cooperating meant not informing on your fellow prisoner. Effectively, this helps your fellow prisoner (because informing on the person

would mean a longer jail term), while costing you little (because if you had informed on your comrade, the police would have to let you go). It is also a bit of a risk because if you cooperate but your fellow prisoner defects, you will be in jail for a long, long time. Among animal behaviorists, cooperation similarly means behaving in a way that helps another, sometimes at a cost to oneself. For this game, cooperation is defined as moving the game piece in a direction that directly benefits one's opponent at an obvious cost to one's own potential success. For example, say you move the penny toward your goalspot, then I cooperate by moving it in the same direction on my turn, and then you move it again into your goalspot. In this case, rather than attempting to win the game for myself, I have cooperated in helping you to win. A game won through **pure cooperation** can be won in three moves (from the center spot straight to a goalspot).

- Competition
  Another strategy for solving the problem of how each player can win the game the most times while losing the least involves competition. In the single-iteration PD, the optimal strategy for a given prisoner is to defect (or "compete" for the prospect of an early release) by informing on the other prisoner. In a situation in which both prisoners essentially "compete" for an early release by informing on one another, neither prisoner benefits because each still gets a jail term. However, that jail term is not as long for the defecting prisoner as it would have been if he or she had cooperated with the fellow prisoner while the fellow prisoner defected. Competition, then (in the sense of competing for the possibility of early release), may yield a maximum payoff (freedom) or an intermediate punishment (an intermediate jail term). For the purposes of this activity, competition might result in a win for a given player (if the opponent cooperates) or it could result in a draw with neither player winning. One might think of a draw as a consequence that is intermediate between winning and losing. **Pure competition** would result in neither player winning. All the moves would be used up, as each player moved the penny back toward their goalspot away from the circle into which the opponent had just moved it.
- Submission
  One player can give up or "submit" after the other directly challenges him or her. In single-iteration (PD) games, this opportunity never arises because each player has only one chance to make a decision about whether to cooperate or to defect. This is unlike our game of 20 moves, in which each move (and, in fact, each game) is like a new decision for a player: "Do I cooperate, or do I defect?" Except with the first move, each player knows what the other player did just before they must make their move or decision. Submitting might be one of those decisions. In the IPD games, submitting could occur not only multiple times within a game but also in multiple games between the same players. Each game in this case represents an opportunity to decide, on the basis of the outcome of the last game, whether to try to win or just to submit. Even if multiple decisions

occur within a game, there is no opportunity to use the outcome of one game (PD) to guide future decisions about whether to try to win or just to submit.

What would submission look like? Imagine I move the penny toward my goalspot and you move it back one. One my next move, I might submit to you and move the penny toward your goalspot yet again so that it may take as few as five moves for you to win the game. Submission would look like moving toward another player's goalspot, rather than to the side or toward their own goalspot.

- Conflict avoidance
  Trying to avoid conflict involves moving the penny in a direction that is neither toward nor away from one's own goalspot but, rather, to the side. Sometimes it is called "staircasing" because it results in the movement of the penny on the game board making a staircase pattern (Fig. 14.2). Again, such a strategy is unique to this game because PD games (unlike IPD games) allow players to make only one decision and do not allow players to interact.
- Combination of strategies ("mixed")
  Sometimes players will attempt a combination of strategies as they consider how the game might be won. A single game of 20 moves might show evidence of submission *and* conflict avoidance.

Data analysis

Each observer should summarize his or her data on the data summary sheet.

1. Calculate how many times (if any) one of your subjects won a game in three moves. How many examples of pure cooperation were there among your 10 games? Enter this in your data sheet.
2. Calculate how many times (if any) your subjects failed to win a game and spent all 20 moves in vain. How many examples of pure competition were there among your 10 games? Enter this in your data sheet.
3. Calculate how many games (if any) show at least one instance of submission among your 10 games. In how many games did at least one player show signs of "giving up"? *In this count do not include games that also show* "conflict avoidance"*; rather, count only those games that show at least one instance of submission.* Enter this in your data sheet.
4. Calculate how many games (if any) show at least one instance of a player trying to avoid conflict by staircasing. How many games of the 10 showed evidence of conflict avoidance? *In this count do not include games that also show submission; rather, count only those games that show at least one instance of conflict avoidance.* Enter this in your data sheet.
5. Calculate the number of games (if any) in which both conflict avoidance and submission were employed as a "mixed" strategy. Enter this in your data sheet.

Data should now be pooled for the class. For both types of games answer the following questions (calculated separately):

1. What was the average number of wins per player observed?
2. What was the total number of draws observed?
3. How many of the games observed show pure cooperation?
4. How many of the games observed show pure competition?
5. How many games observed show mixed strategies?
6. Did anything change in the strategies employed by subjects as they played one another again and again? To examine this create some pie charts in which you compare the number of games that showed each of the strategies for the first, fifth, and tenth games played. Are there any apparent changes in strategy over time?
7. For IPD games: Was there any evidence of a "tit-for-tat" strategy being employed? In other words, did the behavior of one player appear to affect the behavior of the other player in a predictable way? Document the frequency of occurrence of tit-for-tat strategies.
8. If instructed, you may conduct some statistical analyses of your data.
   a. Specifically, you will want to test the null hypothesis that there is no difference between the average number of IPD and PD games won per player.
   b. If you have more data from more than 30 observers for each type of game use a t-test to compare the strategies.
   c. If you have fewer than 30 observers for each type of game use the nonparametric Mann–Whitney U test, which does not make as many restrictive assumptions about normality of the underlying data as the t-test.
   d. Use a chi-square test to determine whether the two types of games (PD vs. IPD) differ in the frequencies of the different strategies they elicit.

Discussion questions

1. Compare the typical outcome of game data between the types of games (PD vs. IPD). Was the average number of wins per player the same? If not, why might that be?
2. Did the subjects appear to play the game differently when they knew they were going to play only once versus when they knew they were going to play the game repeatedly? If so, in what way was the play different?
3. Among the PD games, was there one strategy that seemed to be most common? What about in the IPD games? Were the most common strategies in the two kinds of games the same or different? How do your results compare with predictions from the game theory?
4. In the IPD games, why might one strategy have "worked" better than another for maximizing each player's winnings while minimizing losses?

5. From your examination of the pattern employed by each player during the 10-game bouts, did player A's behavior appear to influence what player B chose to do? If so, how?
6. What do your data tell us about humans and their decisions on when to cooperate with one another? Do you think you might have had different results if you asked humans from a different culture to play this game? If so, why? If not, why?
7. Are there any particular conditions under which you think that the strategies a person employed might be different? Why?

References

Axelrod, R., & Hamilton, W. D. (1981). The evolution of cooperation. *Science, 211*, 1390–1396.

Trivers, R. L. (1971). The evolution of reciprocal altruism. *Quarterly Review of Biology, 46*, 35–57.

Wilkinson, G. A. (1984). Reciprocal food sharing in the vampire bat. *Nature, 308*, 181–184.

Part II. Instructor notes

Classroom management/blocks of analysis

This activity can be done two ways, either collecting data on games played by people outside of class or collecting data on games played by class members. The game board is available in the supplementary materials.

Played outside class

If you want to assign the data collection portion of this activity outside the class, every student will be an observer. The students' mission as an observer is to find volunteers who are not in this class, explain the game to them, and then collect data on the games the volunteers play. You can assign this as either individuals, each needing to gather 5 teams that played 10 PD games and 5 teams that played 10 IPD games, or working in pairs. You can modify how many "teams" you want students to collect data on but remember that there is a limited pool of students. Make sure to provide the students with copies of the game board (or an electronic version). Students will need to do their calculations and answer the questions. If you gather the data from each group, you could then extend the activity to a discussion and additional calculations as a whole.

Played in class

If data are to be collected on class members, then some of the students will be players while others will be observers of the game. The instructor will need to divide the class, and how they are divided depends on how many students you have. One part of the class will follow Procedure 1 (in Part I) and participate in IPD games, while the other portion of the class will follow Procedure 2 (in Part I) for PD games. You will want your data to be somewhat balanced, so decide ahead of time how many "teams" should play IPD games versus PD games. Some suggestions include having students gather in the hallway and then admitting them into the classroom when it is their turn to play. *You do not want to influence the outcomes of the game, so we find it better that students do not know ahead of time what is being done.*

After each team has collected its data, everyone in the class will come together again to share their methods and to compare and discuss their findings.

Teaching the activity

Remember that the purpose of this activity is to determine some possible strategies that humans might use when presented with an opportunity either to cooperate or to behave selfishly. Under most circumstances and in most animals, natural selection favors essentially selfish behavior and does not favor behavior that benefits another at a cost to one's self. The female vampire bat that offers a blood meal to her starving neighbor does so at a cost to herself. If our generous bat made these offers to unrelated neighbors repeatedly, to the point at which she compromised her own health, it is unlikely that her altruistic genes would be passed onto future generations.

Students will need information on altruism and reciprocal altruism. Depending on your method of teaching, you could provide it before or after the students play the game. But it will be helpful for them to understand the material to properly answer the discussion questions.

Data analysis

After the PD and IPD games have been completed, students will need to summarize the results. For each IPD game, they will need to calculate how often each player wins, the winning strategies, and the draws. For PD games, data will be averaged across all the games to calculate how often each player (A or B) wins, the winning strategies, and the draws. The instructor can then collect these data from the students to create a class data set, if they choose.

Determining winning strategies can be difficult. It is worth spending a little time in class going over the four strategies listed in the Student instructions. You may also provide the students with a quick summary of each on a handout. Students will need to evaluate each game board to figure out the strategies that each player used.

You can also expand this activity to include statistical analyses; details are provided in the Student instructions.

Answer key

1. Compare the typical outcome of game data between the types of games (PD vs. IPD). Was the average number of wins per player the same? If not, why might that be?

   Answers depend on the data. You would expect some differences between PD and IPD game outcomes.
2. Did subjects appear to play the game differently when they knew they were going to play only once versus when they knew they were going to play the game repeatedly? If so, in what way was the play different?

Cooperation theory suggests that individuals should only cooperate when provided with the opportunity to play more than once. If individuals only get one opportunity to play with each other, you would expect them to be more aggressive and try to win.

3. Among the PD games, was there one strategy that seemed to be most common? What about in the IPD games? Were the most common strategies in the two kinds of games the same or different? How do your results compare with predictions from the game theory?
Strategies should be different between PD and IPD games. You would expect to see more competition or mixed strategies with PD games. For IPD games, you would expect students to "defect" by using the competition strategy because this move is guaranteed to maximize the payoff of the player no matter what the opponent chooses. However, it can be seen that both players would have been better off choosing to cooperate with each other (hence the dilemma).
4. In the IPD games, why might one strategy have "worked" better than another for maximizing each player's winnings while minimizing losses?
For IPD games, there is no single best strategy against all possible opponents. Therefore the player needs to have a strategy that is capable of doing reasonably well with many different types of strategies. To do this, the player's strategy must reduce negative consequences while trying to induce cooperative behavior. Therefore a good tactic is to be nice (do not defect first), but if the opponent defects, there are repercussions. But at some point the strategy is willing to be forgiving and return to cooperation when the opponent does.
5. From your examination of the pattern employed by each player during the 10-game bouts, did player A's behavior appear to influence what player B chose to do? If so, how?
Answers depend on the data. For IPD games, there is a chance that previous game outcomes and strategies that were used will impact the future games.
6. What do your data tell us about humans and their decisions on when to cooperate with one another? Do you think you might have had different results if you asked humans from a different culture to play this game? If so, why? If not, why?
Culture may affect cooperation between individuals. Students from collectivist cultural traditions would display more cooperative behavior than groups composed of people from individualistic cultural traditions (Cox et al., 1991).
7. Are there any particular conditions under which you think that the strategies a person employed might be different? Why?
Although we have expectations based on altruistic theories, whether or not the students knew each other before playing the game may affect their strategies. If students have a relationship, you would expect to see more cooperation.

Reference

Cox, T. H., Lobel, S. A., & McLeod, P. A. (1991). Effects of ethnic groupb cultural differences on cooperative and competitive behavior on a group task. *Academy of Management Journal, 34*(4), 827–847.

Part III. Supplementary material

Supplementary data related to this article can be found online at https://doi.org/10.1016/B978-0-12-821410-7.00024-8.

CHAPTER

15 Using empirical games to teach animal behavior

Philip K. Stoddard
Department of Biological Sciences, Florida International University, Miami, FL, United States

Chapter outline

Exploring Animal Behavior in Laboratory and Field. https://doi.org/10.1016/B978-0-12-821410-7.00014-5

Part I. Student instructions

Learning goals, objectives, and key concepts

- Understand the value of knowledge in an animal's life.
- Learn when it makes sense to form coalitions for personal (and mutual) advantage.
- Figure out why owners of a resource will spend more defending it than the resource may be worth.
- Learn how animals can use their opponent's psychological mistakes to their own advantage.

Background

Knowledge has value—that's why you committed the time and money to attend college. Members of other animal species likewise benefit from knowledge, whether knowledge of their habitat, knowledge of their predators and competitors, or knowledge of the individuals with whom they might cooperate. Knowledge has two components: information and the ability to use it. Knowing how to use information to personal advantage gives value to that information. By using an intelligent strategy alone, one can do OK at the game of life described below, but combining a smart strategy with up-to-date knowledge of your opponents will allow you to excel. Sharing that information with smart collaborators is almost unbeatable.

A "war of attrition" is a contest with no weapons in which contestants seek to outlast each other, wearing down their opponents with small challenges. The contest ends when one side gives up and leaves. "Symmetry" in this contest occurs when both sides know their own resource levels and the opponent's as well, and thus each contestant can estimate how long both sides can hold out. That symmetry of knowledge makes for a quick and boring game, about as interesting as a game of "Go Fish" played with the cards face up. "Asymmetry" in this knowledge of resources occurs when one contestant knows how much reserve he or she has but the opponent does not, i.e., not showing your cards (Hammerstein & Parker, 1982). This asymmetry can be heightened by observing one's neighbors over time, allowing a contestant to estimate the defenses of various possible opponents. That knowledge makes a much more interesting, intricate, and powerful game. It's like the advantage that accrues to card-counters in poker or blackjack. Little wonder that casinos try to prevent card-counting at the gaming tables; it erases the house advantage and shifts the profit margin statistically to the smart players. Similarly, professional baseball invests in scouts who study opposing teams to decode their secret internal signals. Knowledge has value.

Purpose

As an animal player in this game, your ultimate goal is to produce the most offspring. As a student in this course, your goal in playing this game is to figure out and apply the smartest biologically realistic strategies you can think of to outsmart and outcompete your opponents in the Darwinian game of life. Master the game as a solo player, and then use your knowledge of animal behavior to devise social strategies that will beat the less imaginative soloists. Lessons from studies of cooperation, reciprocal altruism, and kin selection can all be put into play in this game.

Methods

At the most basic level, here's how you play the game.

Your instructor is the game master who keeps time. Every 12 s the timer goes "DING".

Foraging

Players amass resources by foraging at the resource table. On a "ding" a forager can collect five "resource squares" (2-cm square slips of scrap paper). For example, foraging for three "dings" will provide the player with 15 resource squares. A player may choose to keep secret from other players about the number of amassed squares, making the contests that follow into asymmetric wars of attrition.

Reproduction

Any time, a player can move to an open chair, the breeding site.

At the "ding" a seated player may produce one offspring. The player can only reproduce while seated in the chair when the "ding" is given. To reproduce fold one resource square in half and that folded paper is an offspring.

At any time, a player may leave a chair to go back and forage.

Challenge for possession of a breeding site

A standing player without a chair can lay down a square before a resident, and the resident must match it or leave.

The challenge continues until one player gives up or runs out of squares.

A seated player cannot breed while being challenged.

Played squares are considered "detritus" and get returned to the table by the loser.

End of the bout

At the final "ding", players count their offspring (folded squares). This is the player's fitness score. Unfold the squares and return them to the table for the next game.

Results/discussion

After each game, every player reports their offspring count. If you have the technology available, players can quickly report their score with a "clicker". Make a histogram of the class fitness scores and see how your strategy compares to the rest of the population. Is your fitness higher or lower than the mean? The ratio of your fitness to the mean fitness is your "relative fitness." The highest and lowest fitness players should report to the group what strategies they followed. You could keep your strategies secret, but the game is more fun and evolves more quickly if you share your strategies, good or bad, between rounds. In the next game employ the best strategies you can and avoid mistakes made in the prior game.

The group's mean fitness generally rises from game to game, and the variance narrows. It's obvious why group fitness should go up with experience, but why should the variance diminish? The same phenomenon occurs in virtually all games and contests: the massive point spreads common in high-school athletic competitions diminish in college athletics and disappear entirely in professional sports. Reduction of variance with experience (narrowing of the point spread) results from the combination of a couple of factors. But, an innovation or a change in rules can widen the spread again (Hessenius, 1999). Use this game to figure out what factors close the performance gap between contestants and which ones reopen those gaps.

Think also about how this game relates to other aspects of behavioral ecology and psychology you have learned thus far. The following section gives you some hints in this area.

Upping your game by applying lessons of behavioral ecology

This game can be played at multiple levels. It's fun just as it is, and you can definitely improve your score by refining your strategy for allocating your time between foraging and breeding and by figuring out when to hold or fold during the contest for a chair. But in this game, as in life, knowledge is power. No rule says you can't watch the play going on at the chairs while you are foraging at the table. A little discrete spying will greatly improve your performance.

Social strategies change the level of play entirely. The best social strategies tend to run circles around the best solo strategies. Could you work out an arrangement to cooperate in a mutually beneficial manner with nonrelatives (i.e., reciprocal altruism)? Why not choose a trusted partner and share information reciprocally?

Maybe team up to challenge a chair holder. A smart chair holder will give up quickly when confronted by an organized gang.

You can use your knowledge of kin selection to considerable advantage in this game. In nature, sisters often cooperate and brothers sometimes do. Could you and another player declare yourselves to be a coalition of siblings and benefit from indirect fitness by helping each other? The math of kin selection is much easier if you are haploid or cooperating with your monozygotic (MZ) twin (or, like armadillos, MZ quadruplets). Figure out a kin selection strategy and give it a try.

If you want to make the game more interesting yet, then declare your species diploid and require sexual reproduction to reproduce (a fist bump is sufficient for fertilization in this species). Is it better to be male or female? If you are a simultaneous hermaphrodite and can be both sexes at once, will you deploy egg-trading strategies with another simultaneous hermaphrodite (Fischer, 1984)? Can pathogen resistance genes compensate for the 50% fitness cost of sex? Amend the game to introduce pathogens (i.e., pathogen cards) and resistance genes (Major Histocompatability Complex (MHC) trading cards) that can be given to your offspring. You will never run out of interesting, biologically relevant variations of this game.

Part II. Instructor notes

In-class games add life to the teaching and learning of any academic discipline. Animal behavior games are perfect for the class period when a big assignment is due and the students are fried or when a field trip is canceled. Games can always regain attention lost during the "postlunch dip". Games may be theoretical or empirical, and each genre has a specific function in the classroom. Theoretical games (i.e., games theory) must generally be worked out in front of the class, with the students prompted to fill in the matrix. Theoretical games demand a clear statement of the theory's underlying principles and are especially valuable at revealing counterintuitive results. A most delightful introduction to theoretical games in animal behavior can be found in Karl Sigmund's book, *Games of Life* (1993). Empirical games are different in that they reveal the strategies and problems at the animal's level rather than at the theorist's, finding inspiration in the Saki short story, The Schartz-Metterklume Method (Munro, 1914). Empirical games involve students playing the role of animals, each student applying his or her own (more or less thought-out) solutions to the animal's problem as it is defined in the game. Students' understanding of the animal's perspective comes from inductive reasoning, frequently orchestrated by the instructor.

The best empirical games have these attributes:

(i) simple rules that can be explained quickly and remembered easily,
(ii) greater reliance on strategy than chance,
(iii) outcomes that are hard to predict from intuition alone,
(iv) multiple connections to theory and published data.

The keys to get the most pedagogical benefit from an empirical game involve setting up the game and following up afterward. It helps if the game is fun. My chief complaint with instructive games is that we tend to give the students too much instruction while setting up the game (e.g., "Here's the equation, and here's what to do with it ... "). I have found that students get much more from the exercise if I lead *them* to derive principles, perhaps from the lecture and readings, and turn those principles into game rules and parameters. Few students have a clue as to how this can be done, so I ask leading questions and help the students fit their answers into a framework that emerges piece by piece on the board in front of them. Instructors find it far more efficient, of course, to tell the students how the exercise is to work and what everything means. In doing so, however, the opportunity for a deeper understanding is squandered.

My current favorite empirical game is The Asymmetric War of Attrition because it combines the better aspects of behavioral ecology, strategic thinking, and musical chairs. Student players are free to divide their time between foraging for resources and reproducing using their collected resources to produce offspring. Potential breeding territories (chairs) are a limiting resource, so players must compete for chairs by using the resources they collected for breeding to instead fuel the fights for a breeding territory. The basic game can be made more interesting with strategy, social alliances, and familial relations incorporating Hamilton's rule and indirect fitness.

Classroom management/blocks of analysis

A 40-round bout of the game should take 15 min to play and another 5–8 min to analyze. The more bouts you can fit into the time allotted the better because the game gets more interesting with each bout. Students spend the first few bouts of the game getting good at their solo strategies. Subsequent bouts tend to include social strategies and more nuance, where the best lessons of animal behavior can be learned through experience. Three bouts would be the absolute minimum. Six bouts would be better, possibly split across days to allow students the opportunity to devise social strategies with their peers.

Teaching the activity

Preclass preparation

There's no limit to how many students can play at once, provided adequate materials and chairs are available.

Cut paper "resource squares", about 50 per player. These are just scrap paper cut into 2-cm squares.

Set an audible repeat timer to beep every 12 s. I use the "Repeat Timer" app for iPhone, with the phone connected to a portable Bluetooth speaker to be audible in a busy classroom.

If you have use of computer projection in the teaching space, then create a histogram template in Excel or R, into which you can quickly throw and project the students' performance data. If you can keep the previous histograms displayed for comparison with the new data, that will be best.

Chairs: put out (or otherwise designate for play) half as many chairs as players. If the chairs can be moved and the floor is flat, a good arrangement is two rows back-to-back.

Table: spread out resource squares on a table at front of the room some distance from the chairs.

In-class preparation

Rules of play

See the student section.

Duration of play

The first bout should last just 4 min to give students a chance to master the rules. Students make lots of mistakes during the first bout. Catch the mistakes, correct them, and continue to another bout. Subsequent bouts should last 10 min. Announce the number of "dings" remaining in the last 3 min of this bout. Instructors may need to put out more fitness squares midgame if they run low.

Analysis between bouts

On the whiteboard or computer create a quick histogram of the players' offspring counts. One typically gets a bell curve. Checking any outliers in the first couple of bouts usually identifies the few students who didn't understand the rules when first explained. Players on the high end and low end of the main distribution should be asked to explain their strategies to the class. Ask the students to consider the following:

Value of knowledge: How many clever players observe others play, swooping in to displace a resident who has spent a lot of resource squares?

Social behavior: Did any players form coalitions? Mutual defense?

Psychology of the resident: Why does the resident tend to outbid the intruder, even though the game has no inherent resident advantage? Run another bout and encourage students to try a better strategy. Directional selection will ensue before the distribution stabilizes and narrows, so don't erase the histogram between bouts. Rather switch to a different color for scoring the next bout so the changes can be seen and discussed.

The cost of sex: So far, we have considered a haploid game. If you have a more advanced class, you can change the players to diploid and consider the 50% fitness hit from sexual reproduction. A fist bump with another player does the trick, or a player can bump her own two fists together for self-fertilization. How can sexual diploids possibly compete with selfers? Without pathogens and resistance genes, selfing beats sexuals two to one. Try developing a version with pathogens and resistance genes.

Relatedness and cooperation: The application of Hamilton's Rule (degree of genetic relatedness among two individuals) can transform this game into an infinite number of variations. An individual may contribute to his or her own fitness by assisting the reproduction of a relative who has genes in common. Quietly tell a few pairs of students that they are MZ twins ($r = 1.0$), such that the fitness of each is shared by the other. They will quickly come up with cooperative strategies. Create conventional sibling pairs ($r = 0.5$) and see how they compare in fitness to the MZ twin pairs. Help the students work out a haplodiploid version, and then let haplodiploid teams compete against diploids and see which helping strategies work best with each ploidy set.

Postgame analysis: Thoughtful discussion after the play is complete allows you to help students use their empirical experience to construct a theoretical understanding of trade-offs, alternate strategies, applied kin selection, and cooperation. When I run the game, we alternate play with analysis in a recursive manner to allow students to use the discussion to improve their strategy and understand the concepts. I ask the class what they will do differently in the next round and then reflect on how their proposals relate to biology and theory. A few times, I've had the sharpest students innovate by introducing kin selection with no prompting, but I often pull a few students aside and suggest a simple kin selection strategy, e.g., "You two are monozygotic twins. If you work as a team, you can share your fitness points at the end, but no double counting—you split the points in half."

Here is a follow-up quiz that I have used after upper division biology majors play the game for 60–70 min. For the more involved questions 3–5, I'd definitely have them discuss it in groups.

Imagine playing **101** rounds of the "Asymmetric War of Attrition" game with twice as many players as chairs. Remember that you have to forage at least once before you can reproduce so you have *at most* 100 breeding rounds.

(1) What's the **average** number of offspring a player can produce if everybody shares the chairs instead of fighting (i.e., nobody loses squares to challenges)? *This one is easy because the average player has a chair half the time on average.*

(2) Let's get real. Assume half the time is spent fighting over chairs. Now what's the average fitness you'd expect?

(3) In this game, you reproduce asexually, so your offspring are related to you by 1.0, as are your siblings. You decide to improve your inclusive fitness by pursuing indirect fitness instead of direct fitness, i.e., by helping your chair-sitting siblings breed instead of breeding yourself. So you spend your time foraging and feeding your sibling paper squares so she can always reproduce and never lose the chair to a challenger. Determine your indirect fitness by calculating how many **extra** offspring (not total offspring, but additional offspring) one sibling can produce on average when you help that sib to always keep a chair? Justify your answer.

(4) How many of your clonal siblings (r = 1.0) of your MZ twin (r = 1.0) would you have to supply in this way to compensate you with indirect fitness benefits for never having a chair of your own? Explain your logic.

(5) Answer #4 again assuming you were playing the advanced version with sexual reproduction, where siblings are related by 0.5 and assisting those siblings now produces conventional nieces and nephews. Explain.

Answer key

(1) 50. Here's why: After one foraging round, players have 100 rounds left. On average, a player can sit for at most 50 of these (half as many chairs as players), so the most offspring you'd expect from fully cooperative players is 50.

(2) 25. Half of 50 since half the time is lost to challenges. That was easy.
(3) This one is harder. With good provisioning help, you could produce 50–75 more offspring. The best anyone can do is 100 offspring. Someone who sat for 100 rounds and reproduced every time would need an effective way to avoid wasting time with challenges. Showing your (hopefully) massive stash of squares, provided by your provisioner, can intimidate challengers and keep them from wasting your time. A 100 is at least 50 more than you'd do on your own under the best of circumstances, and probably 75 more than you'd do if you had to deal with challenges.
(4) If you spent zero time reproducing, and you'd expect to have 25 offspring of your own if you were fighting for a chair on your own, you'd want to make sure your efforts helping a relative produced 25 or more clonal offspring ($r = 1.0$) from your fully related sibling.
(5) Your sibling is now related to you by a half, so your sibling's offspring are related to you by a quarter ($r = 0.25$). But your own offspring are related to you by 0.5 as well. You need to produce twice as many nieces and nephews to give you the same fitness as having your own offspring. Before you needed 25, so now you need 50.

The particular follow-up you arrange or what evaluation assignments you find appropriate should depend on how long your students played the game and at what level of complexity.

References

Fischer, E. A. (1984). Egg trading in the Chalk Bass, *Serranus tortugarum*, a simultaneous hermaphrodite. *Zeitschrift für Tierpsychologie, 66*, 143–151. https://doi.org/10.1111/j.1439-0310.1984.tb01361.x

Hammerstein, P., & Parker, G. A. (1982). The asymmetric war of attrition. *Journal of Theoretical Biology, 96*(4), 647–682. https://doi.org/10.1016/0022-5193(82)90235-1

Hessenius, C. (1999). Maybe there were giants, or at least outliers: On the .400 batting average myth and the absolute limits of hitting for average in major league baseball. *Journal of Sport Behavior, 22*(4), 514–544.

Munro, H. H. (1914). The Schartz-Metterklume method. In Saki (Ed.), *Beasts and superbeasts*. Available from https://en.wikisource.org/wiki/Beasts_and_Super-Beasts/The_Schartz-Metterklume_Method.

Sigmund, K. (1993). *Games of life: Explorations in ecology, evolution, and behaviour*. Oxford: Oxford Univ. Press.

PART

3

Application of behavior

CHAPTER

Finding food is fun! Location discrimination training

16

Robin L. Foster, Carolyn Loyer

University of Puget Sound, Tacoma, WA, United States; University of Washington, Seattle, WA, United States

Chapter outline

Exploring Animal Behavior in Laboratory and Field. https://doi.org/10.1016/B978-0-12-821410-7.00021-2

Part I. Student instructions

Learning goals, objectives, and key concepts

- Examine how an animal's food search behavior is influenced by recent past experiences.
- Understand how behavior is controlled by its consequences. This process is called operant conditioning and is the foundation of this exercise.
- Explore individual differences and cognitive biases in patterns of food search behavior.

Background

All animals, from bees to primates, are intrinsically motivated to search for food and will remember the locations of reliable food sources. For example, when a honeybee discovers a rich patch of pollen or nectar, it will return to the hive and communicate the precise location using an elaborate symbolic dance so that her nestmates can return to that source (von Frisch, 1950, pp. 55–56). Predators learn and regularly visit locations where their prey gather to eat (Selva et al., 2014), drink, or reproduce (Penney & Lowry, 1967; Rogers & Bryden, 1995), and birds learn the locations of backyard feeders and are more likely to return to feeders that are regularly replenished.

The four quadrants of operant conditioning

Learning is defined as a change in behavior due to experience, and operant conditioning is the process of learning through consequences. The four quadrants of operant conditioning are fundamental principles of learning, including positive reinforcement, negative reinforcement, positive punishment, and negative punishment (Table 16.1). Which quadrant is operating when an animal learns can be determined by answering two questions: (1) is the behavior more or less likely to be repeated in the future? and (2) does the behavior result in something being added to or removed from the environment?

Is the behavior more or less likely to be repeated in the future?

- Behavior will increase if it leads to a pleasant outcome or improves the animal's situation. The behavior is said to have been reinforced.
- Behavior will decrease or stop if it leads to an unpleasant outcome or worsens the animal's situation. The behavior is said to have been punished.

Table 16.1 The four quadrants of operant conditioning.

	Reinforcement: behavior increases	**Punishment: Behavior decreases**
Positive (+) Something is added	**Positive reinforcement (R+)** Behavior increases because it results in something pleasant being added. *Example: The dog approaches the food dish because he gets a treat.*	**Positive punishment (P+)** Behavior decreases because it results in something unpleasant being added. *Example: The rat stops eating the bait because it became ill from the food.*
Negative (−) Something is removed	**Negative reinforcement (R−)** Behavior (escape or avoidance) increases because it results in something unpleasant being removed. *Example: The crow flies into the tree to escape from a cat.*	**Negative punishment (P−)** Behavior decreases because it results in something pleasant being removed. *Example: The dog stops jumping in greeting because its owner turns away.*

Does the behavior result in something being added to or removed from the environment? Some examples can help clarify this question.

- When an animal's search behavior results in finding food, then something (food) was added.
- When a bird flies up into a tree to escape from a predator, then something (a threat) was removed.

The mathematical terms and notation "positive" (+) and "negative" (−) reflect something added or removed, respectively, as a consequence of the behavior. When humans train animals, the terms also refer to the trainer adding or removing something from the environment when the animal does something. It can be helpful to think about "positive" and "negative" as one might refer to medical test results rather than as "good" and "bad" (a positive medical test result can be very bad news!)

Positive reinforcement is the operant conditioning quadrant featured in this lab. When an animal searches for and finds food in a particular location, the consequence of finding food will increase the probability that the animal will search for food in that location in the future.

The following are some additional learning principles and definitions applied in this lab. Additional information about operant conditioning can be found in texts on learning and behavior (e.g., Chance, 2014; Domjan, 2015).

Shaping: Reinforcement of successive approximations to a desired response.
Extinction: Reduction of a response that occurs because it is no longer followed by the reinforcer.

Stimulus discrimination: The tendency for behavior to occur in certain situations but not in others.
Stimulus generalization: The tendency for learned behavior to spread to new situations, which makes learning an adaptive strategy in changing environments.

Learning theory makes predictions about an animal's foraging behavior. According to operant conditioning, whether or not an animal will repeat a behavior depends on its consequences. In this exercise the behavior is "approach a dish", the pleasant reinforcing consequence is finding "food", and the discriminative stimulus is the "dish location". You will train an animal to discriminate between two dish locations. In one location (CS+) the dish has food and approaching it will be *positively reinforced.* In a second location (CS−) the dish is empty; approaches to this dish are not reinforced and will thus be *extinguished.* The notation from learning theory used here might be new to some students. CS stands for "conditioned stimulus" (in the current exercise, CS is the food dish location). The "+" sign means a reinforcer (food) is present and the "−" sign means the reinforcer is absent. To stop responses to the CS−, some research on location discrimination learning has included a punishment condition with unpleasant consequences, such as bitter-tasting food or shock. The use of punishment is not recommended in this exercise because of ethical concerns and unwanted side effects.

After the animal learns to discriminate between the CS+ and CS− locations, you will next observe if and how quickly it will approach a dish placed in a novel location in between CS+ and CS−. This tendency to approach the dish in a novel location can be used as a measure of judgment bias about making decisions to forage in novel locations (Piotti et al., 2018; Roelofs et al., 2016). Judgment bias is defined as a systematic error in cognition that influences decision making and behavior and affects how an individual responds to ambiguous stimuli. Is a food dish placed in a novel location likely to have food or will it be empty? Judgment bias is sometimes compared to the concepts of optimism and pessimism in human psychology, where a positive judgment bias is the expectation of a good outcome in the face of uncertainty and a negative judgment bias is the expectation of a bad outcome.

Purpose

Food is a biological necessity for all living organisms, and animals will invest tremendous effort searching for it. In this exercise, we will explore how an animal's foraging decisions are affected by their recent learning experiences and cognitive biases. Students will use operant conditioning to train an animal to discriminate between two food dish locations. In one location the dish has food and approaching it will be *positively reinforced.* In a second location the dish is empty; approaches to this dish are not reinforced and will thus be *extinguished.* After learning to discriminate between the two locations, you will observe if and how quickly the animal will approach a dish placed in a novel location. Judgment bias is defined as a systematic error in cognition that affects how an individual responds to an ambiguous stimulus. In this experiment, an animal's response to the food dish in a novel location can be used as a measure of judgment bias.

Methods

If you are conducting this investigation during class time, your instructor will set up the testing arena and have everything you need on hand. If you are doing this on your own, please follow these guidelines to test out appropriate setup and food rewards before starting.

Materials needed

1. The **testing arena** should be a large enclosed space, such as an open-field apparatus (Gould et al., 2009), with a **stationing area** where the animal can be secured briefly between trials while the next trial is being set up (Fig. 16.1). The size of the arena can vary but should allow a distance of several body lengths from the release point to the food dish locations for terrestrial species and a greater distance for aerial species. Possible arenas for different animals

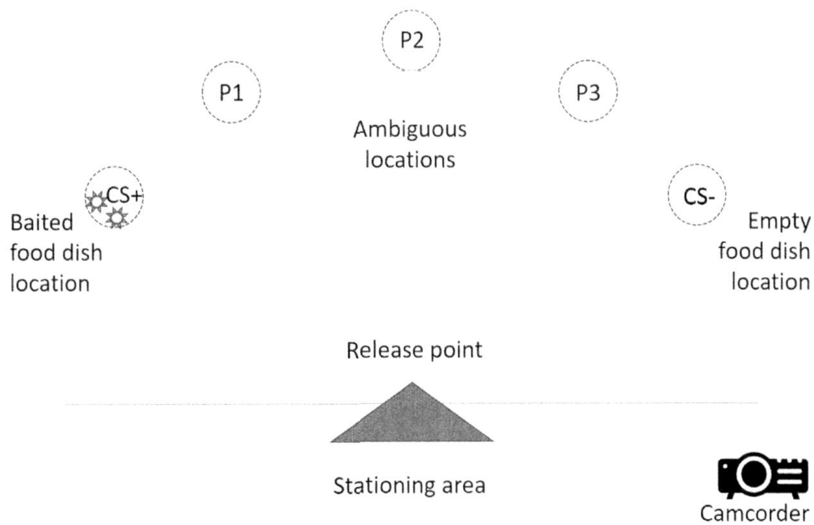

FIGURE 16.1

Experimental setup showing relative positions of dish locations CS+ (baited) and CS− (empty) for discrimination training, as well as ambiguous probe locations P1, P2, and P3 used during generalization test trials. *CS*, conditioned stimulus.

include an open-field enclosure for laboratory animals; a training studio, yard, or driveway for dogs; a flight room for birds; a large tank for fish; a paddock or large box stall for horses and other livestock; and a large open field for wild corvids.

2. **A physical or visual barrier between the arena and stationing area** is recommended so that the animal cannot see where the food dish has been placed for the next trial. In rodents a separate start box may be used as a stationing area and connected to the testing arena by a PVC tube (Barker et al., 2017). If accustomed to being handled, animals may also be removed from the arena between trials and dogs and livestock can be secured on lead line behind a fence covered with a tarp.
3. Choosing a **food dish** style that the animal is familiar with can save time because it will already be comfortable eating from it. If a new dish is used, the animal will need to become comfortable eating from it. The dish should be readily perceived from a distance, but the food should be hidden from view (e.g., under a lightweight object that the animal can easily move aside, partially buried in an inert and inedible substrate, or placed behind the rim of a food dish). In one study with rats (Barker et al., 2017), small treats were buried in scented sand to control for both olfactory and visual cues. Examples of possible food dishes include ceramic or stainless steel cup for rodents, stainless steel coop cup for birds and other small animals, stainless steel bowl on the ground for dogs, food pan on the ground for horses, small dish on the ground for rodents, and a floating ring into which food can be sprinkled for fish.
4. **A food pan** may also be used in the stationing area to encourage the animal to leave the arena and return-to-station between trials in order to set up for the next trial. The food pan in the stationing area should be distinctly different from the food dish used in the training trials.
5. Experiment with different **high-value foods** to see what your animal's preference is. Use small amounts of food on each trial to prevent satiation; when the animal is full, finding food isn't as fun! Possible food reinforcers for rats include milk chocolate baking chips and various cereals such as rice crisps, oat-o's, and fruity cereals. In research using food reinforcement, rats and mice are often maintained at 80% normal body weight to increase their motivation to work for food. If the animal has a healthy balanced diet, an alternative would be to restrict access to food for several hours before each trial.
6. A **stopwatch** is used to measure latency recorded on a **data spreadsheet. Video recording** trials can provide valuable backup as well.

Fig. 16.1 shows five food dish locations in the test arena, but only one dish is set out on any given trial. The CS+ (baited dish) and CS− (empty dish) locations are used during location discrimination training, and the "ambiguous" dish locations, probe trials P1, P2, and P3, are introduced during generalization testing; the probe trial dishes are always empty. Ambiguous location P1 is closest to CS+ and location

P3 is closest to CS−. Fig. 16.1 shows from left to right, CS+, P1, P2, P3, CS−, but some animals will be trained in the reverse order, CS−, P3, P2, P1, CS+.

The five dish locations are the same distance from the animal release point. Exactly what this distance is will depend on the species selected and the size of the open-field arena, but generally three or more body lengths is sufficient. The dish locations are also the same distance from one another, with "P2" centered between CS+ and CS−. Importantly, the greater the degree of separation between CS+ and CS−, the more quickly the animals will learn location discrimination (Graf et al., 2018).

Step-by-step instructions

Step 1. Pretraining

a. Acclimation to the open-field arena (Fig. 16.2)
Many animals are initially distracted by and curious to explore an unfamiliar arena or open-field apparatus. Before beginning discrimination training trials, the animal should be given an opportunity to freely explore the arena. If time allows conduct a few 5-min acclimation sessions each day for 5 days.

b. Shaping approach dish and taking food
To ensure that the next steps go smoothly make sure that the animal is comfortable eating from the food dish and that the food used has high value for this particular species/individual. If possible shape the behaviors of approaching the dish and taking food somewhere other than the testing arena.

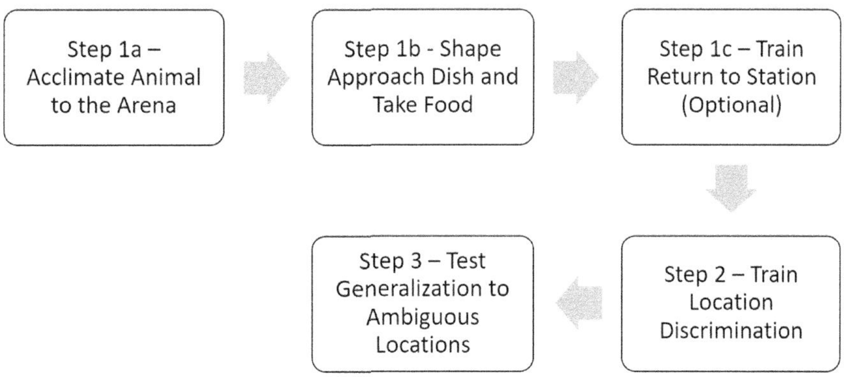

FIGURE 16.2

Flowchart of steps in the pretraining, discrimination training, and generalization testing phases of the lab.

1. Place a small amount of the food in the food dish. Set the dish on the floor near the animal and allow it to eat from it.
2. After it is finished eating, pick the animal up or lure it away with a small treat.
3. Repeat several times, moving the food dish to different locations until the animal approaches the dish and takes food without hesitation.
4. If the dish will be covered or the food will be buried, additional preliminary training trials are needed to introduce that component.

c. Train the animal to return-to-station (optional)
Explicitly training the animal to return to the stationing area is optional but may be recommended when working with confined animals that cannot practically be led or carried from the arena to the stationing area by the experimenter. Return-to-station training is not needed for labs with rats and mice; these species can generally be introduced to and removed from the arena manually between trials.
Place a small bit of food in the stationing area food pan when the animal is close by and "watching". When the animal is finished eating and moves slightly away from the pan place another bit of food in it. Just before putting the food in the pan add a "return-to-station" cue, such as a tap on the pan, a whistle, a flashing light, or a special phrase "go to station". Continue to use the return-to-station cue and reinforcer throughout discrimination training and generalization testing trials.

Step 2. Location discrimination training

This lab uses a procedure called successive discrimination training (Chance, 2014). The animal will learn to associate the dish placed in one location with a food reinforcer (CS+) and that the same dish placed in a second location will be empty (CS−). Discrimination learning is demonstrated when the animal quickly and immediately approaches the CS+ and slowly approaches, or does not approach, the CS−.

a. Training protocol
Flip a coin to decide which side to assign to CS+: heads, right side; tails, left side. Only one dish is placed out on each trial. When the dish is placed in the CS+ location, it is baited with food. When the dish is placed in the CS− location, it is empty. The same dish is used in every trial so that the smell and appearance are held constant. For any one subject the CS+ location, as determined by the coin flip, will be the same for all training and test trials.
Animals learn at different rates, so the total number of training trials will differ across individuals. Trials will alternate CS+ and CS− trials in random order to ensure an equal number of trials at each food dish location and to avoid long stretches of rich (CS+) and lean (CS−) trials, which can affect motivation.

b. Data collection

One student will act as the data recorder and the other student will act as the animal handler. Blank spreadsheets are included with this lab and can be used to record data. If a different order of trials for each subject is needed, the order can be modified using the randomization procedure outlined in Supplement 1.

The independent variable is the location of the dish, either with food (CS+) or empty (CS−). The dependent variable will be go—no go (does the animal approach the food dish?) and/or latency (how long did it take the animal to approach the food dish?). The maximum time limit for a given study will depend on the focal species, for example, 5 s for rats and mice, 20 s for dogs, 30 s for horses, etc.

If "go—no go" is the dependent variable, "1" is recorded if the animal approaches the food dish within the maximum time limit and "0" is recorded if it does not (see Spreadsheets 1a–d).

If latency is the dependent variable, timing starts when the animal is released into the arena and ends when it reaches the food bowl. If the animal does not approach the bowl within the maximum time limit, then that time is recorded on the data sheet (see Spreadsheets 2a–d).

1. The student acting as a data recorder will set up the food dish and record data for each trial, using a stopwatch to measure approach latency.
2. The student acting as handler will make sure the animal is in the stationing area behind the barrier.
3. The data recorder will place a food baited or empty dish, depending on the trial, in the designated location. The data recorder will signal the handler to begin the trial.
4. The handler will release the animal into the arena and the data recorder will start the stopwatch if latency is being recorded.
5. The data recorder will stop the stopwatch either when the animal reaches the food dish or at the maximum trial time and then record the time on the data sheet.
6. If the animal has not voluntarily approached the food dish within the time limit in the first block of four trials, additional pretraining may be needed. With dogs, horses, and livestock the handler may also lead the animal to the dish (for both CS+ and CS− trials) at the end of the trial so it has an opportunity to learn the location—reinforcement association.
7. The handler will return the animal to the stationing area.
8. Trials continue until the animal reaches the learning criterion, defined in the following section.

c. Learning criterion

How do we know when the animal has learned to discriminate between the CS+ and CS− locations? If go—no go is used as the dependent variable, the learning criterion recommended for this exercise is *correct go responses to the baited food dish (CS+) and no-go responses to the empty food dish (CS−) on eight consecutive trials (i.e., two blocks).* If latency is used as the dependent variable, the learning criterion recommended for this exercise is *approaches to the baited food dish (CS+) are faster than approaches to the empty food dish (CS−) on eight consecutive trials (i.e., two blocks).*

Once the animal has met the learning criterion, stimulus generalization testing (step 3) can begin. Most animals will meet the discrimination learning criterion in 30–60 trials, which can be completed in about 1 h of testing.

Step 3. Stimulus generalization testing

After the animal has learned that a dish in the CS+ location has food and a dish in the CS− location is empty, three ambiguous locations are added, P1, P2, and P3. These locations are evenly spaced between CS+ and CS− (Fig. 16.1) and the trials presenting a dish in one of these locations is called a probe trial. P1 is closest to CS+, P2 is centered between CS+ and CS−, and P3 is closest to CS−. These stimuli are ambiguous because the animal does not know if the dish in these locations has food or is empty; by convention, however, these dishes are always empty.

This is a test of stimulus (location) generalization to determine the tendency of learning to spread. How quickly will the animal approach a dish in the new location? If it approaches quickly, CS+ generalization has occurred, and if it approaches slowly or not at all, then CS− generalization has occurred.

a. Testing protocol and data collection

Probe trials will be inserted between blocks of the previously learned CS+ and CS− trials (see Spreadsheets 1b and 2b). Students will maintain their previous roles as data recorder or handler. Other than the addition of probe trials, the procedures remain the same as in discrimination training (step 2).

How an animal responds to the probe trials will provide information about the extent to which its response to CS+ (approaching quickly) and CS− (approaching slowly or not at all) has generalized to the new P1, P2, and P3 locations. An individual that tends to generalize their CS+ response would be said to have a positive judgment bias versus a negative judgment bias for those who generalize their CS− response.

Hypothesis, expected results, and interpretation

Through discrimination training, animals will learn to approach a location that reliably predicts food. Initially they will rapidly approach both the CS+ and CS− locations. After additional trials, animals are expected to quickly approach the baited dish in the CS+ location and to slowly or not at all approach the empty dish in the CS− location.

The pattern of stimulus generalization provides insight into the animals' judgment bias. Stimulus generalization gradients depict an animal's tendency to quickly approach the food dish in a novel ambiguous food location as they would approach the CS+ (positive judgment bias) or to slowly approach the food dish in a novel location as they would approach the CS− (negative judgment bias).

Results/discussion: go—no go as the dependent variable

Discrimination training

1. Data processing: Calculate the proportion of correct responses (i.e., "go" on CS+ and "no go" on CS− trials) for each block of four trials.
2. Summary stats and figures: Use a line graph for discrimination learning (Fig. 16.3A), with trial block on the X-axis and proportion of "go" responses for CS+ and CS− on the Y-axis.

Stimulus generalization testing (i.e., judgment bias)

3. Data processing: Calculate the proportion of "go" responses for all CS+ trials, for all CS− trials, and for each probe location separately. If three probe locations are used, you should end up with five values.
4. Summary stats and figures: Graph the generalization gradient (Fig. 16.4), with the stimulus location, CS+, P1, P2, P3, and CS−, on the X-axis and the proportion of "go" responses on the Y-axis.
   NOTE: Stimulus location order on the figure is CS+, P1, P2, P3, and CS−. Some students will use this order, but others will use the reverse order, CS−, P3, P2, P1, and CS+. In both cases the ambiguous location P1 should be closest to CS+ and P3 closest to CS−.

Results/discussion: latency as the dependent variable

Discrimination training

1. Data processing: Calculate the average latency for the two CS+ and for the two CS− trials in each block.
2. Summary stats and figures: Use a line graph for discrimination learning (see Fig. 16.3B), with trial block on the X-axis and average latency for CS+ and CS− on the Y-axis. Latency for CS+ and CS− are plotted as separate lines.

Stimulus generalization testing (i.e., judgment bias)

1. Data processing: Calculate average latencies for all CS+ trials, all CS− trials, and each probe location.
2. Summary stats and figures: Graph the generalization gradient (see Fig. 16.5), with stimulus location, CS+, P1, P2, P3, and CS−, on the X-axis and average latency on the Y-axis.

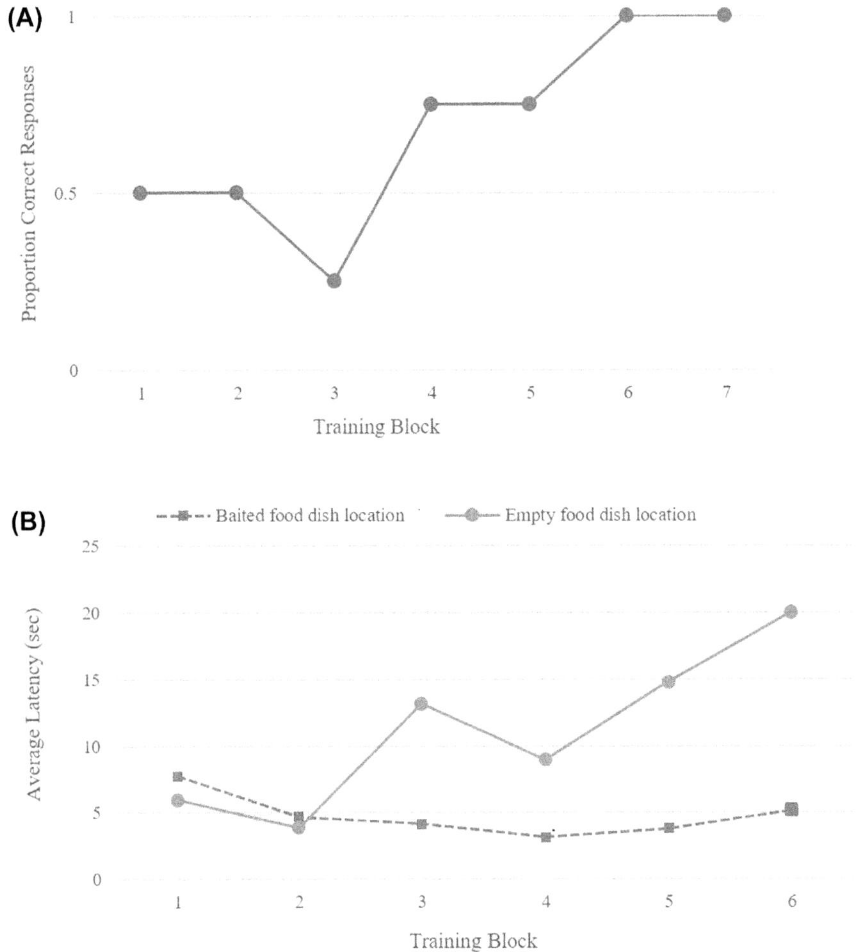

FIGURE 16.3

Location discrimination learning curves plotted using training go–no go (Spreadsheet 2a) and/or latency (Spreadsheet 2b) results. (A) Proportion of correct responses for each block of four trials. The animal initially approached the baited dish at the CS+ and the empty dish at the CS– location at a chance level of accuracy (50%). With additional training the proportion of correct responses increased. The animal met the discrimination learning criterion on trial 28 (block 7). (B) The average latency to approach the baited food dish location (CS+) and the empty food dish location (CS–) for each block of training trials. The animal initially approached both locations quickly. With additional training the latency to approach the CS– location increased and the latency to approach the CS+ location remained low. The animal met the discrimination learning criterion on trial 24 (block 6).

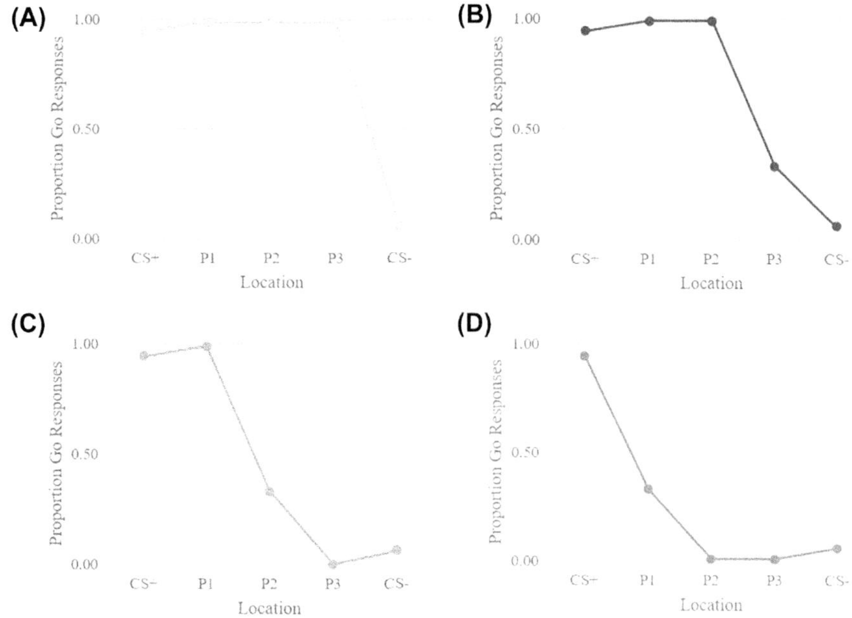

FIGURE 16.4

Examples of different generalization gradient results from testing data with go—no go as the dependent variable. Fig. 16.3A is from sample data in Spreadsheet 1b. As expected, the animals repeatedly approached the CS+ (baited food dish) and rarely approached the CS− (empty food dish). The probability of approaching ambiguous stimuli P1, P2, and P3 depends on the probe location and the animal's judgment bias. (A) High response rates to P1, P2, and P3 shows strong excitatory CS+ stimulus generalization to the ambiguous locations (positive judgment bias). (B) A high rate of P1 and P2 and low rate of P3 responses shows modest CS+ stimulus generalization (somewhat positive judgment bias). (C) A high rate of responses to P1 and low rate of responses to P2 and P3 show weak CS+ stimulus generalization (somewhat negative judgment bias). (D) Low response rates to P1, P2, and P3 shows strong inhibitory CS− stimulus generalization to the ambiguous locations (negative judgment bias).

Conclusions

1. How might using a high-value reinforcer (vs a low-value reinforcer) in discrimination training affect the learning curve (i.e., Fig. 16.3)?
2. What steps should the handler and recorder take to reduce the likelihood of inadvertently or unconsciously "cueing" the subject (i.e., the "Clever Hans effect")?
3. In this exercise an empty pan is used in the CS− location. Why might the animal experience the absence of food as unpleasant or aversive?

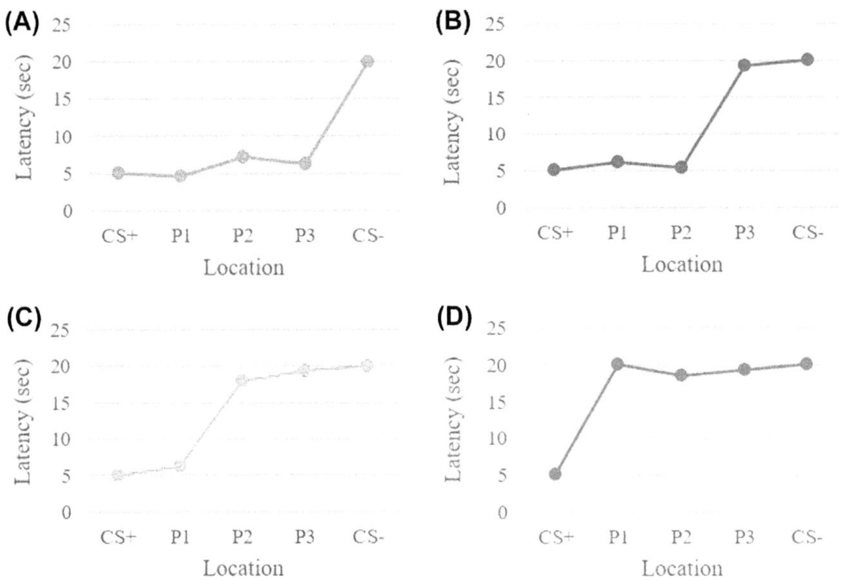

FIGURE 16.5

Examples of potential generalization gradients from test data, with latency as the dependent variable. Fig. 16.4A is from sample data in Spreadsheet 2b. As expected, the animals approach the CS+ (baited food dish) quickly and the CS– (empty food dish) slowly or not at all. Latency to approach P1, P2, and P3 shows different patterns depending on the probe location and the animal's judgment bias. (A) Short P1, P2, and P3 latencies show strong excitatory CS+ stimulus generalization to the ambiguous locations (positive judgment bias). (B) Short P1 and P2 latencies and long P3 latency show modest CS+ stimulus generalization (somewhat positive judgment bias). (C) Short P1 latency and long P2 and P3 latencies show weak CS+ stimulus generalization (somewhat negative judgment bias). (D) Long P1, P2, and P3 latencies show strong inhibitory CS– stimulus generalization to the ambiguous locations (negative judgment bias).

4. How might using a punisher, such as a shock or a loud sound, as the CS– affect the animal's approach to ambiguous stimulus locations?
5. In general, why would an animal be faster or more likely to approach the ambiguous stimulus P1 than P3?
6. What is the potential problem when probe trials at the same location are repeated?
7. In this exercise, we used three probe locations in stimulus generalization testing. What might be an advantage of using more?
8. How would you expect an animal with a negative judgment bias to approach the middle ambiguous stimulus location?

Discussion questions

Students can share results for their animal and, using the collective data, explore individual differences in learning and behavior.

1. *Individual differences in discrimination learning rates.* Some animals learn more quickly than others. The number of trials to criterion and the total number of errors can be compared across animals. The total number of errors includes trials when the animal approaches an empty food bowl in the CS− location (a commission error) and trials when it does not approach a bowl with food in the CS+ location (an omission error). Students can discuss why some animals learn quickly, why some types of errors might be more common than others, and why some individuals make more errors.
2. *Differences in behavior during extinction.* When animals learn that the bowl in the CS− location is empty, they tend to approach more slowly or not at all. What the animal actually *does* during the CS− trial varies and students can discuss the different behaviors observed.
3. *Individual differences in stimulus generalization.* The tendency to approach a novel food location varies across individuals. Students can discuss why some animals will approach ambiguous locations quickly and others approach them slowly or not at all.

References

Barker, T. H., Bobrovskaya, L., Howartha, G. S., & Whittaker, A. L. (2017). Female rats display fewer optimistic responses in a judgment bias test in the absence of a physiological stress response. *Physiology & Behavior, 173*, 124−131.

Chance, P. (2014). *Learning and behavior* (7th ed.) (Cenage).

Domjan, M. (2015). *Principles of learning and behavior* (7th ed.) (Cenage).

Freymond, S. B., Briefer, E. F., Zollinger, A., Gindrat-von Allmen, Y., Wyss, C., & Bachmann, I. (2014). Behaviour of horses in a judgment bias test associated with positive or negative reinforcement. *Applied Animal Behaviour Science, 158*, 34−45. https://doi.org/10.1016/j.applanim.2014.06.006

von Frisch, K. (1950). *Bees: Their vision, chemical senses, and language.* Cornell University Press.

Gould, T. D., Dao, D. T., & Kovacsics, C. E. (2009). The open field test. In T. Gould (Ed.), *Mood and anxiety related phenotypes in mice. Neuromethods* (Vol. 42). Totowa, NJ: Humana Press.

Graf, R., Longo, J. L., & Hughes, Z. A. (2018). The location discrimination reversal task in mice is sensitive to deficits in performance caused by aging, pharmacological and other challenges. *Journal of Psychopharmacology, 32*(9), 1027−1036. https://doi.org/10.1177/0269881118779383

Loyer, C. (2019). *Cognitive bias and intraspecific social behavior in the domestic dog (*Canis familiaris*)* (Unpublished doctoral dissertation). Seattle, WA: University of Washington.

Mendl, M., Brooks, J., Basse, C., Burman, E. P., Blackwell, E., & Casey, R. (2010). Dogs showing separatio-related behaviour exhibit a 'pessimistic' cognitive bias. *Current Biology, 20*. https://doi.org/10.1016/j.cub.2010.08.030

Penney, R. L., & Lowry, G. (1967). Leopard seal predation of Adelie penguins. *Ecology, 48*(5), 878–882.

Piotti, P., Szabó, D., Bognár, Z., Egerer, A., Hulsbosch, P., Carson, R. S., & Kubinyi, E. (2018). Effect of age on discrimination learning, reversal learning, and cognitive bias in family dogs. *Learning and Behavior, 46*, 537–553. https://doi.org/10.3758/s13420-018-0357-7

Riebe, C. J., & Wotjak, C. T. (2012). A practical guide to evaluating anxiety-related behavior in rodents. In A. Szallasi, & T. Bíró (Eds.), *TRP channels in drug discovery. Methods in pharmacology and toxicology*. Totowa, NJ: Humana Press.

Roelofs, S., Boleij, H., Nordquist, R. E., & van der Staay, F. J. (2016). Making decisions under ambiguity: Judgment bias tasks for assessing emotional state in animals. *Frontiers in Behavioral Neuroscience, 10*, 119. https://doi.org/10.3389/fnbeh.2016.00119

Rogers, T., & Bryden, M. M. (1995). Predation of Adelie penguins (Pygoscelic adeliae) by leopard seals (Hydrurga leptonyx) in Prydz Bay, Antarctica. *Canadian Journal of Zoology, 73*(5), 1001–1004.

Sakai, N., & Yamamoto, T. (2001). Effects of excitotoxic brain lesions on taste-mediated odor learning in the rat. *Neurobiology of Learning and Memory, 75*(2), 128–139. https://doi.org/10.1006/nlme.2000.3969

Selva, N., Berezowska-Cnota, T., & Elguero-Claramunt, I. (2014). Unforeseen effects of supplementary feeding: Ungulate baiting sites as hotspots for ground-nest predation. *PloS One, 9*(3), e90740. https://doi.org/10.1371/journal.pone.0090740

Part II. Instructor notes

Classroom management

This exercise can be completed in one 2- to 3-h session or in multiple shorter sessions, depending on the time available and species. Discrimination training with laboratory rodents usually requires multiple shorter sessions. If multiple sessions are used, each session should include a fixed number of trials or blocks of trials determined by the amount of time available. Most dogs and livestock will complete the exercise within an hour.

Teaching the activity

Preclass preparation

The activity is outlined with rats or mice as the subject animals but can be modified for other species. The protocol will vary based on the species of choice. Teams of two to four students are recommended if they will be handing animals and recording data. With multiple observers, the assessment of interrater reliability can be included as part of the data analysis.

Most universities will require prior approval of the protocol from the Institutional Animal Care and Use Committee.

If using a companion animal (dogs, cats, guinea pigs) or livestock (horses, goats, pigs, chickens), the animal's owner should sign a consent form in advance.

The issues to consider are listed in the following.

1. What reinforcers will work best? For this exercise, reinforcers should be discrete (i.e., they can be consumed quickly), small (to avoid satiation), and valuable enough to motivate the animal to keep approaching the CS+ over many trials and for a long time.
2. How and where will the animal be stationed during the interval between trials? Ideally the dish locations in the arena should be out of view while the next trial is being set up and transitioning between the arena and stationing areas should not be stressful. If it isn't possible to set up a blind to prevent the animal from viewing the dish setup, then block with back turned or an umbrella so the animal cannot see if food was placed in the dish.
3. Is latency to approach the best dependent variable for your species? If the animals move quickly, latency may be difficult to accurately measure with a stopwatch, in which case go/no go may be a better dependent variable choice.

4. Are there any ethical concerns with using the species or individual animal for this exercise that might need to be discussed with the students beforehand? Some animals may become frustrated when they approach a food bowl and discover that it's empty. Frustration is usually transient and does not usually constitute an ethical concern.

Materials

When conducting the lab with rats/mice, the following materials are suggested.

1. A square (Fig. 16.6) or circular (Fig. 16.7) open-field arena is commonly used to study behavior in rats and mice. The open field is a walled enclosure with dimensions 60 × 60 cm for mice with at least 30-cm-tall walls and 120 × 120 cm for rats with at least 40-cm-tall walls. Commercial open fields are often transparent plexiglass or PVC. Only one subject is run at a time, so the number of chambers needed will depend on how many animals and teams will

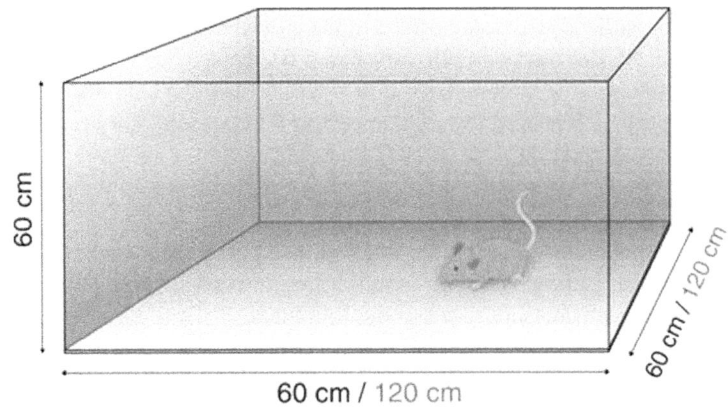

FIGURE 16.6

A square open-field arena. The open-field test is a simple square open arena. The dimensions for mice are shown in black and the dimensions for rats are shown in gray. The apparatus should be at least 60 × 60 cm for mice and 120 × 120 cm for rats. The walls should be about 60 cm high or high enough that animals cannot escape. The entire box is illuminated between 150 and 700 lux.

Figure and caption from Riebe C. J., & Wotjak C. T. (2012) A practical guide to evaluating anxiety-related behavior in rodents. In: A. Szallasi, T. Bíró (Eds.), TRP channels in drug discovery. Methods in pharmacology and toxicology. *Totowa, NJ: Humana Press.*

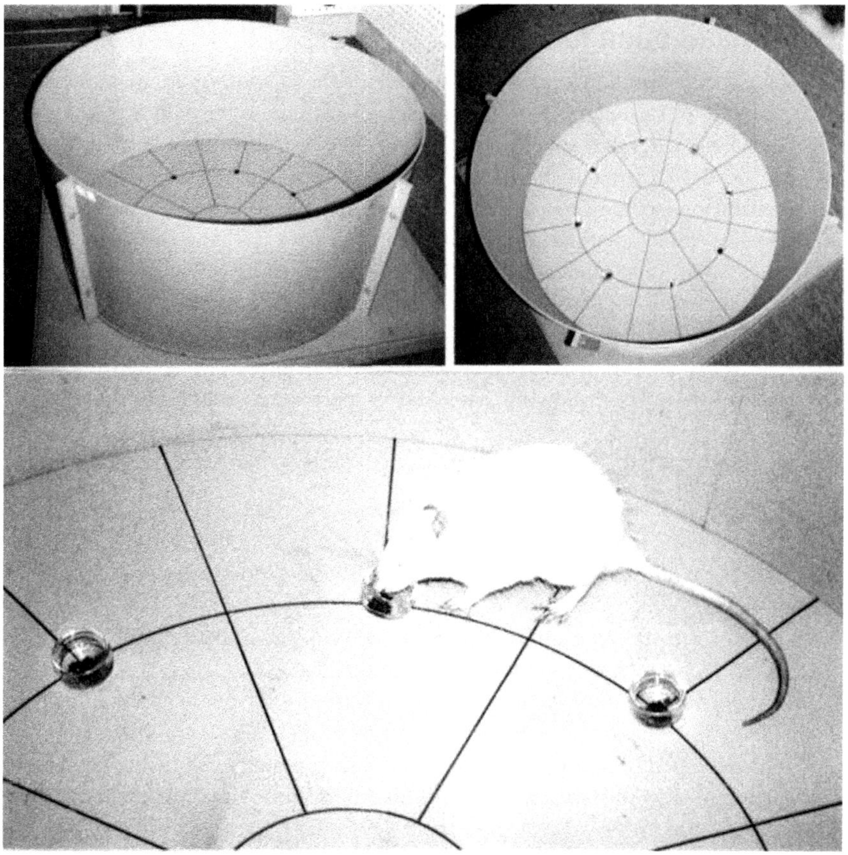

FIGURE 16.7

A circular open-field arena.

From Sakai, N., & Yamamoto, T. (2001). Effects of excitotoxic brain lesions on taste-mediated odor learning in the rat. Neurobiology of Learning and Memory, *75(2), 128–139. https://doi.org/10.1006/nlme.2000.3969.*

be running trials at the same time. If trials can be scheduled such that they are staggered across the day, then fewer will be needed.

2. An overlay that indicates dish locations and animal placement can help standardize these positions across trials and teams.
3. One food bowl for each team. The same food bowl is used on every trial and may be wiped with a clean paper towel between trials. Filter paper trimmed so it is slightly larger than the food dish works well as a cover to block visual stimuli because it is readily available, lightweight, and disposable.
4. Various commercial cereals have been used as food reinforcers for rodents.
5. At least one stopwatch or timer and data sheets per team.

Species selection

With some species, student teams can run trials in the lab with more than one animal at the same time, for example, with rats and mice in an open-field arena or fish in different tanks. With other species such as dogs and horses a single demonstrator animal can be used to teach the process to the whole class, and then after being instructed during lab, students can run trials on their own outside lab and then share data sheets or videotapes during a subsequent lab period. If the instructor plans to repeat a lab with dogs and horses in future class offerings, it may be prudent to compile a library of videos. This way, if it becomes difficult to recruit subjects or time is tight, students can code the trials from video.

The learning and cognition principles featured in this exercise apply to a wide variety of animals. The following are suitable species and suggested modifications for each.

Rats, mice, and other small mammals (e.g., gerbils, hamsters, guinea pigs).

a. The exercise is ideal for laboratory rodents.
b. A circular or rectangular open field can be used as an arena. Distances between release point and food dish locations should be in proportion to the size of the open field.
c. Omit return-to-station training. Rats and mice can generally be introduced to and removed from the arena manually between trials.
d. Rodents are fast! Latency and maximum trial time will be very short and might be challenging to accurately measure using a stopwatch. Go—no go data recording is recommended. If video recording is an option, latency can be captured using playback in slow motion, but this will take more time to complete.

Dogs (Loyer, 2019).

a. Dogs are easy to train and readily available; nearly 40% of households in the United States have a pet dog.
b. The instructor or a colleague may bring their own dog to act as a demonstrator in class in order to show students how to complete the protocol. Outside lab time, students may work individually or in teams with their own or family member dog, or as an enrichment project for university lab dogs.
c. Exclude dogs with a history of aggression or resource guarding against people.
d. If using someone else's pet dog, the owner should be present in the stationing area where the dog may interact with the owner between trials.
e. Discrimination learning and generalization has been studied in companion dogs and dogs living in shelters. See Supplement 2 for video examples of discrimination training with dogs.
f. Most dogs meet the learning criteria within 20 trials and rarely require more than 30 trials.

g. If using an arena that is not enclosed allow the dog to freely explore the area while on leash.
h. All locations are the same distance from the animal release point (approximately 6–8 m). The dishes are also the same distance from one another (approximately 3 m apart), with location "P2" centered between CS+ and CS−.
i. A small amount of high-value food works well for repeated trials of dog training. Food includes small pieces of boiled chicken breast, a spoonful of canned dog food, slices of hot dog, or small pieces of string cheese.

Horses (Freymond et al., 2014) and livestock.

a. Horses may be an option for schools with a stable on campus. Students should work in pairs or teams when training horses.
b. Only someone with experience should handle horses.
c. If working in an arena that is not enclosed allow the horse to freely explore the area while on lead line.
d. A round pen or paddock familiar to the horse and with few environmental distractions can function as the arena.
e. Following the protocol as written, horses require about 30 trials (60 min) to meet the criterion for discrimination. See Supplement 2 for video examples of discrimination training with horses.
f. A small amount of high-value food works well for repeated trials of horse training. Food includes small pieces of carrot or apple, a handful of pelleted commercial feed, or small pieces of commercial horse treats.
g. The exercise may be completed in one long session, or in several sessions, with each session consisting of 20 trials (five blocks of trials).
h. If a horse becomes anxious or distressed during the exercise (social isolation and environmental distractions being the most common reasons), it will lose interest in food and the exercise should be stopped for both ethical and practical reasons.
i. All locations are the same distance from the animal release point (approximately 6–8 m). The dishes are also the same distance from one another (approximately 3 m apart), with location "P2" centered between CS+ and CS−.

Fish.

a. Aquarium fish including sticklebacks, goldfish, and zebrafish have been subjects in location discrimination studies.
b. Use a floating ring to contain food flakes to a small area on the surface of water instead of using a food dish. Different ring colors can be used for feeding in the arena and in the stationing area.
c. Use an eyedropper in place of a food dish. When lowered into the water at the CS+ location, it will release bloodworms or other high-value food when the fish arrives. At the CS− location, the eyedropper is empty and will not release food.

In-class preparation

Analyzing data

Discrimination analysis

Latency between CS+ and CS− approaches can be compared using a paired t-test. Difference scores between CS+ and CS− can also be analyzed against the null hypothesis that there will be no difference in latency between the stimulus locations.

Identifying a population-level judgment bias can be done using an ANOVA (analysis of variance) of average latency to arrive at each location, with post hoc t-tests. Use only group data from the stimulus generalization testing for the CS+ and CS− locations, compared to the ambiguous (probe) locations.

Differences in judgment bias can be analyzed using an adjusted judgment bias score used by Mendl et al. (2010) in their research:

$$\text{Adjusted judgement bias score} = \frac{(\text{latency to probe location} - \text{mean latency to [CS+]})}{(\text{mean latency to [CS−]} - \text{mean latency to [CS+]})}$$

Scores closer to zero indicate a more positive judgment bias and scores closer to one indicate a more negative judgment bias. Some individuals move more quickly than others, and compared to a raw latency score, the adjusted judgment bias score accounts for variation in animals' baseline speed.

Areas of potential confusion or difficulty for students

1. The animal will not eat from the food dish.
   a. Use shaping. Place food around the dish, try the food dish upside down (so the animal eats from the flat bottom surface and doesn't have to place its head inside), and gradually build to having food inside the food dish. This should be done outside the arena.
   b. Try a different food reinforcer or increase the animal's motivation to eat by increasing the time since its most recent access to food.
   c. Modify the food dish. Try a nonreflective surface or select a shallower dish.
2. The animal will not leave the owner or handler (for dogs and other animals on lead).
   a. Use shaping. The owner and/or handler can accompany the animal to the food dish in the first four trials. In subsequent trials, the owner/handler decreases their involvement by stopping before reaching the food dish and then by not accompanying the animal at all.
3. The animal will not leave the release point when placed in the arena. If it occurs early in the discrimination training the animal may not have an opportunity to learn the task.

a. Pretraining may be needed. If the animal does not approach the food dish voluntarily the handler should lure or lead the animal to the food dish after the maximum time has elapsed.
b. Use shaping to encourage the animal's approach to the dish. Shaping can take two forms. For most animals, the food dish can be placed close to the release point, gradually moving the dish closer to the final location over several trials. For animals being led, the handler may walk halfway or more to the food dish for a few trials and then walk one-third of the distance to the feed dish for a few trials. The handler may need to alter the distance they approach or the number of trials they use to shape the animal's approach to the food dish in the arena.

4. Timing.
   a. If latency will be a dependent variable start the timer as soon as the animal is released and free to approach the food dish.
   b. Agree upon a clear criterion for when to stop the timer, e.g., when the animal's nose crosses into the food dish the first time.
5. Dealing with interrupted/compromised trials. Issues may arise for various reasons. The data recorder may accidently fail to start the timer or stop it too soon. Alternatively, the animal may be distracted by external stimuli (sounds, visual distractions, etc.). Discuss these possibilities in advance and decide how compromised trials will be dealt with.
   a. If sessions are video recorded flag the trial by immediately displaying a written note in front of the camera lens to mark the time and issue. This will make using the video to correct an issue faster and more accurate (it avoids someone having to scan and count each trial when searching through the video to review the trial in question).
   b. If the trial is a learning trial, you may be able to simply continue in the discrimination training. Make a note on the data sheet regarding the issue and discuss whether that trial should be excluded from the count of the number of trials to reach criterion, if the latency should be excluded from the final data.
   c. If the trial is a testing trial, but is not a probe trial (i.e., the food dish is not placed in one of the ambiguous locations), it may be appropriate to repeat the trial.
   d. If the trial is a probe trial (i.e., the food dish was placed at one of the ambiguous locations), it may be necessary to continue the trial sequence and add a repeat of the missed/interrupted probe trial at the end.

Recommendations for extensions or continuations for more advanced classes

1. Expand the stimulus generalization test.
   a. Replicate probe trials and compare changes in go—no go responses and/or approach latencies with successive trials.

   b. Include additional ambiguous locations for a more precise generalization gradient.
2. Develop alternative or additional behaviors of interest.
   a. Code and describe behaviors other than direct approaches to the food dish, such as refusals to leave the stationing area, exploration of the arena, etc.
   b. Examine approach patterns. For example, when animals approach the CS−, do they first detour to the CS+ location? When they do not approach the CS− within the allotted time, is it because they did not leave the stationing area or because they spent time exploring the rest of the arena?
   c. For dogs, horses, and other animals being led by a handler look at the animals' tendency to engage with the handler. Do they use social referencing (glancing between the arena and the handler)? Does this differ across individuals? Does this differ based on the stimulus presented? Does this change over time?
3. Evaluate factors that might influence the variables in the study, such as the number of trials to reach criterion and the number or types of errors.
   a. Individual differences in animals' age, sex, or history.
   b. Differences in how the animal has been trained/managed in its daily life (e.g., the number of other animals it is housed with, how often the animal interacts with people, etc.)
4. Add reversal learning trials. Aging and disease can impact cognitive flexibility, which has been revealed through impaired reversal learning (Graf et al., 2018).
   a. After reaching criterion on the stimulus discrimination training trials switch the CS+ and CS− locations and look at reversal learning performance.
   b. Reversal learning performance is assessed by comparing the total trials to reach criterion and looking at the number of perseverative (continuing to approach the now empty food dish in the previous CS+ location) and regressive errors (failing to maintain the correct response set after making a reversal).
5. Increase the complexity of statistical analyses.
   a. Instead of using the adjusted judgment bias scores, students may apply survival analysis to evaluate approach time based on stimulus location.
   b. Compare all probe trials to CS+ and CS− and then break down by probe location.

Answer key

1. How might using a high-value reinforcer (vs a low-value reinforcer) in discrimination training affect the learning curve (e.g., Fig. 16.2)?
   High-value reinforcers may increase the number of trials to meet criterion (i.e., slower discrimination learning). Animals trained with a high-value reinforcer

may be more motivated to approach the food dish, and therefore it may take longer to extinguish approaches to the CS–.

2. What steps should the handler and recorder take to mitigate the "Clever Hans effect"?
   The Clever Hans effect refers to a person's unconscious and unintentional cues that the animal picks up on and which influence behavior. To mitigate this effect, handlers should maintain a neutral facial expression and consistent body posture. The handler and data recorder should remain hidden from view during trials, if possible. Blind observations are the gold standard, but they are not easily accomplished with this protocol.
3. In this exercise an empty pan is used in the CS– location. Why might the animal experience the absence of food as unpleasant or aversive?
   Animals will stop approaching the pan in the CS– location because it is not reinforced. This process is called "extinction". At face value, extinction seems neutral, but it involves an unmet expectation of reinforcement and thus frustration is commonly seen. Individuals differ in whether and how intensely they experience frustration during extinction.
4. How might using punishing approaches to the CS– with a shock or a loud sound affect the animal's approach to ambiguous stimuli during tests of stimulus generalization?
   This exercise used an empty food dish—the absence of a reinforcer—as the CS–. A punisher as the CS– would create a much more unpleasant consequence that would inhibit approach behavior. The animal would be more likely to avoid the CS– location. The potential effect of a punisher on stimulus (location) generalization might be that animals will be more wary of approaching the novel locations at greater distances from the CS– location (this is called an inhibitory generalization gradient) and show a more negative judgment bias. When an unpleasant punisher is used as the CS– an animal may stop participating at all, resulting in an incomplete data set.
5. In general, why would an animal be faster or more likely to approach ambiguous stimulus P1 than P3?
   P1 is closer to the CS+, meaning that it is perceived as more similar to the CS+. The animal therefore is more likely to generalize its CS+ behavior and quickly approach the food pan at the P1 location.
6. What is a potential problem when probe trials at the same location are repeated?
   Animals will quickly learn that approaching the food dish is not reinforced on probe trials (the dish is empty). A behavior that is not reinforced tends to decrease, the latency to approach the food dish on subsequent probe trials at the same location is expected to increase, and the probability of approaching is expected to decrease.

7. In this exercise, we used three probe locations in stimulus generalization testing. What might be an advantage of using more?
Including the approach latency information from more than three probe trial locations would provide a finer generalization gradient, showing more precisely at what point the animal switches from generalizing the CS+ to generalizing the CS−.
8. How would you expect an animal with a negative judgment bias to approach the middle ambiguous stimulus location?
Even though the middle ambiguous location seems "neutral" an animal with a negative judgment bias will likely generalize the CS− to all ambiguous locations. Thus the animal is expected to slowly approach the food dish in the middle ambiguous location or not approach it at all.

Part III. Supplementary material

Supplementary data related to this chapter can be found online at https://doi.org/10.1016/B978-0-12-821410-7.00021-2.

CHAPTER

Using natural behavior as a guide for welfare

17

Malini Suchak

Department of Animal Behavior, Ecology, and Conservation, Canisius College, Buffalo NY, United States

Chapter outline

Exploring Animal Behavior in Laboratory and Field. https://doi.org/10.1016/B978-0-12-821410-7.00019-4

Part I. Student instructions

Learning goals, objectives, and key concepts

At the end of this exercise, students will be able to

- use an animal-based approach to quantify natural and abnormal behaviors in a captive animal,
- identify the strengths and limitations of using natural behavior as a guide for the welfare of captive wildlife.

Background

Over time, caging for zoo animals has morphed from concrete and bars to naturalistic enclosures. These changes reflect both an evolving perspective on animal welfare and changing visitor expectations. However, a natural-looking environment alone does not necessarily satisfy an animal's needs. David Fraser, an animal welfare scientist, recommends evaluating the welfare of an animal based on how well its adaptations match up with its current environment (Fraser, 2008). Captive wildlife evolved for millenia in the natural environment, and their adaptations are likely to have a significant amount of overlap with their wild counterparts. Fraser highlights two potential areas of concern: adaptations that the animal possesses that are not needed in its current environment and challenges that the animal faces that are unique to captivity; therefore animals do not possess any adaptations through their evolutionary ancestry to help them meet those challenges.

This activity will focus on the first concern, adaptations that the animal possesses that are not needed in the current environment. Examples of this commonly found in zoos are adaptations related to finding and catching food (as food is provisioned), finding a mate (as breeding is often, although not always, controlled), and ranging behavior (as the animal is contained within an enclosure). The inability to express these adaptations is sometimes referred to as behavioral restriction, that is, when an animal is prevented from expressing its full behavioral repertoire. In cases where behavioral restriction causes suffering, the animal may develop abnormal behaviors or experience the negative physiological effects of chronic stress.

Enrichment is frequently designed to allow animals to use adaptations no longer needed in the current environment. For example, false termite mounds for "fishing" are common features of chimpanzee exhibits, animals who spend time in the water might have pools or streams for swimming, and arboreal animals should have trees for climbing. Provisioning these items in an enclosure is a part of a *resource-based approach* to ensure positive animal welfare (Whitham & Wielebnowski, 2013). While resource-based approaches are a starting point, it is important to remember that the environment can only be designed to provide the potential for good welfare; whether or not an animal experiences good welfare depends on how they are experiencing it.

For that reason, many researchers emphasize *animal-based* approaches to assess welfare (Whitham & Wielebnowski, 2013). Animal-based approaches require collecting behavior and/or physiological data on individual animals and monitoring those data for concerning changes or patterns. Of key interest to researchers is how animals spend their time, or allocate their activity budgets, and comparing those data to the activity budgets of their wild counterparts.

It is important to note that although the wild may be used as a standard against which to compare captive activity budgets, differences between the wild and captivity should not necessarily be assumed to indicate a welfare problem in captivity (Hosey et al., 2013). Some activities that wild animals do may simply not be important when resources are provided in captivity. Furthermore, it is important to note that "the wild" can refer to many different environments, with a wide range of human impacts. Therefore wild animals may also be experiencing behavioral restriction that can cause suffering. However, when interpreted appropriately, differences in the activity budgets of wild and captive animals can serve as an important starting point to identify if an animal is experiencing behavioral restriction and is unable to use important adaptations in its current environment.

Purpose

In this activity, students will research the natural behavior and adaptations of a species and compare the behavior observed in the zoo (or other captive wildlife) setting to the behavior in the wild. The value of animal-based approaches to assess welfare will be highlighted.

Methods

Species selection

Any species of captive wildlife will work for this exercise. Ideally, observations will occur in a zoo or an aquarium; however, other settings, such as wildlife rehabilitation or a sanctuary, may also work.

Materials needed

- Ethogram worksheet
- Observation sheet
- Cell phone or stopwatch

Step-by-step instructions

1. Visit the zoo/aquarium. Select a species that is on exhibit. Watch your species for 15–30 min, noting down all the behaviors you see. Read the signs in the exhibit, looking for any species-typical behaviors described on the signs. Use this information to construct an ethogram using the sheet provided. For each behavior on your ethogram note the definition and categorize your behaviors into the following categories: locomotion, resting/inactivity, foraging/feeding, social behavior (if applicable), and any other relevant categories for your species. It is strongly recommended you create a category for abnormal behaviors, which include stereotypies (unvarying, repetitive behaviors including pacing, swaying, etc.), self-injurious behavior (such as hair or feather plucking, skin picking, etc), and other anomalous behavior (such as coprophagy, which is the eating of feces).
2. Fill out the top of the observation sheet and identify the resources present that encourage natural behavior or the use of adaptations. Add the behaviors from your ethogram to your observation with one behavior in each column.
3. Complete your observations.
   a. Use focal animal sampling to record the behavior of one individual of your species using 1-min scan samples for at least one 30-min observation period. If you are studying a group-housed species, you should randomly choose an individual to observe. Do this by numbering the individuals. Use a random number generator (such as random.org) to choose which individual to observe.
   b. Use your cell phone to time 1-min intervals. Most cell phones have a stopwatch feature, but an interval timer app, such as SIT timer, can be helpful so you do not have to keep looking down at your phone. Each minute place a tally mark in the corresponding box of any behavior you see. If time allows, take a break and return for a second 30-min observation period later in the day or on another day. Fill out a new sheet for each observation, keeping in mind that resources can change from one observation period to the next.
4. When you return to campus research the wild behavior and natural history of that species using reputable sources (i.e., books or peer-reviewed journal articles). If you are making more than one trip to the zoo, you should do this research in between visits.

Results/discussion

Analytical approach

The first step in an animal-based approach to assess welfare is to calculate your animal's activity budget. First, categorize your behaviors into larger categories, such as locomotion, resting/inactivity, foraging/feeding, social behavior, abnormal behavior, etc. The specific categories you use should be tailored to your species and should be mutually exclusive. For each category of behavior add up the total number of times you observed that behavior and divide by the total number of scans you observed to create a percentage. The percentages should add up to 100 across all your behaviors. Once you have this activity budget compiled, you can use it to get a sense of how individuals of your species spend their day in the zoo. This will then be compared to the sources you found about individuals of this species in the wild to assess how "natural" the behavior of your species is in captivity, which is one way researchers can assess welfare. However, there are strengths and weaknesses to this approach. The guiding questions should help you think critically about ways we should and should not use activity budget comparisons to evaluate welfare.

Conclusions

Questions to guide students

1. What was the behavior you most frequently observed? What behavior is described as being particularly important in the wild? Do these behaviors align? Why or why not?
2. Were there any behaviors that arc particularly important to the wild animal that you did not observe? Do you think this was a problem for your individuals? Why or why not?
3. What resources were available to facilitate natural behavior in your exhibit? Did any individuals interact with those resources? Do you feel they were effective in eliciting natural behavior?
4. What time of day did you observe? How might that have impacted the animal's activity budget?
5. How many times did you observe? How might that impact your perception of the animal's activity budget?
6. What reasons, other than poor welfare, could explain why a zoo animal is not performing its natural behavior?
7. Did you observe any abnormal behaviors? Sometimes abnormal behaviors are expressions of natural behaviors that cannot be performed in the animal's current setting. Using information you found about the wild counterpart of your species form a hypothesis about why your species might perform abnormal behaviors.
8. If an animal is performing a natural behavior, is that sufficient to know it has good welfare? What other types of information might you need to gain a better understanding of the physical/mental health of the animal?

References

Fraser, D. (2008). *Understanding animal welfare: The science in its cultural context*. Oxford: Wiley-Blackwell.

Hosey, G., Melfi, V., & Pankhurst, S. (2013). *Zoo animals: Behaviour, management, and welfare* (2nd ed.). Oxford: Oxford University Press.

Whitham, J. C., & Wielebnowski, N. (2013). New directions for zoo animal welfare science. *Applied Animal Behaviour Science, 147*, 247–260.

Part II. Instructor notes

Classroom management/blocks of analysis

Time needed: This activity could be done during a standard 3-h lab period, which would allow the students to do several 30-min observation blocks. Because of the time needed to travel to the zoo, a 1-h class period might not be feasible. If only an hour is available, it is highly recommended that the students return on a second day (perhaps on their own time) to do another observation. Regardless of the time available for observation, it is critical that the observation sheets are set up in advance. After the trip to the zoo, the instructor may want to allocate class time to guide research into the behavior of the wild counterparts of the species. If students are advanced, they may be able to do that research on their own.

Teaching the activity

Preclass preparation

- No Institutional Animal Care and Use Committee (IACUC) approval is needed. It should be stressed to students that they should be as unobtrusive as possible and should not deliberately alter the behavior of the animal in any way.
- Any species of captive wildlife can be used; however, large, easily identifiable species will be easier to observe. Students should also keep in mind the likelihood of finding information on the species in the wild—charismatic megafauna will be easier to compare between captivity and wild than fish or invertebrates.
- It is possible to translate this activity to other settings, such as a sanctuary, farm, wildlife rehabilitation, or laboratory. Some sanctuaries and wildlife rehabilitation centers are not open to the public and many do not allow research. Instructors should expect the necessary permissions to take longer in these settings. If there is a local farm or agricultural school, observations could be done there. However, domesticated animals do not have appropriate wild counterparts. Instead, students could compare between more intensive farm settings and free-range settings or between confined farm animals and their feral counterparts. If the university has a laboratory that allows class use, students could also do observations there. Students could still compare species to their "wild" counterparts, as there are many studies of "natural" behavior in free-living mice and rats; however, the discussion should include some analysis of the impact of inbreeding on natural behavior.

Modifications to the activity

If time allows, an excellent modification is to repeat observations on different days and/or different times. Comparing the same time across different days will give a consistent look at that individuals' pattern (as one single day could represent an anomaly). Multiple times per day gives a better representation of their overall activity budget, as individual activity patterns vary throughout the day.

In-class preparation

The analysis for this is straightforward. To calculate an activity budget, simply take the behaviors, group them into biologically meaningful categories, and then calculate the overall percentage of scans the animal was doing that behavior. Altogether the categories should add up to 100% of the animal's activity budget.

Areas of potential confusion or difficulty for students

- It is possible that the animal the students choose has limited information available about the behavior of its wild counterpart.
    - How to address and/or modify: Students should be advised to consider this when they choose their animals at the zoo. Students may also want to broaden their search strategy to reputable internet sites when they return to the campus.
- The animal only performs one behavior (likely inactivity), or the behavior does not seem representative during the observation period.
    - How to address and/or modify: This provides a good area for discussion about "snapshot" assessments of welfare. One cannot get a holistic impression of welfare without detailed knowledge of the individual animal. This is a good teaching moment to explain why one cannot walk up to a zoo animal, or into any other setting that houses animals, and know everything they need to know to make an assessment.
- Problems identifying individuals: Assessments need to be done on individual animals or small groups. If a student picks a species where there are too many animals and/or they are difficult to identify, the student may not be able to complete the activity.
    - How to address/modify: Work with the student to see if there are any identifying features that can be used (e.g., sometimes fish have individualized markings that are clear upon closer inspection or sometimes one individual is bigger or smaller than the others and size can be used as an identifier). If not, one individual can typically be followed as long as the observer does not have to look back and forth from paper note-taking to the animal. Dictation (which is readily available via an app on most smartphones) can help.

- Boredom/lack of variation in behavior: Students may become bored if their animal sits or sleeps for the full 30 min. While it may be difficult to motivate good data collection, this is a normal part of observing animal behavior!
  - How to address/modify: This presents a nice opportunity to analyze whether inactivity in the species is a sign of boredom or poor welfare. The species' wild counterpart would likely spend much of its day foraging or moving. In the zoo, these needs are taken care of. But what does that mean for the animal?

Recommendations for extensions or continuations for more advanced classes

This could become an extended independent study project for an advanced study requiring repeated trips to the zoo for observations. In this case, observations should be spread equally across the day and throughout the week to fully capture the animal's behavioral repertoire.

Sample answers to the guiding questions

1. What was the behavior you most frequently observed? What behavior is described as being particularly important in the wild? Do these behaviors align? Why or why not?
   Sample answer: My species spent most of the time sleeping. Wild counterpart reports show this species foraging or traveling most of the time. These behaviors do not align, but this is likely because animals in the zoo are fed by keepers and do not have to spend their days finding food.
2. Were there any behaviors that are particularly important to the wild animal that you did not observe? Do you think this was a problem for your individuals? Why or why not?
   Sample answer: There were a lot of studies in the wild about social behavior, but I did not observe any social behavior because they were all sleeping the whole time. I think this could be a problem if they are never social, but it might have been the time of day I was there.
3. What resources were available to facilitate natural behavior in your exhibit? Did any individuals interact with those resources? Do you feel they were effective in eliciting natural behavior?
   Sample answer: I noted the presence of a puzzle feeder, which would require them to spend more time and energy foraging. There was also a pool of water because my species likes to swim. I noticed one individual using the feeder, but no one in the pool. I think these items were not very exciting or interesting to the animals. They might need to switch it up a bit or make it more rewarding for them to use it.

4. What time of day did you observe? How might that have impacted the animal's activity budget?
Sample answer: We were there in the afternoon, which was the hottest part of the day. I was not surprised they were mostly sleeping. We probably should have come in the morning or maybe the late afternoon instead.
5. How many times did you observe? How might that impact your perception of the animal's activity budget?
Sample answer: We observed the animals twice, once earlier in the day and once later. I think it had a big impact because the animals were doing different things each time. So I don't really know what is "normal" for them, as they didn't behave consistently across sections.
6. What reasons, other than poor welfare, could explain why a zoo animal is not performing its natural behavior?
Sample answer: The animal could be sick or maybe it just doesn't want to do that behavior. Maybe in the wild the animals do that behavior because they have to, but in the zoo, they have the option not to do it. It's also possible they never learned how to do the behavior maybe because they don't need it in captivity.
7. Did you observe any abnormal behaviors? Sometimes abnormal behaviors are expressions of natural behaviors that cannot be performed in the animal's current setting. Using the information you found about the wild counterpart of your species form a hypothesis about why your species might perform abnormal behaviors.
Sample answer: I saw one of my individuals pacing for a part of the time. I read that in the wild, they range over 2 km per day, so maybe they are motivated to walk a lot. I would hypothesize that they are pacing because they need to roam.
8. If an animal is performing a natural behavior, is that sufficient to know it has good welfare? What other types of information might you need to gain a better understanding of the physical/mental health of the animal?
Sample answer: I think it could be a good sign but maybe not a complete picture. I would want to know about its physical health, which could probably be determined from vet records (but unless it was seriously injured, you might not be able to tell from observing them). I would also want to know about its history. As I said earlier, maybe some individuals never learned to do a natural behavior because they were born in captivity. So if some individuals were doing the behavior and some weren't, it might not indicate that the ones that do it have better welfare, maybe they just have a different life experience.

Part III. Supplementary material

Supplementary data related to this chapter can be found online at https://doi.org/10.1016/B978-0-12-821410-7.00019-4.

CHAPTER

Conservation behavior: effects of light pollution on insects

18

Brett Seymoure[1], Elizabeth K. Peterson[2,3], Rachel Y. Chock[4]

[1]*Living Earth Collaborative, Washington University in St. Louis, St. Louis, MO, United States;* [2]*Communities to Build Active STEM Engagement, Colorado State University-Pueblo, Pueblo, CO, United States;* [3]*Department of Biology, Colorado State University-Pueblo, Pueblo, CO, United States;* [4]*Recovery Ecology, San Diego Zoo Wildlife Alliance, Escondido, CA, United States*

Chapter outline

Exploring Animal Behavior in Laboratory and Field. https://doi.org/10.1016/B978-0-12-821410-7.00001-7

Part I. Student instructions

Learning goals, objectives, and key concepts

- Observe, describe, and quantify behavior
- Ethically and safely handle animals
- Effectively develop research questions, hypotheses, and defend predictions
- Provide feedback on hypotheses and defenses
- Apply behavioral ecology to solve conservation questions

Introduction

Conservation behavior is the application of knowledge of animal behavior to solve wildlife conservation problems (Blumstein et al., 2010). Conservation behavior integrates multiple fields: (1) conservation biology, focused on the conservation of biodiversity; (2) animal behavior, the study of nonhuman behavior; (3) behavioral ecology, the study of interactions between an organism and its environment in an evolutionary context; and (4) wildlife management, the application of scientific principles to conserve and manage wild populations (Blumstein et al., 2010). There are many aspects of animal behavior, particularly those related to fitness, which can be used by wildlife practitioners to help guide conservation management and policy. These can include empirical research in areas such as foraging, habitat selection, antipredator behavior, communication, animal personality, and sexual selection. The most effective application of conservation behavior, however, often comes from an understanding of problems or difficulties in wildlife management and applying an animal behavior approach to identify potential solutions. Utilizing conservation behavior may not be the only way to solve conservation problems and may not provide the complete solution, but it is an important approach to have in the toolbox.

Conservation behavior approaches have been particularly valuable in increasing the success of captive breeding, improving outcomes for reintroduction and translocation, and understanding how to mitigate anthropogenic impacts. For example, conservation behavior has resulted in successful captive breeding programs of giant pandas, elephants, and many other endangered species that are historically difficult to breed in captivity (Swaisgood et al., 2000; Zhang et al., 2004). Relocation of threatened species has also been aided through conservation behavior, including the success stories of Stephens' kangaroo rats (*Dipodomys stephensi*) (Shier & Swaisgood, 2012). Conservation behavior is often employed to mitigate human changes to the natural habitat of many species. Wildlife crossings over busy freeways can reduce animal mortality on roads, but an understanding of behavior is important to make these structures effective (McDonald & St. Clair, 2004).

As humans are causing the sixth mass extinction event, referred to as "the Anthropocene" (Barnosky et al., 2012; Crutzen & Stoermer, 2000; Vince, 2011; Zalasiewicz et al., 2010), the more conservation behavior becomes necessary to protect the biodiversity on the planet from which humans receive vital ecosystem services. The Anthropocene has been induced by climate change, pollution, deforestation, habitat loss and fragmentation, exploitation of resources, and conversion of forest to cropland or savanna (Vince, 2011; Zalasiewicz et al., 2010). Although much attention on pollution is focused on chemical toxins in the environment, less attention has been focused on light pollution, which is contributing to the Anthropocene.

Light pollution

Animals have evolved under consistent light cycles, including day and night, seasonality, and lunar cycles. Animals rely upon these natural light changes to entrain circadian rhythms, which determine the timing of an animal's behavior, e.g., rest, foraging, reproduction, etc. (Branch, 2006; Gaston et al., 2013). However, after the invention of the light bulb over 150 years ago, the Earth is losing these natural light cycles because of light pollution, or as it is referred to in the scientific community, ALAN (artificial/anthropogenic light at night) (Falchi et al., 2016; Hölker et al., 2010). ALAN is growing in intensity and geographically at a rate of 2% per year, and its importance in ecosystem health, human health, and energy use is becoming very clear (Kyba et al., 2017; Sanders & Gaston, 2018; Stevens et al., 2013).

All environments are composed of light ranging from a lot of photons of many wavelengths (i.e., bright daylight) to very few photons of only longer wavelengths (i.e., starlight). Animals have evolved eyes and other photoreceptor cells to detect the photons that comprise light environments and visual information. The specific composition of light in the environment provides cues to animals that influence activity patterns and behavior. However, owing to the spectral nature of light, light pollution alters these natural light environments because artificial light sources produce only certain wavelengths of light at times when there would naturally be little light available (e.g., twilight and night). Furthermore, there are numerous types of light sources contributing to light pollution, including high-pressure sodium (HPS), mercury vapor, metal halide, incandescent, halogen, compact fluorescent lamp (CFL), and light-emitting diodes (LEDs). Each type of light source has a unique spectral output (see Fig. 18.1A).

Interestingly, and ecologically important to note, most animal eyes see different colors than human eyes. While humans are trichromatic (having three different types of cones: blue, green, and red), many animals are dichromatic (two classes of photoreceptors, in most mammals and many insects including cockroaches, see

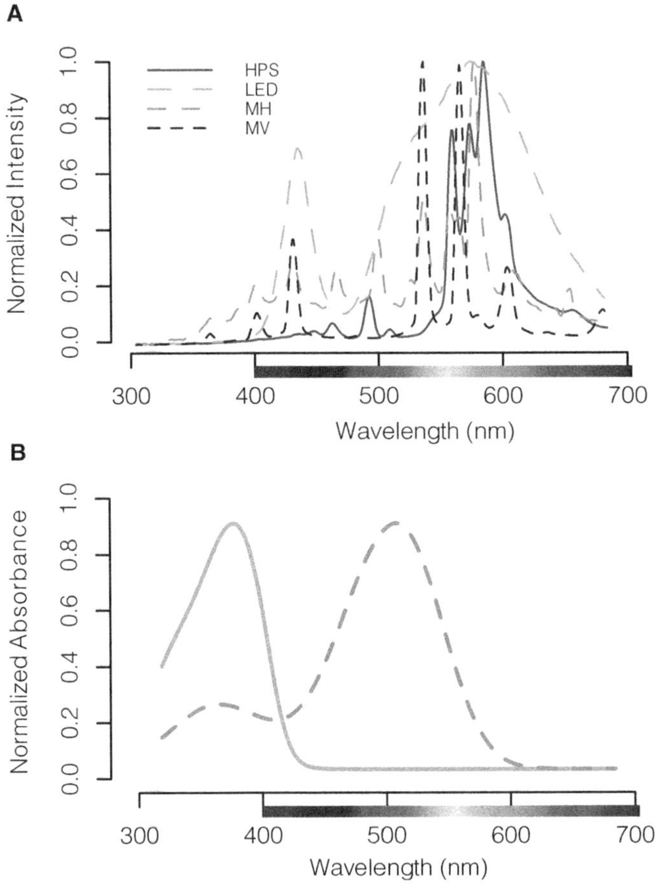

FIGURE 18.1

Color spectra of different artificial lights and the visual sensitivity of cockroaches. (A) Different artificial light sources that are commonly used by municipalities: high-pressure sodium (HPS), white light-emitting diode (LED), metal halide (MH), and mercury vapor (MV). Notice that all the spectra are quite different and very "peaky," meaning that only certain wavelengths are very abundant. For example, the MV spectrum, represented in blue, only has four main peaks with one near 420 nm, one near 530 nm, another at 570 nm, and lastly one around 605 nm. Although the MV spectrum has only four main peaks, we see this color as white because the MV spectra stimulates each of our three cone photoreceptor cells. (B) American cockroach spectral sensitivities. Cockroaches have two different photoreceptor types: a UV photoreceptor (peak at 365 nm) and a long photoreceptor (peak at 507 nm). In both figures, a color bar is included that matches the human perceived color of each respective wavelength, for example, 400 nm corresponds to violet perception and 650 nm corresponds to red perception.

Fig. 18.1B). It is important to realize that although humans see a certain visual scene with specific colors, our perception is likely very different from most animals that are sharing that visual scene (Seymoure, Buxton, et al., 2019, Seymoure, Linares, & White, 2019). This is especially important when studying light pollution effects on animals, as the light polluted world is composed of numerous types of lights.

Just as animals vary in their color vision, they also vary in their response to differently colored lighting (Cronin et al., 2014). Many species exhibit phototaxis, which is the movement in response to light. Positive phototaxis is movement toward light and negative phototaxis is movement away from light. Furthermore, many animals have phototaxis to only certain colors of light. In the context of light pollution, previous research has shown that many animals are more sensitive to shorter wavelengths of light (i.e., ultraviolet and blue) and less sensitive to longer wavelengths of light (i.e., orange and red) (Davies et al., 2013; Donners et al., 2018). For example, sea turtle hatchlings become misoriented when exposed to blue and green light sources on the horizon (Karnad et al., 2009), while numerous insects fly around blue lights more than red lights (van Langevelde et al., 2011). However, migrating birds are attracted to red lights at night and many fireflies don't flash their courtship signals under the illumination of longer wavelengths of light (Firebaugh & Haynes, 2016; Poot et al., 2008). So as is the case in conservation biology, there are trade-offs to what lights to use to reduce the ecological consequences.

Light pollution has been shown to affect eight classes of organisms across numerous biological functions ranging from predation to courtship to navigation (Seymoure, Buxton, et al., 2019, Seymoure, Linares, & White, 2019). One group of animals that is a large contributor to the ecosystem services appears to be very susceptible to light pollution—insects (Owens et al., 2020). Unfortunately, insects are in grave shape globally, as 33% of insect taxa are declining due to ALAN as well as pesticides, climate change, habitat destruction, and other pollutants (Dirzo et al., 2014; Sánchez-Bayo & Wyckhuys, 2019). It is estimated that 30% of insects attracted to light at night won't survive through the night! Numerous species of insects, especially nocturnal species, are attracted to light sources and research has shown that phototaxis is species specific and spectra specific (Donners et al., 2018; Owens & Lewis, 2018; van Langevelde et al., 2011). As the effects of light pollution increase for species around the globe, the ability to use a conservation behavior framework to address the consequences of light pollution becomes more and more important.

Often, an organism's response to a stimulus is not composed of a single response behavior but a compilation of behaviors. For instance, foraging behavior is not a single isolated behavior. Foraging behavior includes additional behaviors acting in unison to produce the response, such as neural responses and decisions about what food to eat, sensory behaviors to detect the presence of food or to locate food sources, hunting behaviors to acquire food sources, and sometimes communication with other members of the population. One behavior that is critical for foraging behavior is activity, which is mediated by an individual's circadian rhythm and thus the light environment. Light pollution may be especially disruptive to activity patterns, as light pollution can modify normal circadian rhythm patterns, attract or deter species from light sources, and change predation pressure, particularly for species that are vulnerable to predators who hunt using visual cues. Even a change in perceived predation pressure, regardless of an actual increase in predation, can impact activity, and many nocturnal animals alter their activity during the full moon when the night is brighter (Gaston et al., 2017). It's possible that behavioral responses to light pollution are tightly linked with changes in activity patterns induced by exposure to artificial light.

Purpose

In this experiment, we will study the impact of light pollution on the activity and foraging behaviors using Madagascar hissing cockroaches as the model system. You will develop your own hypotheses using the background information provided. You will then test your hypotheses, evaluate the results of your experiment, and revise and retest your experiment up to three times.

Methods

Study subject: Madagascar hissing cockroaches

Cockroaches are infamously known as household pests (Darmo & Ludwig, 1995), particularly in the southern states of the United States. However, very few cockroach species (<1%) are actually nuisance species to humans (Darmo & Ludwig, 1995). Today, there are approximately 3500 known cockroach species ranging throughout a wide variety of habitats across the globe (Darmo & Ludwig, 1995).

The Madagascar hissing cockroach (*Gromphadorhina portentosa*) is one species of cockroach that is commonly used in research, teaching laboratories, and even kept as pets (see Fig. 18.2). They live on the island of Madagascar, act as scavengers in the ecosystem, and are usually found in rotten logs in savannas (Darmo & Ludwig, 1995). However, little is known about this species from observations in the wild, and our knowledge of the Madagascar hissing cockroaches is limited to research and observations in teaching laboratories and to owners (Darmo & Ludwig, 1995).

Madagascar hissing cockroaches are quite large (approximately 2–3 inches long and 1 inch wide) and they do not have any wings (Darmo & Ludwig, 1995). They are suitable species to use in the classroom because they do not bite, are docile, do not produce odors, and move much slower than other species (Darmo & Ludwig, 1995), not to mention they have some very unique and interesting behaviors!

Reproduction and life cycle: Madagascar hissing cockroaches are unique from many other insects because they bear live young, have parental care, and have a much longer life cycle (Darmo & Ludwig, 1995). Madagascar hissing cockroaches gestate for approximately 60–70 days then give birth to live young, which emerge

FIGURE 18.2

A picture of a juvenile Madagascar hissing cockroach on a bark.

Image taken by Almabes at English Wikipedia; transferred from en.wikipedia to Commons by Sreejithk2000 using CommonsHelper, Public Domain.

from the ootheca (the egg case); these newborns are actually nymphs, and they resemble miniature adults. Female Madagascar hissing cockroaches have higher levels of parental care than most insects and nymphs stay close to their mothers for protection. It takes approximately 5–10 months for nymphs to reach full maturity, during which time they will molt (shed their exoskeleton) 6 times (Darmo & Ludwig, 1995).

Sexual dimorphism: Sexual dimorphism is defined as differences in form and structure between the sexes as a result of sexual selection (Nordell & Valone, 2014). In Madagascar hissing cockroaches, there are several morphological differences between adult males and females. (1) The most obvious difference between males and females is the pronatal humps on the thorax (see Fig. 18.3). These humps resemble horns and are much larger on the male (Darmo & Ludwig, 1995). In females, the horns are either absent or extremely small. (2) The female's thorax is much smaller than the males. Although males have larger pronatal humps, females weigh 1.4 times the males' weight (Carrel & Tanner, 2002). (3) Both males and females have a "sternite" on the last segment of their abdomen. In females, the sternite is quite large and wider than the male's sternite (Durrant et al., 2016).

Social behavior: Madagascar hissing cockroaches communicate via an audible hissing sound (Nelson & Fraser, 1980). In order to produce this hissing sound, air has to be pushed through openings in the abdomen, called spiracles (Nelson & Fraser, 1980). When an adult produces this hissing sound, it is trying to communicate a warning of danger or it can be a defensive behavior in the presence of a predator (Petrie, 2012). Females and late-stage nymphs rarely hiss, but when they do, it's usually due to a threat or disturbance of some kind (Petrie, 2012). Female Madagascar hissing

FIGURE 18.3

Sexual dimorphism in Madagascar hissing cockroaches. A male (left) and female (right) Madagascar hissing cockroach sitting side by side. Notice the sex differences in the pronatal humps where the hump in the male is much larger than the female's hump.

Images taken by Brett Seymoure.

cockroaches are also much more sociable than males; their interactions with other females or with males are typically friendly (Petrie, 2012). Males, on the other hand, form dominance-based territories (Carrel & Tanner, 2002). Males violently defend their territories by hissing, posturing, and even ramming into each other with their pronatal humps (Clark & Moore, 1994). Hissing plays key roles in courtship and mating as well (Clark & Moore, 1994). Previous research has found that males will produce five distinctive hissing sounds depending on territoriality, defensive signaling, or if they are courting a female (Fraser & Nelson, 1984).

Activity and foraging behaviors: Unfortunately, like other behaviors, not much is known about Madagascar hissing cockroaches in the wild, such as their diets, microhabitats, or activity (Carrel & Tanner, 2002). Like many species, Madagascar hissing cockroaches are nocturnal and are only active at night. During this laboratory, we will be exploring their foraging and activity behaviors under artificial light.

Handling instructions

- Although the Madagascar hissing cockroaches that you will be working with are captive populations, artificial environments and human handling can be severely stressful. Increasing the stress of experimental subjects is not only unethical but also can alter the results of the experiment. In order to minimize stress, we will emulate natural habitats and handle the adults in a gentle manner.
- Adults are strong individuals and will wander and escape if not monitored. Therefore adults need to be watched carefully to avoid losing them.
- Although you will be working with adults, there may be nymphs in the colony's habitat. Nymphs are able to escape through very small spaces. A piece of cheesecloth should be placed under the lid of each habitat and carefully placed back under the lid covering the entire top of the habitat.
- When handling the cockroaches, it is essential to wear gloves to ensure that any oils on your skin are not transferred to the adult. We need to be very gentle to not hurt or injure them. The adults should be picked up gently on each side of the thorax; the thorax is the segment of the body below the head. Adults use hooks and sticky pads to climb, so it's optimal to allow the adult to crawl onto your hand (or from your hand to another person's hand) instead of picking them up.
- The laboratory room will be maintained between 72 and 76°F at their optimum temperature.
- As Madagascar hissing cockroaches are nocturnal, we will keep the lights off during testing.

Distinguishing nymphs from adults

- Prior to molting, there is a distinguishable line lengthwise on the back of the nymph. This is because the exoskeleton has to crack and split to allow for the larger nymph or adult to emerge (Darmo & Ludwig, 1995).
- After molting, a white, black-eyed nymph emerges from the exoskeleton. Until the exoskeleton hardens (it will darken to normal black/brown coloration), the nymph or adult is very fragile. Therefore any nymph or adult that has just molted will not be used in this experiment and it should be left alone in the tank and not handled.

Sexing adults

- Use either the size of the pronatal humps or sternite to sex females (see Fig. 18.3).

Materials needed for this activity

- Adult Madagascar hissing cockroaches
- Two tanks/habitats for the cockroaches (one for females, one for males)
- Two recovery tanks for "used" cockroaches (one for females, one for males)
- Gloves
- Food for the cockroaches (shredded lettuce, sliced bananas and apples)

- Timers
- Clamp light with aluminum reflector (diameter larger than width of the arena)
- LED lights of different colors or Bluetooth-controllable LED bulb for each arena
- Black construction paper or dark curtains to block other light sources in the arena
- Arena—aquaria or terraria that have width smaller than the aluminum reflector diameter
- Screen cover for cockroaches in the arena
- Lightly colored graph paper cut to fit the arena
- Cut egg cartons or other shelter for cockroaches in the arena

Step-by-step instructions

Let's pretend that the American Samoa National Park has hired you, a conservation behaviorist, to determine what lights to use during a new project to conserve insects inside the park. The American Samoa National Park is updating lighting in the park from HPS to LED to save energy. They will also be lighting specific trails at night for hikers. The American Samoa National Park is working with the Natural Sounds and Night Skies Division of the National Park Service to reduce the effects of these trail lights on the ecosystem and the natural resources of the park. American Samoa has important vertebrate species that need to be protected and many of these vertebrates feed upon insects, especially cockroaches (why not? lots of protein!). Thus you have been hired to figure out what LED lighting will be the best to not disturb cockroaches so that the main food source of many important vertebrate species, as well as the main insect decomposer, is not disturbed in the park.

Next, you do a thorough literature search to see if there has been any previous research on the effect of light pollution on foraging and activity in cockroaches. To your surprise, no one has studied this problem, which means your work will be novel!

1. Given the lack of research conducted on this topic, you will need to use the background information provided to develop intelligent research questions, hypotheses, and predictions based upon what you already know.
2. Complete worksheet 1 with your research question, hypothesis, and prediction.
3. Design a protocol to test your hypothesis and record on worksheet 2. When you are finished designing your experiment, it's time to run your experiment. You will have 45 min to test your hypothesis, compile your results, and report your results to the class or team.
   Some things to remember:
   - Collect and organize all the materials you will need for your experiment before proceeding.
   - Put on a new pair of disposable latex gloves before handling a cockroach. It is important to change your gloves every time you handle a cockroach or a food item. If you forget to change your gloves, this could alter your results

because the cockroaches may detect other cockroaches and the scent from the food could be transferred to the cockroach (Carrel & Tanner, 2002).

4. Record your results on your worksheet.
5. Once you have finished your first test discuss your research questions, hypotheses, predictions, design, and results. How would you improve your experiment? Rewrite your research questions, hypotheses, predictions, and experimental design.
6. When you are finished redesigning your experiment, it's time to rerun your experiment. Test your hypothesis, compile your results, and report your results to the class or team.
7. Clean up. When you have completed your experiment
    - rinse your arenas with water and lay on the counter to dry,
    - return all materials according to the instructions on the side lab bench,
    - spray your lab bench with ethanol and wait for it to air dry,
    - once you have finished cleaning up, you may proceed to the worksheet.
8. Interpret your results and report your findings and recommendations to the American Samoa National Park by completing worksheet 3.

Part II. Instructor notes

There is a movement in education pedagogy to throw out the traditional "cookbook" model for teaching laboratories in favor of inquiry-based or research-based approaches (Gallet, 1998; Peterson & Carrico, 2015; Russell & French, 2001; Taylor, 2014; Volkmann & Abell, 2003). This is because this cookbook model for teaching laboratories is a procedural approach, where students perform a series of highly detailed instructions without forethought into, for example, hypothesis development, experimental design, or problem-solving. Inquiry-based approaches have been a successful model used in both Science, Technology, Engineering, and Mathematics (STEM) lectures and laboratories across many fields (Gallet, 1998; Luckie et al., 2012; Peterson & Carrico, 2015; Russell & French, 2001; Stone, 2014; Taylor, 2014; Volkmann & Abell, 2003; Wilson et al., 2010).

The format for this laboratory is based off of the pedagogy outlined by Taylor (2014). In Taylor's (2014) pedagogical design, students are provided with the necessary background on a topic (i.e., principles of osmosis), equipment, and supplies. Students work collaboratively in groups to design their own experiment to test the principles of osmosis; they then implement their experiments, which realistically result in failures. In Taylor's (2014) design, classroom discussion about the trials and tribulations of their experiments results in students redesigning their experiments, then implementing them again. Classroom discussion again allows students to fine-tune their experiments and replicate them a third time, usually with great success. Using this model, students are able to practice the following skills: asking and defining problems, developing and using models, planning and carrying out investigations, analyzing and interpreting data, and using computational thinking (Taylor, 2014). But what is unique to this pedagogical design are the following skills: forming explanations and problem-solving, argumentation using scientific evidence, critical thinking, and oral communication (Taylor, 2014). Problem-solving, in particular, is an important skill both used and developed that produces well-formulated experiments by the third "run" of their experiments.

Therefore in this laboratory, students are able to practice creating hypotheses and experimental designs by developing them first on their own and then collaboratively in groups. They then practice problem-solving through the process of making mistakes when they run their experiments, discussing and reevaluating their approach, and repeating their experiment multiple times. We recommend having students repeat this process three times if time permits; in our experience using Taylor's (2014) method, it takes three trials for students to utilize their problem-solving skills to develop sound hypotheses and experimental designs and accumulate reliable data.

Classroom management/blocks of analysis

This laboratory can be completed within a standard 2-h period (see Table 18.1). We recommend allowing pairs of students to run through their experiment once, then discuss their results in larger teams to solicit feedback. Each pair of students can utilize that feedback to rewrite their hypotheses and retest their experiment again after a brainstorming period. This process of running through their experiment twice should take approximately 2 h.

If time permits (i.e., if it's a 3-h lab), we recommend allowing students to discuss their results one last time in their teams to solicit feedback then rerun their experiment for a third time before evaluating whether or not their hypotheses are supported or rejected with the data collected. By allowing students to rerun their experiment a third time, students are able to use problem-solving skills to achieve the best results

Table 18.1 Phases of 2-h lab exercise, including prelab reading and postlab reflection to be done outside of class.

Phase 1: Preclass preparation (at home), worksheet 1
• Students read the provided worksheet, background information, and procedures. • Students independently come up with a research question, hypothesis, and prediction.
Phase 2: Run experiment round 1, worksheet 2
• Students work in pairs to discuss and revise research questions, hypotheses, and predictions. Students design a protocol to test their hypothesis (~15 min). • Students run their experiment and record data (~30 min). Provide at least two cockroaches per pair, or enough for a team so that they can rotate animals. Cockroaches respond to light very quickly, so 1- to 5-min trials will likely be sufficient, and students have an opportunity to change their experimental design after the first round based on their initial observations. • In teams (5–7 students), students report results to team and solicit feedback for each experiment (~15 min).
Phase 3: Run experiment round 2, worksheet 2
• Student pairs change or refine their research questions, hypotheses, predictions, and experimental design based on team feedback (~15 min). • Students run their experiment and record data (~30 min). • Students sketch graphs and discuss with partner whether hypothesis is supported or rejected (~15 min). • If there is time, then allow students to report to their teams one last time then run their experiment for a third and final time.
Phase 4: Reflection (in class or at home), worksheet 3
• Students answer the worksheet questions and reflect on their results and experimental design. • Students write up their findings and recommendations for the National Park.

because they have improved their hypotheses and experimental designs. This inquiry-driven method could also be used for additional lab periods to maximize student growth in the learning objectives.

1. Teaching the activity
   a. Preclass preparation (see Table 18.2)
      i. *Student organization:* We recommend organizing students into two types of groups: (1) each student should be paired with one other student to run their experiments and (2) each pair of students should be paired with other students in teams (teams should be composed of five to seven students, if possible) where they will discuss and practice argumentation and providing feedback.
      ii. *Arena preparation:* To streamline the laboratory experiment, we suggest that the arenas are prepared in advance for the students to use; this will also allow for more control by the instructor. See Fig. 18.4 for our suggested setup. Depending on the time and level of independence, students could design and set up their own arena or could be given the

Table 18.2 Directions for prelab setup.

Cockroach colony	
>1 month in advance	Acquire Madagascar hissing cockroaches. Separate males and females. Place in terrariums with food and water, and cover with cheesecloth and tight-fitting lid. Switch photoperiod (dark during the day, bright at night) in an incubator or a closet with light on a timer.
2 weeks in advance	Remove food from cockroaches. Continue to provide water and change regularly.
Lab setup	
Step 1	Set up test arenas following diagram and place at workstations for pairs of students (recommended). Alternatively, set out materials if you want students to design their own arenas.
Step 2	Set out a variety of colored LED light bulbs in aluminum reflectors (students will likely use two to three different colors in each experiment). Alternatively, provide one smart LED bulb in an aluminum reflector for each pair, along with the app installed on an iPad or instructions for students to download and use the app.
Step 3	Set out latex gloves and small containers for students to transport cockroaches from the colony to their lab bench and hold them for the duration of the experiment.
Step 4	Make the lab space as dark as possible to minimize other light sources (e.g., close the blinds or temporarily cover the windows). Keep cockroach colonies covered until you make the room dark for the experiment.

LED, *light-emitting diode.*

FIGURE 18.4

Arena setup with a 5-gallon aquarium. (A) The setup includes a 5-gallon aquarium with all side walls covered with opaque paper to shield any external light. (B) The same paper is used on the bottom of the aquarium, along with an inch-wide grid for being able to determine where the cockroach is at any given time in the experiment. Furthermore, the arena has food on one side and shelter on the other. Romaine lettuce and egg cartons are shown here.

full arena, including the graph paper substrate for measuring the location of cockroaches.

iii. *Light bulb selection:* We suggest that teachers either have several different fixed colors of LEDs that screw into the aluminum reflector fixtures OR have one smart LED bulb for each arena that can be controlled with an app. However, using an app may take a few minutes for students to set up their devices and requires a reliable Wi-Fi signal in the classroom. If iPads are available, we suggest having each bulb paired with an iPad running the respective app. Although smart LEDs will take more prep time, the number of hypotheses that can be tested increases exponentially, as the students can control brightness, color, and saturation (e.g., pink vs. red).

iv. *Study species:* We recommend using Madagascar hissing cockroaches (*G. portentosa*) for this experiment, as they are easy to acquire and maintain, are slow moving, and show variable response to different light treatments. However, other invertebrate species (such as vinegar flies or crickets) can be used instead. We do not recommend *Periplaneta* or *Blattella* cockroaches, as they are quite fast and prone to escaping.

Obtaining cockroaches

Madagascar hissing cockroaches can be obtained from captive breeding populations from the Carolina Biological Supply Company (Madagascar Hissing Cockroach, Item #143668, pack of 3, $28.50) or from Honeybees100 via Amazon.com (Madagascar Hissing Cockroach, 12 Adult Starter Colony, $41.99). They should be purchased and delivered so that arrival occurs at least 1 month in advance, if possible. Once delivered, adult males and females should be separated so they don't continue to mate; be aware that it's possible that females arrive already gravid. Two colonies should be established to avoid reproduction: one female colony and one male colony.

Caring for cockroaches

Adults should be immediately placed in habitats designed for the Madagascar hissing cockroaches (Carolina Biological, Madagascar Hissing Cockroach Habitat, Item #143660, $53.20); other options for tanks include fish tanks or similar tanks, as long as they have a lid so that nymphs cannot escape. It's recommended to provide wood chips as bedding, in addition to multiple tree branches and inverted cardboard flats for climbing (Carrel & Tanner, 2002; Darmo & Ludwig, 1995). The habitats can be closed with cheesecloth (to keep nymphs from escaping) and a tight-fitting lid. Water can be provided in a small petri dish placed at the bottom of the habitat. Dry dog food, supplemented with fresh vegetables and fruits, can be provided weekly as needed. The optimum temperature range for Madagascar hissing cockroaches is between 72 and 76°F (Petrie, 2012); therefore we recommend keeping them in an incubator at this temperature if possible.

Preparing cockroaches for the experiment

Madagascar hissing cockroaches are nocturnal and typically inactive during the day (Petrie, 2012), which can be problematic in a laboratory run during the day aimed at observing their behavior. In order to observe their behavior during the daytime, their photoperiod needs to be reversed. To accomplish this, they can be placed in an incubator with the lights off during the day to simulate night and the lights turned on at night to simulate daytime (Darmo & Ludwig, 1995). If no incubator is available, a closet with good circulation and a light on a timer to be on during the night and off during the day will suffice.

To test activity via foraging behaviors in this experiment, it helps if food is not provided to the cockroaches for 2 weeks prior to experimentation, but it's important to continually provide water. Removing food standardizes the baseline motivation for foraging, and students are more likely to see changes in activity under different lighting regimes than recently fed cockroaches, which are more likely to hide regardless of the light source. As cockroaches are able to survive for 4 weeks without food and water, they can retain body moisture for this period and it is not unethical to remove food (Yoder and Grojean, 1997).

Disposing of cockroaches

If cockroaches are obtained only for this laboratory, the best disposal practices are to freeze the cockroaches in a container for at least 24 h and then place dead individuals in a biohazard bag for the institutions' health and safety department to dispose of (likely through incineration). Although the cockroaches can also be euthanized via alcohol, we do not recommend this because disposal of insects in alcohol will require disposing as hazardous waste owing to the flammability of alcohol. Please consult with your Environmental and Health Safety Office for specific details of disposal. Of course, we suggest keeping the cockroaches alive, as they are very low maintenance and allow for more use in teaching labs as well as for outreach activities.

Modifications

As mentioned earlier, other insects and arthropods could be used in this laboratory. However, we have only vetted the Madagascar hissing cockroach for this experiment. Crickets are likely to alter their behavior under ALAN, including changing in song rate and activity patterns. Vinegar flies may show attraction or avoidance to light sources depending on species. Wolf spiders are also likely to show changes in behavior under artificial light.

Another possible modification is altering the intensity of light (different wattages of lamps) instead of altering color. We don't recommend this, as the lab is written for altering colors of light. However, if conditions constrain the use of differently colored light bulbs, this lab could be run by studying different brightness levels of light.

In-class preparation

As each pair of students will be writing their own hypotheses and using different experimental designs, analyzing data for the entire class will not be possible. Depending on the level of the class, student pairs could conduct their own analysis (e.g., chi-square tests). At any level, students can graph their results by hand and report on the trends that they observe. Ensuring that students are comfortable creating simple graphs of independent and dependent variables will help the lab run smoothly. We have provided examples of graphs and data tables in the supplementary materials; however, we strongly encourage instructors NOT to share these materials with students to avoid biasing their own study designs.

Part III. Supplementary material

Supplementary data related to this chapter can be found online at https://doi.org/10.1016/B978-0-12-821410-7.00001-7.

References

Barnosky, A. D., Hadly, E. A., Bascompte, J., Berlow, E. L., Brown, J. H., Fortelius, M., Getz, W. M., Harte, J., Hastings, A., Marquet, P. A., Martinez, N. D., Mooers, A., Roopnarine, P., Vermeij, G., Williams, J. W., Gillespie, R., Kitzes, J., Marshall, C., Matzke, N., … Smith, A. B. (2012). Approaching a state shift in Earth's biosphere. *Nature, 486*(7401), 52–58.

Blumstein, D. T., Fernández-Juricic, E., & Others. (2010). *A primer of conservation behavior.* MA: Sinauer Associates Sunderland.

Branch, L. C. (2006). Ecological consequences of artificial night lighting, 65.00 (hardcover); 29.95 (paper). xx 458 pp.; ill.; index. ISBN: 1–55963–128–7 (hc); 1–55963–129–5 (pb). 2006. In C. Rich, & T. Longcore (Eds.). *The quarterly review of biology* (Vol. 81 (4), p. 414). Washington (DC): Island Press. https://doi.org/10.1086/511599.

Carrel, J. E., & Tanner, E. M. (2002). Sex-specific food preferences in the Madagascar hissing cockroach *Gromphadorhina portentosa* (Dictyoptera: Blaberidae). *Journal of Insect Behavior, 15*(5), 707–714.

Clark, D. C., & Moore, A. J. (1994). Social interactions and aggression among male Madagascar hissing cockroaches (*Gromphadorhina portentosa*) in groups (Dictyoptera: Blaberidae). *Journal of Insect Behavior, 7*(2), 199–215.

Cronin, T. W., Johnsen, S., Justin Marshall, N., & Warrant, E. J. (2014). *Visual ecology.* Princeton University Press.

Crutzen, P. J., & Stoermer, E. F. (2000). Global change newsletter. *The Anthropocene, 41*, 17–18.

Darmo, L., & Ludwig, F. (1995). Madagascan giant hissing roaches. *Carolina TIPS, 58*(2), 5–6.

Davies, T. W., Bennie, J., Inger, R., de Ibarra, N. H., & Gaston, K. J. (2013). Artificial light pollution: Are shifting spectral signatures changing the balance of species interactions? *Global Change Biology, 19*(5), 1417–1423.

Dirzo, R., Young, H. S., Galetti, M., Ceballos, G., Isaac, N. J. B., & Collen, B. (2014). Defaunation in the Anthropocene. *Science, 345*(6195), 401 406. https://doi.org/10.1126/science.1251817.

Donners, M., van Grunsven, R. H. A., Groenendijk, D., van Langevelde, F., Bikker, J. W., Longcore, T., & Veenendaal, E. (2018). Colors of attraction: Modeling insect flight to light behavior. *Journal of Experimental Zoology Part A, Ecological and Integrative Physiology.* https://doi.org/10.1002/jez.2188.

Durrant, K. L., Skicko, I. M., Sturrock, C., & Mowles, S. L. (2016). Comparative morphological trade-offs between pre- and post-copulatory sexual selection in giant hissing cockroaches (Tribe: Gromphadorhini). *Scientific Reports, 6*, 36755.

Falchi, F., Cinzano, P., Duriscoe, D., Kyba, C. C. M., Elvidge, C. D., Baugh, K., Portnov, B. A., Rybnikova, N. A., & Furgoni, R. (2016). The new world atlas of artificial night sky brightness. *Science Advances, 2*(6), e1600377.

Firebaugh, A., & Haynes, K. J. (2016). Experimental tests of light-pollution impacts on nocturnal insect courtship and dispersal. *Oecologia, 182*(4), 1203–1211.

Fraser, J., & Nelson, M. C. (1984). Communication in the courtship of a madagascan hissing cockroach. I. Normal courtship. *Animal Behaviour, 32*(1), 194–203.

Gallet, C. (1998). Problem-solving teaching in the chemistry laboratory: Leaving the cooks. *Journal of Chemical Education, 75*(1), 72.

Gaston, K. J., Bennie, J., Davies, T. W., & Hopkins, J. (2013). The ecological impacts of nighttime light pollution: A mechanistic appraisal. *Biological Reviews of the Cambridge Philosophical Society, 88*(4), 912–927.

Gaston, K. J., Davies, T. W., Nedelec, S. L., & Holt, L. A. (2017). Impacts of artificial light at night on biological timings. *Annual Reviews, 48*: 49–68.

Hölker, F., Wolter, C., Perkin, E. K., & Tockner, K. (2010). Light pollution as a biodiversity threat [review of light pollution as a biodiversity threat]. *Trends in Ecology & Evolution, 25*(12), 681–682.

Karnad, D., Isvaran, K., Kar, C. S., & Shanker, K. (2009). Lighting the way: Towards reducing misorientation of olive ridley hatchlings due to artificial lighting at Rushikulya, India. *Biological Conservation, 142*(10), 2083–2088.

Kyba, C. C. M., Kuester, T., Sánchez de Miguel, A., Baugh, K., Jechow, A., Hölker, F., Bennie, J., Elvidge, C. D., Gaston, K. J., & Guanter, L. (2017). Artificially lit surface of Earth at night increasing in radiance and extent. *Science Advances, 3*(11), e1701528.

van Langevelde, F., Ettema, J. A., Donners, M., WallisDeVries, M. F., & Groenendijk, D. (2011). Effect of spectral composition of artificial light on the attraction of moths. *Biological Conservation, 144*(9), 2274–2281.

Luckie, D. B., Aubry, J. R., Marengo, B. J., Rivkin, A. M., Foos, L. A., & Maleszewski, J. J. (2012). Less teaching, more learning: 10-yr study supports increasing student learning through less coverage and more inquiry. *Advances in Physiology Education, 36*(4), 325–335. https://doi.org/10.1152/advan.00017.2012.

McDonald, W. R., & St Clair, C. C. (2004). The effects of artificial and natural barriers on the movement of small mammals in Banff National Park, Canada. *Oikos, 105*(2), 397–407. https://doi.org/10.1111/j.0030-1299.2004.12640.x.

Nelson, M. C., & Fraser, J. (1980). Sound production in the cockroach, *Gromphadorhina portentosa*: Evidence for communication by hissing. *Behavioral Ecology and Sociobiology, 6*(4), 305–314.

Nordell, S. E., & Valone, T. J. (2014). *Animal behavior: Concepts, methods, and applications*. Oxford University Press.

Owens, A. C. S., Cochard, P., Durrant, J., Farnworth, B., Perkin, E. K., & Seymoure, B. (2020). Light pollution is a driver of insect declines. *Biological Conservation, 241*, 108259. https://doi.org/10.1016/j.biocon.2019.108259.

Owens, A. C. S., & Lewis, S. M. (2018). The impact of artificial light at night on nocturnal insects: A review and synthesis. *Ecology and Evolution, 8*(22), 11337–11358.

Peterson, E. K., & Carrico, P. (2015). Laboratory exercise in behavioral genetics using team-based learning strategies. *Bioscene: Journal of College Biology Teaching, 41*(2), 32–39.

Petrie, K. (2012). *Madagascar hissing cockroaches*. ABDO.

Poot, H., Ens, B. J., de Vries, H., Donners, M. A. H., Wernand, M. R., & Marquenie, J. M. (2008). Green light for nocturnally migrating birds. *Ecology and Society, 13*(2). http://www.jstor.org/stable/26267982.

Russell, C. P., & French, D. P. (2001). Factors affecting participation in traditional and inquiry-based laboratories. *Journal of College Science Teaching, 31*(4), 225.

Sánchez-Bayo, F., & Wyckhuys, K. A. G. (2019). Worldwide decline of the entomofauna: A review of its drivers. *Biological Conservation, 232*, 8–27.

Sanders, D., & Gaston, K. J. (2018). How ecological communities respond to artificial light at night. *Journal of Experimental Zoology Part A, Ecological and Integrative Physiology, 329*(8–9), 394–400.

Seymoure, B., Buxton, R., White, J., Linares, C., Fristrup, K., Crooks, K., Wittemyer, G., & Angeloni, L. (2019). *Anthropogenic light disrupts natural light cycles in critical conservation areas*. https://doi.org/10.2139/ssrn.3439670.

Seymoure, B. M., Linares, C., & White, J. (2019). Connecting spectral radiometry of anthropogenic light sources to the visual ecology of organisms. *Journal of Zoology, 308*(2), 93–110.

Shier, D. M., & Swaisgood, R. R. (2012). Fitness costs of neighborhood disruption in translocations of a solitary mammal. *Conservation Biology: The Journal of the Society for Conservation Biology, 26*(1), 116–123.

Stevens, R. G., Brainard, G. C., Blask, D. E., Lockley, S. W., & Motta, M. E. (2013). Adverse health effects of nighttime lighting: Comments on American Medical Association policy statement. *American Journal of Preventive Medicine, 45*(3), 343–346.

Stone, E. M. (2014). Guiding students to develop an understanding of scientific inquiry: A science skills approach to instruction and assessment. *CBE-Life Sciences Education, 13*(1), 90–101.

Swaisgood, RR, Lindburg, DG, Zhou, X, & Owen, MA (2000). The effects of sex, reproductive condition and context on discrimination of conspecific odours by giant pandas. *Animal Behaviour, 60*(2): 227–237. https://doi.org/10.1006/anbe.2000.1464

Taylor, J. R. (2014). A simple inquiry-based lab for teaching osmosis. *The American Biology Teacher, 76*(4), 265–269.

Vince, G. (2011). Geology. An epoch debate. *Science, 334*(6052), 32–33, 37.

Volkmann, M. J., & Abell, S. K. (2003). Rethinking laboratories. *The Science Teacher, 70*(6), 38.

Wilson, C. D., Taylor, J. A., Kowalski, S. M., & Carlson, J. (2010). The relative effects and equity of inquiry-based and commonplace science teaching on students' knowledge, reasoning, and argumentation. *Journal of Research in Science Teaching: The Official Journal of the National Association for Research in Science Teaching, 47*(3), 276–301.

Yoder, J. A., & Grojean, N. C. (2008). Group influence on water conservation in the gial Madagascar hissing-cockroach, Gromphadorhina portentosa (Dictyoptera; Blaberidae). *Physiological Entomology: 22*, 79–82.

Zalasiewicz, J., Williams, M., Steffen, W., & Crutzen, P. (2010). The new world of the Anthropocene. *Environmental Science and Technology, 44*(7), 2228–2231.

Zhang, G, Swaisgood, RR, & Zhang, H. (2004). Evaluation of behavioral factors influencing reproductive success and failure in captive giant pandas. *Zoo Biology, 23*(1): 15–31. https://doi.org/10.1002/zoo.10118

CHAPTER

19 Animal enrichment: creating functional and stimulating enrichment for captive animals. Observing and assessing their use and impact

Clara Voorhees

Department of Biology and Mathematics, D'Youville College, Buffalo, NY, United States

Chapter outline

Exploring Animal Behavior in Laboratory and Field. https://doi.org/10.1016/B978-0-12-821410-7.00028-5

Part I. Student instructions

Learning goals, objectives, and key concepts

- To design meaningful and stimulating enrichment for captive animals.
- To observe animal(s) with and without enrichment and discuss any changes in behavior.
- To understand the vital importance of enrichment for animal welfare.

Background

Enrichment is an animal husbandry method used to increase the quality of life of captive animals in zoos, aquaria, animal rescues/preserves, research laboratories, and home by identifying and providing the environmental stimuli needed for both physical and mental welfare (Sherpherdson et al., 1998). Enrichment can be beneficial in terms of decreasing aggression, increasing activity, and decreasing abnormal behavior, as well as increasing biological functioning such as increased lifetime reproductive success (Carlstead & Shepherdson, 2000; Newberry, 1995). Environmental enrichment can be social or nonsocial; nonsocial enrichment includes physical, sensory, food, and cognitive elements. These forms of environmental enrichment can result in increased behavioral diversity (Baer, 1998).

Lack of a stimulating environment can lead to poor welfare of captive animals. Welfare of an individual is the individual's state as regards to its attempts to cope with its environment (Broom & Fraser, 2007). Captive animals typically live in conditions in which their dietary needs are met on a daily basis, predation is low to nonexistent, and their movement is highly controlled (Newberry, 1993). These conditions are at times unintendedly stressful for the animals. Sources of stress in captive environments that are identified and addressed with enrichment can improve their welfare (Carlstead & Shepherdson, 2000). Sensory stimulation has been well studied as a potential method of environmental enrichment in captive animals (Wells, 2009). Environmental enrichment that encourages expression of natural behaviors while decreasing abnormal behaviors can improve both the physiological and psychological welfare of captive animals (Claxton, 2011).

Providing an animal with toys, music, or other stimuli that have little functional relevance is a common mistake when enriching animals (Newberry, 1995). Enrichment is more than just providing a toy or new scent to a cage or habitat; enrichment should address one or more of the five categories (social, physical, sensory, food, and cognitive). Environmental enrichment can include the improvement of enclosure design, the provision of feeding devices, novel objects, appropriate social groupings, and sensory stimuli (Claxton, 2011). It is important to understand your animal's species-specific behavioral repertoire in order to determine which category of enrichment should be used. Understanding your animal's behavioral repertoire will enable you to choose the enrichment that will properly stimulate your animal and avoid the enrichment that could negatively impact its welfare. Fernandez and Timberlake (2019) researched the use of foraging devices as enrichment in captive walruses (*Odobenus rosmarus*). Previous work on captive walruses examined the relationship between walrus stereotypies and species-typical foraging behaviors of wild walruses. Stereotypies in captive animals are defined as repetitive, invariant behavioral patterns with no obvious goal or function. Fernandez and Timberlake (2019) studied two types of walrus stereotypies: the pattern of swimming in their pool and repetitive sucking of some item (usually their own flipper). The researchers tested two enrichment devices to stimulate the use of flippers and mouth suctioning of small food in the devices. The walruses with the devices that contained food increased their movement in the pools and decreased repetitive sucking of items. Fernandez and Timberlake (2019) concluded that providing stimuli that supported completed foraging sequences can reduce stereotypies and increase locomotion in captive walruses. Enrichment that encourages natural behaviors can lead to good welfare by decreasing stereotypies and other abnormal behaviors (Corcoran, 2015).

Purpose

This activity will introduce you to the four nonsocial enrichment categories of animal enrichment, with the goal of increasing the activity of your animals. Working with a partner or in a small group (two to four students), you will observe your species in its "habitat" and determine which category of enrichment you would plan on creating and assessing. You should be able to determine which type of enrichment is appropriate for your species and create an activity budget for your animal's activity level before and after the enrichment is given. These activity budgets will then be compared to one another to see if the enrichment causes an increase in activity.

Methods

Species selection

Depending on the species, singly housed animals can be used; however, social animals should be kept in social groups if used for this activity. Both vertebrate and invertebrate species are appropriate for this activity; however, very small invertebrates may be hard to observe. Your instructor will determine which species are appropriate. If possible, you may be able to observe animals at your local zoo or aquarium. Be sure to follow instructions of the animal care staff at the institution. Your instructor will coordinate with the zoo or aquarium for your visit. Examples of enrichment are provided in the supplementary materials.

Materials needed

The materials needed for this activity will vary based on the species available to work with. For physical and cognitive enrichment, you should have items such as paper bags, large and small; pieces of wood ranging in size from very small to large; paper egg cartons; string and rope of varying thickness; carabiner clips; supplies for papier mache (balloons, newspaper, flour); and plastic toys of varying size and color. For sensory and food enrichment, you should have items such as body mists and sprays, cooking spices, sugar-free jello, pasta in a variety of shapes and thicknesses, peanut butter, honey, and fresh fruits or vegetables. You can use standard timers or cell phones (timing and recording video) for this activity and should have a notebook for observations. You may also wish to have materials available for construction such as hot-melt glue, superglue, duct tape, etc.

Step-by-step instructions

1. You will work in groups of two to three individuals. Each group will first decide which animal they will be working with; the instructor will provide a list of the available animals.
2. You will research or review instructor-provided material on species-specific behavior repertoires. You will then observe your animal(s) for 15–20 min to obtain a baseline of activity level and assess how they interact with their environment. You should create a rough draft of an ethogram for your animal's behavior. An ethogram is a catalog of an animal's behavior; your ethogram should include event and state behaviors.
3. After initial observations are completed, you will create a data sheet of your focal animal's behavior, recording the type of behavior and duration. You may create your own data sheet or use the blank data sheet provided.
4. Each group will then decide which category of enrichment is appropriate for its animals.
5. Once the category of enrichment has been chosen, you will be given access to the materials available in the lab for that category. Each group will design and build its enrichment for its animal(s). Before presenting your enrichment to your animal(s), your instructor should inspect the enrichment to make sure it does not pose a danger to the animal(s).
6. After approval, you can place the enrichment in the animal's habitat. Before you begin make sure your partner(s) is clear on which individual he or she is observing, if more than one is in the habitat. You will be making continuous observations of your animals. Use a stopwatch or your phone to time the length of time for each behavior in seconds. Record the type and duration of the behavior in your lab notebook. The minimum observation duration for this activity is 15 min, but if time permits, 30–45 min is preferred. Some animals may be neophobic (fearful of new objects) and may need more time to acclimate

to the new additions to their habitat before engaging with it. Be sure to provide the appropriate time for your species to engage with the enrichment.

7. Ideally, you will videotape your animal's interactions with the enrichment while you record the type of behavior and duration. You can then review your tape and compare observations. Review of recordings can be done outside the lab if needed.

Results/discussion

Each group should now review its recordings of its animal's interactions with the enrichment. Compare the observations made in real time to the video. To facilitate this, observations should be translated to data sheets; you can use the data sheets provided or create your own. Behaviors should be grouped into categories such as social, asocial, active, inactive, play, aggression, etc. for comparison of before and after enrichment presentation. Using the duration of behaviors observed calculate the percentage time spent in each behavior. Create an activity budget graph comparing behaviors of your animal(s) before and after enrichment is given. Graphs can be done by hand on a graph paper or created using Excel or a similar program, if available. You should create pairs of bars for each category, for example, "Social before" paired with "Social after", this will enable you to visually compare the level of activity that the animals engaged in before and after the enrichment is presented (Fig. 19.1).

Now that you have a visual representation of your results compare the activity levels of your animal(s) before and after the enrichment is given. Was there a

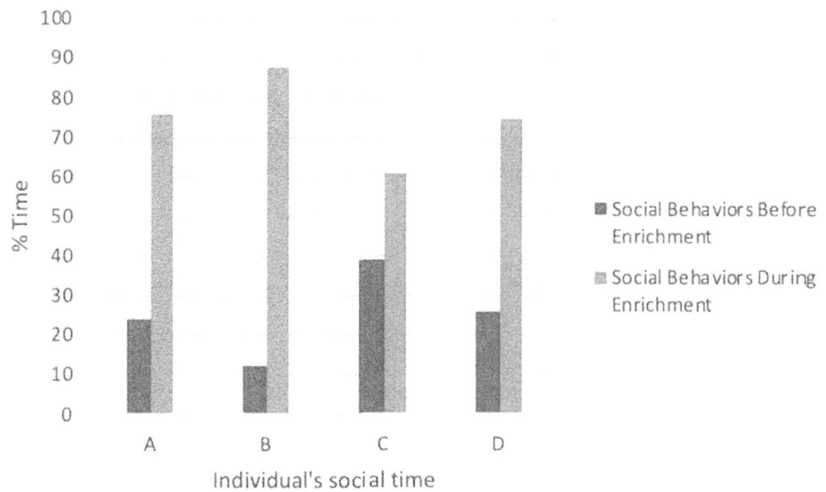

FIGURE 19.1

Amount of time spent engaging in social behaviors before and during enrichment.

difference in activity levels in any of the categories? Did your group's choice of nonsocial enrichment (physical, sensory, food, and cognitive) increase activity levels in some categories more than others? Which categories of behaviors saw the greatest increase? What might you conclude about this? Did you create enrichment that was appropriate for your animal(s)?

Analytical approach

If time permits conduct a two-tailed paired *t*-test to compare individuals' activity levels before enrichment and during enrichment treatments. A two-tailed paired sample *t*-test is a statistical procedure used to determine whether the mean difference between two sets of observations is zero. In a paired sample *t*-test, each subject or entity is measured twice, resulting in pairs of observations.

The paired sample *t*-test has two competing hypotheses, the null hypothesis and the alternative hypothesis. The null hypothesis assumes that the true mean difference between the paired samples is zero. The alternative hypothesis assumes that the true mean difference between the paired samples is not equal to zero.

How to calculate a two-tailed paired sample *t*-test in excel

Spreadsheets have a built-in function to perform paired *t*-tests. To perform this test in Excel put the "before" numbers in one column and the "after" numbers in the adjacent column, with the before and after observations from each individual on the same row. Then enter =TTEST (*array1, array2, tails, type*), where *array1* is the first column of data, *array2* is the second column of data, *tails* is normally set to 2 for a two-tailed test, and *type* is set to 1 for paired *t*-test. The result of this function is the *P*-value of the *t*-test. $P \leq .05$ is considered statistically significant. It indicates strong evidence against the null hypothesis, as there is less than a 5% probability the null is correct (and the results are random). Therefore we reject the null hypothesis and accept the alternative hypothesis. The alternative hypothesis states that the independent variable (enrichment treatment) did affect the dependent variable (behavioral response), and the results are significant.

Questions

- How do the activity levels differ before and after enrichment is presented?
- Did reviewing the videotapes of your animals confirm your original observations or did it necessitate changing your observations?
- Did your enrichment increase the activity level of your animal(s)?
- Did your enrichment increase the activity level of your animal(s) in the category element you predicted (physical, sensory, food, and cognitive element)?

- What surprised you by how your animal(s) reacted, or failed to react, to the enrichment?
- What is the goal of animal enrichment? What are the benefits of providing enrichment to animals?
- Why did we focus on only the nonsocial enrichment categories for this lab?
- When creating enrichment for your animal(s), what things did you and your group take into consideration concerning safety, behaviors, resources available, and location of enrichment? How did this shape what enrichment was created?
- Was the enrichment created by you and your group appropriate for your species?
- If you were to do this activity again, what would you do the same and what would you do differently?
- How does enrichment impact an animal's welfare?
- Why is it important to consider animal welfare in designing enrichment?

References

Baer, J. F. (1998). A veterinary perspective of potential risk factors in environmental enrichment. In D. J. Shpherdson, J. D. Mellen, & M. Hutchins (Eds.), *Second nature: Environmental enrichment for captive animals* (1st ed., pp. 277–301). Washington, D.C.: Smithsonian Institution Press.

Broom, D. M., & Fraser, A. F. (Eds.). (2007). *Domestic animal behaviour & welfare* (4th ed.). Cambridge, UK: Cambridge University Press.

Carlstead, K., & Shepherdson, D. (2000). Alleviating Stress in zoo animals with environmental enrichment. In G. P. Moberg, & J. A. Mech (Eds.), *The biology of animal stress* (1st ed., pp. 337–354). New York, NY: CABI Publishing.

Claxton, A. M. (2011). The potential of the human-animal relationship as an environmental enrichment for the welfare of zoo-housed animals. *Applied Animal Behaviour Science, 133*, 1–10.

Corcoran, M. (2015). Environmental enrichment for aquatic animals. *Veterinary Clinics of North America: Exotic Animal Practice, 18*, 305–321.

Fernandez, E. J., & Timberlake, W. (2019). Foraging devices as enrichment in captive Walruses (*Odoberius rosmarus*). *Behavioural Processes, 168*. https://doi.org/10.1016/j.beproc.2019.103943.

Newberry, R. C. (1993). The time-space continuum, and its relevance to farm animals. *Etologiá, 3*, 219–234.

Newberry, R. C. (1995). Environmental enrichment: Increasing the biological relevance of captive environments. *Applied Animal Behavior Science, 44*, 229–243.

Sherpherdson, D. L., Mellen, J. D., & Hutchins, A. (Eds.). (1998). *Second nature: Environmental enrichment for captive animals*. Washington, DC: Smithsonian Institute Press.

Wells, D. (2009). Sensory stimulation as environmental enrichment for captive animals: A review. *Applied Animal Behaviour Science, 118*, 1–11.

Part II. Instructor notes

Classroom management/blocks of analysis

This activity can be completed in a standard laboratory period (3 h) or in two lecture class sessions, depending on the number of students. If done in lecture sessions, analysis of video and comparisons of activity levels can be done as homework. If students are researching their animal's species-specific behaviors, you may want to have this done as a homework assignment before lab. Alternatively, instructors could provide information about the species-specific behaviors of the animals available in the lab for the students.

Teaching the activity

Preclass preparation

No Institutional Animal Care and Use Committee (IACUC) approval is needed for this activity if working with invertebrates in the lab. However, if vertebrate species are being studied for this activity, you will need to complete an IACUC application at your institution. If you plan on working with your local zoo or aquarium to observe enrichment at the institution, you would require an IACUC review. If so, I would suggest the lab be done over two lab periods or 2 days, 1 day to observe the animals for a baseline and 1 day to observe the animals with enrichment. The students may not be able to create enrichment for the zoo or aquarium but may use enrichment that the animal care staff have created. I would recommend working with invertebrate species that are easy to maintain in the lab and do not require IACUC approval, such as scorpion (*Pandinus imperator*), bull's head cockroach (*Archimandrita tesselata*) or Madagascar hissing cockroach (*Gromphadorhina portentosa*), blue death-feigning beetle (*Cryptoglossa verrucosa*), and giant vinegarroon (seasonal variation only, *Mastigoproctus giganteus*). *P. imperator* can be kept in groups of three or more as long as enough food is provided and can be purchased from a variety of online reptile pet retailers, for example, https://www.joshsfrogs.com and https://www.backwaterreptiles.com. The bull's head cockroach (*A. tesselata*) can be housed in large numbers and can be purchased from a variety of online reptile pet retailers such as https://roach-ranch.com and https://www.roachcrossing.com/. The Madagascar hissing cockroach (*G. portentosa*) can be housed in large numbers and can be purchased from retail pet stores such as Petco. The blue death-feigning beetle (*C. verrucosa*) can be housed as three to five individuals per habitat (in a 5-gallon tank) (can be purchased online at https://shop.bugsincyberspace.com/Blue-Death-Feigning-Beetle-bic977.html).

This lab is well suited for inside the laboratory, especially if working with insects or small mammals. If animals are unavailable in the lab or the budget does not permit their purchase, animal videos could be substituted. However, if using videos, the students would only be able to observe the behavior of the animals without their enrichment. Students could still build enrichment for the animals observed via video but would be unable to observe the animal's interaction with the enrichment. Follow-up for a video substitution could include the "panel discussion" or "gallery walk" activities described later.

In-class preparation

Students may be confused or may find it difficult to create enrichment for their animals. Sometimes having too much material available can be overwhelming. Start by showing the students different ways the materials could be used to create enrichment in the nonsocial categories and encourage them to be creative, or have examples of enrichment built for animals not being used in the lab for them to look at; this will avoid copycats. Sample results are in the supplementary materials. To expand upon this activity, students could do a "panel presentation" or "gallery walk." For a panel presentation, the students will create a "panel" based on their enrichment experiment. Students will develop an expanded expertise on their topic either in class or out of class for homework.

Steps: (1) Assign students to groups based on their research group or group similar projects together. (2) Ask students to complete fairly extensive individual research on their own topics. (3) Ask them to complete some research on other students' topics, i.e., research sufficient enough to come to class with a list of potential questions. (4) In class arrange five (less or more, depending on group sizes) at the front of the room and ask panelists from the first group to take their positions. (5) Ask the moderator to introduce the panelists and, following a brief presentation of 2 or 3 min per person to state their major findings, to take questions.

For a gallery walk, the instructor will write questions on large pieces of paper hung on the wall around the room. Students in their research group will move from question to question, writing responses to each question on the wall.

Steps: (1) Instructor adds questions at different locations around the room on sheets of paper hung on the wall. (2) Students will visit the different stations and add ideas to each sheet. (3) Discuss the ideas shared and any questions the students had about the ideas. Prompts or questions on the sheets could be possible enrichment experiment design scenarios or dilemmas.

If time permits have the groups complete statistical analysis of the results to compare activity levels before and during enrichment using a two-tailed paired sample t-test.

Statistical procedure is used to determine whether the mean difference between the two sets of observations is zero. In a paired sample t-test, each subject or entity is measured twice, resulting in pairs of observations.

Answer key

1. *What is the goal of animal enrichment? What are the benefits of providing enrichment to animals?* The goal of animal enrichment is to increase the quality of life of captive animals. Benefits of providing enrichment are good welfare, increased activity levels, lack of boredom, etc.
2. *Why did we focus on only the nonsocial enrichment categories for this lab?* This lab focused only on the nonsocial enrichment categories because depending on the species available to you in the lab, there may not be enough novel individuals to introduce.
3. *When creating enrichment for your animal(s) what things did you and your group take into consideration concerning safety, behaviors, resources available, and location of enrichment? How did this shape what enrichment was created?* Answers will vary by groups/lab sections.
4. *Was the enrichment created by you and your group appropriate for your species?* Answers will vary by groups/lab sections.
5. *If you were to do this activity again, what would you do the same and what would you do differently?* Answers will vary by groups/lab sections.
6. *How does enrichment impact an animal's welfare?* Enrichment can reduce aggression and fear and enable animals to be calmer in their environment. Enrichment encourages natural behaviors and this leads to good welfare.
7. *Why is it important to consider an animal's normal behavior in designing an enrichment?* It is important to understand your animal's species-specific behavioral repertoire in order to determine which category of enrichment should be used. Understanding your animal's behavioral repertoire will enable you to choose the enrichment that will properly stimulate your animal and avoid the enrichment that could negatively impact its welfare.

Part III. Supplementary material

Supplementary data related to this chapter can be found online at https://doi.org/10.1016/B978-0-12-821410-7.00028-5.

CHAPTER

A nonverbal test battery for evaluating physical and social cognition

20

Malini Suchak, Abigail L. Hines
Department of Animal Behavior, Ecology, and Conservation, Canisius College, Buffalo NY, United States

Chapter outline

Exploring Animal Behavior in Laboratory and Field. https://doi.org/10.1016/B978-0-12-821410-7.00010-8

Part I. Student instructions

Learning objectives

- To be able to describe a species-appropriate response to a given cognitive test.
- To be able to compare and contrast responses to cognitive tests across taxa.
- To be able to explain limitations in cognitive test design.

Background

Cognition tests seek to understand what an animal is thinking about or how it is using information obtained from the environment to make decisions. Cognitive processes are internal thought processes that are unobservable, and therefore tests must be carefully designed to elicit observable behavioral responses, which are then used to make inferences about the underlying cognition. This is similar to intelligence assessment tools used in primary schools, such as an IQ test or gifted student assessment. Nonhuman cognition tests are historically designed from a human perspective with the abilities of primates in mind. This often results in other species being compared to humans, with humans being used as the benchmark for "advanced" cognition. However, the anthropocentrism, or human-centered view, of this approach hinders the ability of scientists to properly assess the skills being tested. This is similar to a person thinking a dog is acting "guilty" when they make a mess or eat something they should not have, when, in fact, the dog might be showing the look in response to being scolded (Horowitz, 2009). We, as humans, interpret the situation in the way our minds are accustomed to in that context. This is particularly problematic in tests of animal cognition, which often fail to take into account the *umwelt*, or the sensory perception of the species being tested, and often use the result that is "correct" for humans as a measure of how smart another species is. Sometimes this can work in primates, who largely share our sensory perception. However, intelligence and cognitive abilities expand beyond this small portion of the animal kingdom. This activity will allow students to explore the different sensory perspectives and physical capabilities of animals that are frequently considered intelligent as they are tested with a cognition battery developed for primates.

This activity will utilize a modified Primate Cognition Test Battery, loosely based off of one designed by Herrmann et al. (2007). In the original study, a battery of 39 tests was given to more than 100 chimpanzees and human children. You will participate in five of these tasks.

Purpose

The purpose of this activity is to experience a simulation of cross-species comparative cognition tests by participating in a subset of the Primate Cognition Test Battery. In addition to completing the task as a primate, you will be asked to explore these tasks from the perspective of other species and develop an understanding of how difficult it is to assess cognition in other species using human-centered tasks.

Methods

Groups of students will be assigned a species with certain physical abilities. Each student in the group will rotate through five stations using the physical limitations of their species. After each station, the group will reflect on the objective of the test. Please keep in mind, you may not start at Station 1.

Your group will be assigned to one of four species and is expected to use the mindset and physical limitations of that species: crow (beak and claw), rat (mouth and hand, with limited vision), primate (hands), or dog (mouth and paw).

You will need a pencil, a worksheet, and to gather with your group. Tools to imitate animal abilities will be provided, if any are needed. An open mind is useful in this exercise as well.

Steps

1. Meet with your assigned group and look over the worksheet.
2. Proceed to the first station, remembering that it is important to remain as quiet as possible. (Nonhumans cannot talk!)
3. No communication with experimenter; remember to act as your species.
4. After each group member has completed the test, quietly review the questions on the worksheet together.

Questions

1. Which species had the easiest time with each test?
2. How significant were the physical or sensory limitations?
3. Did you come up with alternative solutions to better match your capabilities?
4. How does morphology hinder or enhance a species' testing ability as shown by these tests? How does this impact our perception of its cognitive ability?
5. How do animals know what the right answer is for a cognitive test? Was this easier for some stations than others? Why?
6. Which taxa had the hardest time overall, across all the stations? Why do you think that is?
7. In what ways can each of the five stations be modified to account for the physical ability of the test subject?

Part II. Instructor notes

Classroom management/blocks of analysis

Time needed: This activity can be done in one standard 1-h class period by having each student group assigned to a single taxon and rotating through the five stations one time as that taxon. After a full rotation, the groups can come together to share their experiences from their perspective (thus the students would learn about other taxa from discussion and not from experiencing it themselves). If a longer block is available (such as a 3-h lab period), the groups can rotate through all the stations four times, once as each taxonomic group.

In addition to the in-class time, at least 15–20 min is recommended for setup and another 15 min for takedown of the activity.

Teaching the activity

Preclass preparation

No Institutional Animal Care and Use Committee (IACUC) approval is needed for this activity.

This activity can be conducted with students in the class as the subjects. The students will be assigned to represent different taxonomic groups and expected to mimic the movements of those groups in their solutions (i.e., when acting like a corvid, the student can use three fingers of one hand as a "claw" and cooking tongs in the other hand as a "beak"). It is best conducted in a room with tables or larger desks, as it can be difficult to properly set up the materials on a standard-size desk.

It is strongly recommended the instructor familiarize him or herself with the Primate Cognition Test Battery, a cross-species battery of tests developed by Herrmann et al. (2007). The goal of cross-species comparisons is typically to develop hypotheses on how cognition evolved. However, it is extremely challenging to make sure the test is "fair" for all species, and results often look like humans have exceptional cognition, when really humans were advantaged by the testing method. Boesch (2007) published a detailed analysis of discrepancies in the methods of tests for humans versus nonhuman primates that are thought to bias results to favor human subjects. In this activity, we will further expand beyond humans and nonhuman primates to other species with vastly different physical abilities and sensory perception to further explore these challenges.

Materials and setup

The following materials are required:

- four paper plates,
- one cardboard tube,
- three large cups,
- two strips of paper,
- small cups to hold rewards,
- candy rewards (Starbursts and M&Ms or Skittles recommended),
- one set of cooking tongs is needed as a corvid "beak",
- two socks can be used to create "paws" for the dog,
- ski or safety goggles lined with red cellophane or tissue paper can be used to create rat vision

It is strongly recommended to have extra cardboard tubes if available—several can be baited in advance to speed up the trials. In general, extra materials are recommended in case the subject gets "creative" when solving the problem and destroys the material in the process. If materials are not available to create rat vision, sunglasses can be used to generally impair their vision slightly (the darker the better). We do not recommend using blindfolds to impair vision for safety reasons.

Five tables should be spaced around the room as different stations. They should be labeled with the numbers 1–5. Do not label the stations with the corresponding cognitive abilities. Materials should be placed at each station as in the following.

1. Station 1 (numerosity): Two small plates placed in front of the experimenter and two small plates off to the side next to the experimenter. Small candies (M&Ms) recommended to reduce cost, but not required.
2. Station 2 (social learning): Cardboard tube, paper towel or tissue paper, pen or other "poking" tool, and Starburst or similar-sized candy. Candy should be inside the cardboard tube; use the paper towel balled up to hold it in place.
3. Station 3 (transposition): Three cups; Starburst or similar-sized candy.
4. Station 4 (gaze following): No materials required.
5. Station 5 (causality): Two strips of paper, approximately 30 inches long, and one is cut across approximately 4 inches from the top. Starburst or similar-sized candy.
6. Upon entering the classroom, students should be given a worksheet, a card with information about their taxon on it, and a cup for "rewards"
7. It is extremely helpful to have other "experimenters" to staff the extra stations. Teaching Assistants (TAs) or students who have already taken the course work well in this role.

In-class preparation

Students should be divided into groups of three to four. Each group is assigned a taxon (by giving them one of the taxon cards, which can be printed out and provided to the class) and is assigned a station to start. Make sure each group starts at a different station; this will spread them out and prevent crowding at station 1. All members of the group will take turns completing a given station before rotating to the next station. Groups should move through the stations in order, circling back to station 1 after they complete station 5, and until they complete a full rotation. At that point, if the class is doing the version of the activity where every student participates as every taxon, the group should be assigned a new taxon and should start rotating through the stations again.

For all stations the experimenter should be seated across from the student (the "animal") and the trial begins when the "animal" is seated and paying attention. Please see the videos in supplementary materials for a demonstration of the methods.

Station 1: Two small plates are placed in front of the experimenter. The "animal" sits opposite. The experimenter places five candies on one plate (one at a time) and three candies on the other plate (one at a time). The experimenter covers the plates with the two extra plates, so the candy is now out of sight. Next, the experimenter adds three extra candies to one of the plates, again, one at a time, while blocking the "animal's" view using the extra plate. The experimenter can vary which plate gets the additional candy and which side each number is on from trial to trial. Plates are recovered fully and the animal gets to choose. Do not instruct the "animal" to choose a plate, as the "animal" should try this on his or her own. If the "animal" does not choose a plate within 30 s, the experimenter should end the trial and say "next". Do not allow the "animal" to grab both plates, but if this happens pull them both out of the way. If the "animal" successfully selects one plate within 30 s, the experimenter should end the trial and say "next" after the "animal" gets their reward.

Station 2: The candy is placed inside the cardboard tube using the paper towel to hold it in place. The experimenter first demonstrates using the pen to poke out the candy and unwrapping it. Then the tube is rebaited, and the tube and pen are laid on the table in front of the "animal". Do not instruct the "animal" to complete the task, as the "animal" should try it on his or her own. If the "animal" does not solve the problem within 30 s, the experimenter should say "next" and end the trial.

Station 3: Three cups are set in front of the experimenter. When the "animal" sits down, the experimenter places a candy under one of the cups. As quickly as possible, the experimenter moves the cups in the style of "three-card monte". More moves and faster moves make the task more difficult for the "animal" to track. After the experimenter is done moving the cups, the experiment sits back and allows the "animal" to pick. Do not instruct the "animal" to choose a cup, as the "animal" should try this on his or her own. If the "animal" does not choose a cup within 30 s, the experimenter should say "next" and end the trial. Do not let the "animal" choose multiple cups. If this happens pull them away. If the "animal" selects a cup within the 30-s window, the experimenter should end the trial and say "next" after giving the "animal" their reward.

Station 4: The experimenter first glances down toward the table briefly (2 s) and then very obviously moves his or her head and gazes to the side or up. This position is held for five full seconds. At the end of 5 s, the experimenter returns his or her gaze to the "animal" and says "next".

Station 5: The experimenter places a candy on each strip and sits back. Do not instruct the "animal" to choose a strip, as the "animal" should pick one on their own and pull it toward themselves. If they do not pick a strip within 30 s, or they pick the broken strip, the experimenter should end the trial and say "next." Do not let them pick both strips. If this happens pull them both away. If the "animal" successfully gets the food within 30 s, the experimenter should end the trial and say "next" after the "animal" gets their reward.

At the end of the simulation, it is strongly recommended to provide ample time for a debrief. Each station can be discussed using the worksheet as a guide for the questions. In the sample answers to the worksheet (provided), there are notes for the instructor of points that could foster critical thinking in the students. For example, although the answer to quantity may seem obvious to us, in other tests, dogs do not always consistently pick the larger quantity. How might that impact their performance on this test?

The debrief can be particularly effective when paired with the videos from the original Primate Cognition Test Battery study. These can be found in the supplemental materials of Herrmann et al. (2007): https://science.sciencemag.org/content/suppl/2007/09/06/317.5843.1360.DC1?_ga=2.100587608.791245804.1586789553-2038281503.1539979281. You have to download the zip file of all the videos and select the ones that match the stations as follows:

Station 1 (numerosity): children, video 1146282s11; chimpanzees, video 1146282s12
Station 2 (social learning): children, video 1146282s21; chimpanzees, video 1146282s22. (Note: for some reason this task was originally run quite differently in children and chimpanzees. Our format follows the chimpanzee version.)
Station 3 (transposition): children, video 1146282s7; chimpanzees, video 1146282s8
Station 4 (gaze following): children, video 1146282s29; chimpanzees, video 1146282s30
Station 5 (causality): children, video 1146282s19; chimpanzees, video 1146282s20

When watching the videos, the students' attention can be focused on identifying ways in which the test was run differently across the two species (e.g., the children had verbal instructions, their parents present, etc.) as well as how the two different species might have perceived or approached the task. This can be compared with their own experience as other taxa to discuss the difficulty of creating a fair task that can be used to compare cognitive abilities of different species.

Areas of potential confusion or difficulty for students

The primary area of confusion is students not knowing what is being asked of them because there are no verbal instructions. This is normal and part of the experience. It is a good opportunity to remind students that animals engaged in cognitive tests do not get verbal instructions and have to figure things out on their own.

There is also a chance that a student or group of students might not be engaged in the simulation and/or might not faithfully try to represent the way their taxon would complete the task (e.g. by "cheating" and using their hands as any of the nondextrous species). This is more likely to happen after the first rotation (as they will be repeating the work) and it is advised to end the experiment and spend more time on the debrief, particularly to spend more time analyzing the videos of the chimpanzees and humans doing the task.

Recommendations for extensions or continuations for more advanced classes

An advanced class could use this as an opportunity to design a better, more species-appropriate version of the task. If animals are available, this could be extended to test their designs on living members of the species; however, it should be advised that IACUC approval would be required at that point.

Answer key for worksheet

It is highly recommended that the worksheet be used as the basis of a discussion, rather than a graded assignment. As the point of the activity is to introduce the difficulty of assessing cognitive abilities in other animals, there is typically a lot of variation in answers, which provides good fodder for discussion.

Station 1

1. Numerosity (the ability to estimate quantities). Addition or math would also be appropriate answers.
2. Animal's perspective: pick the larger quantity. Experimenter's perspective: same answer.
   Instructor's note: This is a great place to note, however, that in studies of dog quantity estimation, dogs do not always reliably pick the highest quantity. So although many species are food motivated and will naturally gravitate toward picking a larger quantity of food, this is not as universal as people might think.
3. This one is pretty universal. As long as a species can either point or choose, usually they can solve this task. Although rats may have had trouble with visual impairment, the candy should make noise when being placed, allowing them to count the rewards (and a true rat would probably be able to smell the difference).

Station 2

1. Social learning.
   Instructor's note: They might also say tool use, but the social part is essential here because it's not just tool use, it is learning how to use the tool by watching another.
2. Animal's perspective? Anything that enables the animals to get the reward. Experimenter's perspective: Imitation (copying actions to get a reward).
   Instructor's note: This is a critical point. The experimenters expect the animal to correctly imitate its actions. However, there are other ways to get the reward (e.g., pulling it out instead of poking it out, shaking it, banging it against the desk, ripping the tube in half). While the animal might have succeeded, they would not be considered "correct" from the experimenter's perspective.
3. This task heavily advantages primates. It is possible for the dog and corvid to solve the task, but coordination for the corvid will be challenging. The rat may have difficulty due to impaired vision. For any taxa other than primates, true imitation (copying the actions to get the reward) may be impossible.

Station 3

1. Transposition or object permanence (tracking objects).
2. Animal's perspective: find the reward. Experimenter's perspective: same answer.
3. This task should be fine for all visual species as long as they can point or choose a cup. This should be harder for rats, as they could not really see the cup being moved, but the class should be reminded that a true rat could probably smell the location of the reward.

Station 4

1. Gaze following.
2. Animal's perspective: following the experimenters gaze. Experimenter's perspective: same answer.
   Instructor's note: Students often have no idea about this one because it is unrewarded. This is a good place to discuss how rewards communicate to an animal what they are doing right or wrong (and relate outcomes in animal cognition tests to basic operant conditioning).
3. Nonvisual species (rats) would likely find this impossible. Other species should be fine.

Station 5

1. Causality, cause and effect (some might say tool use if they view the paper as a tool).
2. Animal's perspective: get the food. Experimenter's perspective: choose the connected piece of paper.
3. Nonvisual species may not be able to see which paper strip is solid; dogs may have difficulty with the dexterity required.

References

Boesch, C. (2007). What makes us human (*Homo sapiens*)? The challenge of cognitive cross-species comparison. *Journal of Comparative Psychology, 121*(3), 227–240.

Herrmann, E., Call, J., Hernández-Lloreda, M. V., Hare, B., & Tomasello, M. (2007). Humans have evolved specialized skills of social cognition: The cultural intelligence hypothesis. *Science, 317*, 1360–1366.

Horowitz, A. (2009). Disambiguating the "guilty look": Salient prompts to a familiar dog behaviour. *Behavioural Processes, 81*(3), 447–452.

Part III. Supplementary material

Supplementary data related to this chapter can be found online at https://doi.org/10.1016/B978-0-12-821410-7.00010-8.

PART

Communicating behavior

4

CHAPTER

21 Learning from the primary literature of animal behavior

Rebecca Burton

Department of Biology, Alverno College, Milwaukee, WI, United States

Chapter outline

Exploring Animal Behavior in Laboratory and Field. https://doi.org/10.1016/B978-0-12-821410-7.00025-X

Part I. Student instructions

Learning goals, objectives, and key concepts

- To accurately identify key results and methods in primary literature.
- To evaluate the validity of methods and conclusions.
- To synthesize the findings of a study with what is already known and identify future avenues of research.

Background information

Primary research articles are written by the person or team who conducted the research described. Articles that summarize papers by other authors are *secondary* sources or review articles. *Peer-reviewed* articles are published only after they have been evaluated by professionals with expertise in the field. Reviewers use criteria such as validity of experimental design, clarity of communication, completeness of background information, and reasonableness of conclusions to determine whether an article is suitable for publication. Peer-reviewed literature makes advances in animal behavior possible by giving researchers credible information about the results of other studies.

Reading the primary literature is a skill that requires analysis, evaluation, and problem-solving. Primary literature is written using specific *conventions*—organization, format, and terminology—that make the style fairly consistent. Practice will help you become an effective reader of articles in a variety of scientific journals.

Typical organization of a primary research paper

Abstract

Briefly summarizes methods and main findings.

Introduction

Sets context about the behavior and species of interest, states reasons for the study, and lists hypotheses and predictions.

Methods

States what was done clearly and completely enough so that another researcher could evaluate the validity of the study and replicate it.

Results

Describes the findings, usually including figures and/or tables.

Discussion

Evaluates the findings in relation to the findings of previous studies. Discusses possible reasons for the findings and often identifies future research possibilities.

References

List of cited sources in alphabetical order.

Purpose

You will read a peer-reviewed research article on your own. Working in a group, you will analyze the different sections of the paper to identify key concepts, processes, and findings. You will evaluate which hypotheses were supported or refuted by the findings.

Methods

1. Before class read *How to Read a Technical Article* and take notes.
2. Before class read the assigned technical article (allow plenty of time). Be sure to take detailed notes on the article, questions you may have, and so on.
   - Highlight important points
   - Make note of questions
   - Look up unfamiliar terms
   - Use any additional materials from your instructor, such as the *Technical Paper Summary and Analysis* worksheet
3. During class discuss the paper and the questions provided by your instructor, such as the *Technical Paper Summary and Analysis* worksheet
   - Share the "airtime;" don't dominate or stay out of the discussion.
   - Stay on task and help your team stay on task.
   - Don't pretend to agree if you do not.
   - Do not attack people; challenge ideas using evidence.
   - If you are answering questions as a team, then agree on how you will decide on your answer.
   - Be aware of any time limits and deadlines.
   - Consult with your instructor/teaching assistant (TA) if you get stuck.
   - Optional: review another group's answers.

Making a methods flow diagram

Methods can be difficult to follow in text. Using a methods flow diagram as a visual model will help you clarify and communicate what the authors did. To illustrate this process consider an experiment to test whether fish preferred conspecific fish that had not been exposed to risk compared to fish that had been exposed to risk. This example experiment is represented by a flow diagram (Fig. 21.1).

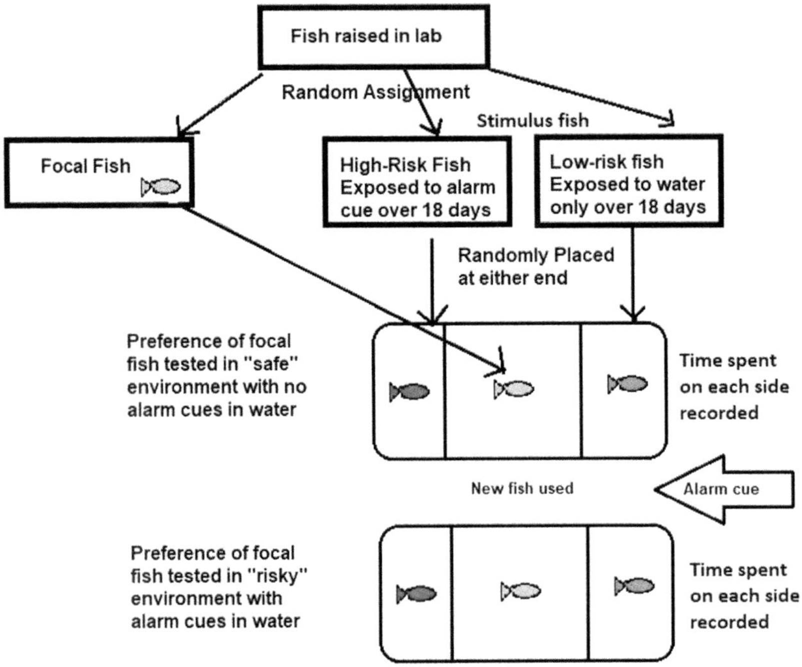

FIGURE 21.1

Example of a methods flow diagram (based on McGhee 2019)

To make your own diagram start by identifying the source of the animals (such as a wild population or a tank in a lab). Place a box or label for that at the top of your diagram. If the animals were randomly assigned to groups show this on your diagram.

Next identify the independent variable. It will be the variable that the investigators hypothesized as the cause of a behavior. In an experimental study, it will be the variable that the investigators manipulated. In an observational study, the researchers may not manipulate the independent variable. For example, experimenters might be comparing behavior in different seasons. Add boxes (or something similar) with labels for each category of the independent variable to your diagram. In the example experiment, the independent variable was the presence of stimulus fish that had been exposed to "risky" or "safe" environments. The diagram shows how some stimulus fish were exposed to an alarm chemical and some were exposed to plain water.

Identify the dependent variable that the investigators were measuring. This will be the behavior that the investigators have hypothesized to be influenced by the independent variable. In the example experiment, the dependent variable was which stimulus fish the focal fish would spend more time near. Sketch the experimental setup in your diagram. Your sketch should be as simple as possible, with just the key details. Use labels to make the meaning clear.

If the paper describes a series of tests, add the later steps to your diagram. In the example experiment, there were two stages to the experiment. The researcher wanted to know whether the preference of the focal fish would depend on the environment that the focal fish was in. One condition exposed the focal fish to an alarm chemical. The other did not. The diagram has a notation that new fish were used for the second phase. This sort of detail is critical to include.

Analysis and summary questions

1. According to the abstract, what did the authors do and find (briefly)?
2. According to the introduction, what did previous researchers discover about this type of **behavior**?
3. According to the introduction, what do we know about **this species**?
4. What **question** was this experiment designed to answer and why is it important to answer this question?
5. What did the authors **do** to answer the question?
6. What were the **most important** findings?
7. According to the discussion, what do the data mean?
8. Do you agree with the authors' interpretation of the data? Is there other evidence that might be important in interpreting the data?
9. What would you do differently **or** what experiment would you do next?
10. List two or three vocabulary words and define them in your own words.

Part II. Instructor notes

Classroom management/blocks of analysis

The actual discussion of the article in class will usually take about 45–60 min, but the reading and initial analysis of the article by students will take much longer, depending on the length and complexity of the article and the preparedness of the students. For students with perceptual and learning differences, who may need to have the article and figures translated into an accessible format, work with your institution's support services.

Students will generally improve with each iteration. If they are reading an article a week, then expect them to provide more and higher-level questions in later weeks or shorten the time scheduled for the discussion. Students may be able to read articles independently later in the semester. The ability to read animal behavior primary literature should transfer well to other scientific literature.

Scaffolding

Article selection

The amount of scaffolding needed will vary depending on both the students and the article selected. A list of potential articles is included in the Supplementary materials. Aspects to consider when selecting an article are

- length,
- complexity of language (technical and nontechnical),
- centrality of the concept (key concept vs. an exception),
- complexity of experimental design,
- complexity and familiarity of figures,
- familiarity of statistics and models.

Making a flow diagram of the experiment tends to cause students the most difficulty but gives them the clearest idea of what was actually done. The first time they do this task you may want to walk them through it or provide a diagram that is partially completed. You can fade this out with later iterations.

Vocabulary, technical or nontechnical, tends to be an issue for many students. Teaching them how to analyze words, showing them resources for definitions, and providing them with an article-specific glossary are all ways to assist them in both the long and short term.

Students may be able to handle more challenging articles

- after a few iterations with other articles,
- when provided with a glossary and/or reading guide specific to the article,
- when provided with simple rules when statistical tests are beyond their experience (e.g., "$P < 0.05$ indicates a statistically significant difference between the treatments"),
- when the class works together on tasks such as diagramming the experimental method.

Materials needed, including variations

- *How to Read a Technical Article* (available in Supplementary materials).
- Worksheets (link or print, such as *Technical Paper Summary and Analysis* available in Supplementary materials).
- If students are working online, then multiple-choice quizzes on your course management system (CMS) (examples in Supplementary materials).
- Optional: assignment cards (in Supplementary materials).
- Optional: materials for students to collaborate in diagramming experiment (large paper, markers, etc.).

Preclass preparation

- Success starts with the selection of an appropriate article (see Supplementary materials for suggestions).
- Discuss the structure of technical articles at least a week in advance of the first time you do the first activity.
- How to Read a Technical Article be italicized as it is in the rest of the chapter.
- Do not flirt with copyright infringement. **Link** to the article on your CMS.
- Students can work in groups to review and discuss a paper. Depending on the class size, groups of about 10 students work well. In a larger class, multiple groups may be assigned the same paper. If you choose to assign different papers to each group, you must assign groups and papers in advance. If you wish students to present different papers to the rest of the class, it will probably only be possible to share two papers in a 50-min class period.
- Decide how you will assign students to groups (using color-coded assignment cards, randomly, determined by you, etc.) and how they will answer questions (online, using the *Technical Paper Summary and Analysis* worksheet, using clickers, etc.) and prepare accordingly. If using the *Technical Paper Summary and Analysis* worksheet make sure that students print or upload their completed work before class.

In-class preparation

At least for the first time, provide additional instruction. You can skip some of this in later iterations.

- Vocabulary can be a barrier to student understanding. Explicitly teach how to approach unfamiliar words (parts of a word, selecting the correct definition, using a pronunciation tool, such as Google).
- Model how you would annotate an article or graph.
- Identify areas that seemed to be challenging and provide assistance.
- Identify misconceptions (anthropomorphism, "good-of-the-species" arguments, teleology, etc.).
- Connect article to past/future course topics.

In-class activity

- Students should have read the article and answered the questions.
- Use Assignment Cards (Supplementary materials) at the beginning of the class to assign student responsibilities. Students can get these randomly or as first come, first choice.
- Have students report on the question or vocabulary on their card.
- Discuss the student responses and related topics.
- If the class is small enough, the discussion and completed worksheets may be sufficient for assessing student performance. See the next section for modifications based on class size and mode.

Modifications to the activity

- For classes meeting online, students can have their discussion in a CMS forum.
- Large classes can be broken into smaller groups for discussions, whether in class or in CMS forums.
  - If there are multiple groups in an in-person course, cards can be color-coded to divide students into groups.
  - Online, students can be randomly assigned a number and then referred to the table of Assignment Cards (Supplementary materials) to find their role.
- If using multiple groups, groups could compare answers, such as their flow diagram of the experiment or ideas for improving or extending the study.
- If groups are smaller, some cards/questions can be taken by instructor, TA, or class as a whole.
- For large courses, students can be randomly assigned on the day of class as the "panel of experts" to explain specific sections of the article to the entire class.
- For large/online courses, multiple-choice questions can be used instead of worksheets to decrease instructor workload.
  - If not using a worksheet, multiple-choice questions could be online, clicker, paper, or the IF-AT (Immediate Feedback Assessment Technique; http://www.epsteineducation.com/home/about/).

- To decrease difficulty, partial diagrams of experimental methods can be used to scaffold the skill of diagramming experiments.
- To make complex figures easier for students to understand, marking them up or providing a simplified version can be helpful.

Part III. Supplementary material

Supplementary data related to this chapter can be found online at https://doi.org/10.1016/B978-0-12-821410-7.00025-X.

CHAPTER

The fine print: process and permissions for behavioral research

22

Susan W. Margulis[1], **Heather Zimbler-DeLorenzo**[2]

[1]*Department of Animal Behavior, Ecology, and Conservation, Department of Biology, Canisius College, Buffalo, NY, United States;* [2]*Department of Life and Earth Sciences, Perimeter College at Georgia State University, Decatur, GA, United States*

Chapter outline

Exploring Animal Behavior in Laboratory and Field. https://doi.org/10.1016/B978-0-12-821410-7.00009-1

Part I. Student instructions

Learning goals, objectives, and key concepts

- To review the requirements for research permissions.
- To understand the difference between experimental and observational research.
- To complete an application for permission to conduct research.

Background

Behavioral research studies generally fall into two broad categories: *observational* and *experimental*. In an experimental study, the researcher alters specific factors, called independent variables, to determine how they influence a measured factor, or dependent variable. In an observational study, the researcher does not control any variables, but rather observes the dependent variable in the context of normal variation in the independent variables. In behavioral research, the dependent variable is some component of behavior.

All experimental research involving vertebrate species must be reviewed and approved by an *Institutional Animal Care and Use Committee* (IACUC) according to the Federal animal welfare regulations and the U.S. Public Health Service Policy (National Institutes of Health, 2017). If you are using human subjects, you will need to obtain approval from an *Institutional Review Board* (IRB).

An IACUC is required by federal regulations for the use of nonhuman vertebrate animals in research, teaching, and testing. Animal use activities must be reviewed and approved by the IACUC before a project/study can begin. It is, therefore, important that you understand the process in your institution, including the timeline and deadlines for protocol review. Some IACUCs require review for observational studies, but this is not always the case. Experimental studies are often conducted on invertebrates (and therefore IACUC review is not required) or in research labs that have experience with the IACUC proposal and review process (National Institutes of Health, 2017).

Observational studies may be able to bypass IACUC review. However, depending on where the research is conducted, permits or approvals may be needed. For example, observing wildlife may require state or federal permits. Observing people and their pets (at a dog park, for example) may require approval by an IRB, which evaluates research that involves human subjects. Zoos and aquariums represent another excellent locale for conducting observational studies. Here too, approval may be needed.

Zoos represent an excellent venue for behavioral research. Many species that would be entirely inaccessible in the wild may be viewed up close and consistently. Recognizing individual animals can often be accomplished easily. There is no need to painstakingly habituate animals to human presence. Of course, one needs to be mindful of the captive environment and how this might influence behavior. Nevertheless, zoos may provide you with a chance to conduct authentic research.

Most zoos are quite amenable to permitting students to conduct observational research studies. An observational study is one in which the researcher does not manipulate specific variables, but rather takes advantage of natural variation in the environment to ask questions. For example, a group may contain individuals of different ages and sexes; does behavior vary in any predictable way? Does weather or crowd alter behavior? Many questions can be posed in this way.

Anyone may pay admission (or purchase a membership) and observe animals in a zoo. As long as you are observing during normal zoo hours and from the public area, no formal permission is *required,* but it is *highly recommended.* Most zoos have a process by which you can submit a proposal for research, gain approval, and thus have the support of the zoo for conducting your research. Although you may not *require* such assistance, it is highly recommended that you go through the research proposal process. With an approved proposal, you may be able to get valuable assistance—help in identifying individuals, information about husbandry practices, and other details that can be extremely helpful for your study.

Purpose

This activity will introduce you to the process of getting permission for the conduct of research. You will complete either a zoo research proposal for an observational study or an IACUC proposal for an experimental study.

Methods

Option 1: observational study

If you choose an observational research proposal, you will complete a research proposal form for a possible study at your local zoo. Depending on the class, you may conduct a full research project at the zoo or you may develop a proposal only. In either case, it's important to be clear and concise! You can find the standard Association of Zoos and Aquariums (AZA) research proposal form on the AZA website: https://www.aza.org/research-and-science. The proposal has seven sections:

1. Basic information and approval signatures
2. Abstract
3. Research project
4. Permits and support

5. Impact on holding institution
6. Biomaterials
7. Supplemental materials

For the purposes of this activity, you will complete sections 1, 2, 3, and 5.

Your instructor will work with you to develop appropriate research questions. Remember, you cannot manipulate the animals or their environment. Your study proposal will be observational. Pay special attention to section 5, impact on the holding institution. This section is unique to zoo research proposals. How will your study impact the zoo—its staff or animals? Although you may think it will have no real impact, you should remember that if you are planning to ask the zoo for access, then that is an impact. If you ask for contact information for a keeper, then that too will impact the institution.

Step 1. Research the zoo. What species does it have? What species are you interested in?

Step 2. If possible visit the zoo and spend some time observing the species you will propose a study for. If you are unable to visit, then observe some zoo webcams if these are available.

- **a.** Identify individuals, if possible. Note that this will help you determine how much assistance you may need from the zoo.
- **b.** Determine what research question you will propose.

Step 3. Finalize your methods.

- **a.** Prepare your ethogram (what behaviors will you observe?).
- **b.** Specify how much data you intend to collect.
- **c.** Explain how you will analyze your data.

Step 4. First complete Section 3 of the research proposal form (Fig. 22.1). This is where you will clearly describe what your research question is and how you will answer it. Be sure that each section is completed appropriately.

Step 5. After you complete Section 3 write your abstract, a summary, in Section 2.

Step 6. Now complete Section 5, impact on the hosting institution (Fig. 22.2). This section is particularly important for studies conducted at a zoo. How will your proposed research impact the animals, the visitors, or the staff? Justify any possible impacts. Think about all the ways that your research might have an impact. Note that for section C, your answers to the first three questions should be "Yes". For the final question, although you may think this too is a yes, if you are likely to ask the keepers any questions, no matter how minor, you should answer "no" and explain what you are likely to need.

Step 7. Lastly, complete Section 1, the cover page.

Step 8. Submit your draft as indicated by your professor.

Step 9. Based on your instructor's feedback, you can revise your proposal and submit the final version.

Section 3 – Research Proposal Form, AZA – Research Project

This section has a 5-page limit, and is the most important – here is where you explain your project

A. Brief Description – include literature review, objectives, significance, and contributions
B. Study Subjects – sample size, justification for using these subjects
C. Methodology – include experimental design, research methods, analysis plans.
D. Health, Safety, and Welfare Considerations –
    i. Describe any potential health/safety/psychological risks for animals or humans involved.
    ii. Describe how you will monitor health/safety and psychological wellbeing for this study.
    iii. For human studies, under what conditions will a subject be able to choose to withdraw?
    iv. For animal studies, describe the final disposition of subjects at the completion of the study.
E. Dissemination of results (check all that apply)

☐ Undergraduate thesis	☐ Non-peer reviewed publication
☐ Graduate thesis or dissertation	☐ Peer-reviewed publication
☐ Scientific meeting presentation	☐ Public presentation
☐ Website	☐ Other:________________

FIGURE 22.1

Overview of section 3 of the AZA research proposal form. The complete, fillable form can be found at https://www.aza.org/assets/2332/research_proposal_form_current.docx.

Section 5, AZA Research Proposal Form AZA – Impact on Hosting Institution

This section is unique to zoo-based research.

A. Please describe the PI/and or Co-Pi's prior experience working or conducting research in a zoo/aquarium setting and with the target species (1500 characters)
B. Identify the zoo/aquarium time and effort needed to assist with project completion.
C. Role of zoo/aquarium staff in this project:
    i. Is the project purely observational? Y/N
    ii. Can the study be conducted from public areas? Y/N
    iii. Can the study be conducted during operating hours? Y/N
    If you answered "no" to any of these questions, please describe what is requested of the hosting institution:
D. Describe any records or archival materials that would be required from the institution to support the study:
E. What are the benefits of participation for the hosting institution?
F. How will the zoo/aquarium and staff be acknowledged in dissemination of results?
G. Attach specific protocols for activities and procedures that will be implemented at the hosting institution in conjunction with this study not included in other sections of this application.
H. What are the anticipated costs and reimbursements (if any) to the institution for participation?

FIGURE 22.2

Overview of section 5 of the AZA research proposal form. Your answers to section C should all be "Y". The complete, fillable form can be found at https://www.aza.org/assets/2332/research_proposal_form_current.docx.

Option 2: experimental study

If you choose an experimental project, you will complete your school's *Institutional Animal Care and Use Form*, sometimes called an Animal Subject Approval Form. Every higher education institution has an IACUC. Although forms will vary among institutions, there are some common types of information required.

1. Responsible person or lead/principal investigator (there may also be a student section)
2. Project details, such as project title and dates
3. Project funding sources
4. Purpose of the study
5. Background and rationale for the study
6. Characteristics of the study population
7. Housing
8. Description of experimental design and justification
9. Procedures to be performed on the organism
10. Potential hazards
11. Conclusion of the study (what will happen to the organism)

Your instructor will work with you to develop appropriate research questions. For the purposes of this activity, you will complete sections 1, 2, 4, and 5.

Step 1. Develop your research question and determine what organism is best to test your hypothesis.

Step 2. Finalize your methods.
- **a.** Prepare your experimental design and what materials you will need.
- **b.** Specify how much data you intend to collect.
- **c.** Explain how you will analyze your data.

Step 3. Complete the purpose of your study section of your IACUC form. This is where you will clearly describe what your research question is and how you will answer it. Be sure that each section is completed appropriately.

Step 4. Now, you need to find some scholarly articles to provide information on your question for the background and rationale section of the IACUC form. Make sure to gather details about the species, as this will help you determine what you need for your study subjects. Complete the "study population" part of the IACUC form.

Step 5. As you think about your experimental design consider how you will keep your organism and complete the "housing" portion of the IACUC form. What materials do you need to do this properly?

Step 6. Next, you will need to provide a detailed experimental design for your project. This will also need to include a justification, or why, of the process you have chosen.

Step 7. Lastly, consider your organism and the procedures you are performing and what the consequences may be to yourself/others and the animal. Also, what will happen with the animal once the project is over?

Step 8. Submit your draft as indicated by your professor.

Step 9. Based on your instructor's feedback, you can revise your proposal and submit the final version.

Results/discussion

1. What sorts of questions can be studied using an observational method? What sorts of questions can be studied using an experimental approach?
2. What are some advantages of conducting an observational study at your local zoo? What are some disadvantages? What are some advantages of conducting an experimental study? What are some disadvantages?
3. How much data should be collected for a research study (either observational or experimental)?

Reference

National Institutes of Health, Office of Laboratory Animal Welfare. (2017). *The IACUC*. Retrieved from https://olaw.nih.gov/resources/tutorial/iacuc.htm.

Part II: Instructor's notes

Classroom management/blocks of analysis

This activity can be done with minimal in-class time if the endpoint is a proposal only. Allow one regular class period for explaining what an IACUC is and the types of research that require IACUC review. If you choose to have students conduct the research (rather than just propose it), this activity can and should encompass the entire semester.

The instructor should use the IACUC form from his/her institution. Make sure to download the forms so they are available for the students. Some institutions have required training that investigators should also complete. This could be an interesting addition to the activity and allows students to learn more about animal safety.

Teaching the activity

It is advisable to check with your local zoo to determine what process or forms are needed. Many zoos will accept the standardized research form that is available on the website of the AZA, but it is worth confirming this, as some zoos have their own forms or addendums. If students will actually be completing a research study, as part of a semester-long project, they should complete the form in its entirety. If this is an exercise in proposing a study, you may decide to omit certain parts of the proposal as indicated earlier. Furthermore, you need to check on the timing of the proposal review at the zoo. A strictly observational study is not going to impact the zoo animals and although approval is not guaranteed, it is likely. If students are preparing a proposal, but not actually doing the research, then do not send the proposals to the zoo.

Reach out to the local zoos in order to determine their process for research approval. How often do they review proposals? What do they require for student proposals? For example, some zoos may require a letter from an advisor or a supervisor. Will students be permitted to enter the zoo at no cost, and what entrance should they use? If funding is available, we recommend purchasing memberships for students who will be visiting the zoo multiple times. Alternatively, if you are permitted to charge a "lab fee" for the course, then consider using these funds to purchase zoo memberships for the class. If the class will be visiting the zoo as a group, field trip options may be less expensive. If students will need to communicate with zoo staff (e.g., for advice on how to identify individuals), it's important for them to realize that they need to be respectful of keeper time. Have specific questions, keep it brief. Make sure contact phone numbers and emails are obtained (with the approval of the zoo).

Once you have established a relationship with your local zoo or aquarium, cultivate this relationship. Even if students are only completing proposals vet these proposals, and if some seem to have potential for a full-fledged research study, then encourage the students to do so. It will benefit the zoo and the students, and you may generate some valuable research.

Answer key

1. What sorts of questions can be studied using an observational method? What sorts of questions can be studied using an experimental approach?
The most important thing for students to recognize is that they cannot manipulate any aspects of the animals' environment for an observational study. They can ask questions that relate to the normal variation in the environment and in the animals—differences based on age, sex, season, and crowd level—these are all suitable questions for observational studies. For an experimental study, the best questions manipulate or vary only a single independent variable (holding all others as constant as possible). This permits the researcher to draw conclusions about the influence of that independent variable on the dependent variable.
2. What are some advantages of conducting an observational study at your local zoo? What are some disadvantages? What are some advantages of conducting an experimental study? What are some disadvantages?
There are of course many possible answers to this question. Students have an opportunity to observe species that they almost certainly would not be able to observe in the wild. They may learn something that will benefit the zoo. They also gain experience at designing a study and evaluating methodology, and if they are actually conducting the study, they will gain experience at data analysis and interpretation as well. Disadvantages include an inability to control any aspect of the environment and being limited to days and times that the zoo is open to the public. Advantages of an experimental study include the ability to tightly control the environment and to set up experimental and control groups. Disadvantages include the challenge of getting IACUC approval (for vertebrates) and the need to provide all animal care.
3. How much data should be collected for a research study (either observational or experimental)?
There is no single correct answer for this question. It is important to recognize, however, that the more data one collects, the more reliable the information will be. A few hours of observation or an experiment with a small sample size and limited replication may be heavily biased by (for example) environmental conditions at that specific time. As more data accumulate, more informative patterns begin to emerge. If students are collecting data as part of a semester-long project, we recommend requiring a minimum of 25–30 h of data.

CHAPTER

Writing science for the general public

23

Erin A. Weigel[1], Carol M. Berman[2]

[1]*Program in Evolution, Ecology and Behavior, University at Buffalo, Buffalo, NY, United States;* [2]*Department of Anthropology, University at Buffalo, Buffalo, NY, United States*

Chapter outline

Exploring Animal Behavior in Laboratory and Field. https://doi.org/10.1016/B978-0-12-821410-7.00004-2

Part I. Student instructions

Learning goals, objectives, and key concepts

- To identify the big picture from a research study and its importance in a broader context.
- To learn how to identify, replace, and/or eliminate jargon.
- To compose a story for the general public using engaging journalistic techniques.

Background

The moment a scientist is asked "So, what do you study?" some scientists delight in the opportunity to discuss their research with curious nonscientists, whereas many adopt the ever-charismatic "deer-in-the-headlights" expression. How can the intricacies of the work be summed up in a few sentences? Simplifying the research risks underselling its value or the level of effort and time involved, whereas droning on with too much detail may lead to glazed eyes and empty nods. The main reason these questions are so difficult is because scientists are trained to communicate in detail about the merit of their work with other scientists using very technical language.

Yet there are countless circumstances in which the findings need to be conveyed to a more general audience. In addition to simply sharing the wealth of knowledge currently inaccessible to the general public, it's important to remember that poor communication can lead to widespread mistrust or misunderstandings of scientists and their findings. This is especially true when dealing with socially charged issues, such as stem cell therapy, use of animal models, or climate change (Brownell et al., 2013). Scientists need to effectively communicate their work to the media or the general public directly. This can be of vital importance when dealing with issues, such as personalized medicine or vaccinations, that have the potential to directly impact the quality of individual lives. Of course, it is also essential that researchers are able to effectively communicate about their work to policy-makers who influence governmental decisions about regulation and funding because policy-makers often lack a deep foundational understanding of the science (Brownell et al., 2013). In the field of animal behavior, information about findings may need to be communicated to zoos, animal technicians, or even pet owners, who all want to better understand the animals they work with and how to improve their care.

The number of channels for communicating science with a general audience is increasing in this digital age. While some individuals seek out a profession in writing science articles for news and magazines (online and in print), there are many less formal ways of informing the public about ongoing research, such as through social media, blogging, or YouTube videos. Often referred to as "popular science", these forms of writing tend to use common language, analogies, humor, and a more journalistic style than technical writing (Rakedzon et al., 2017). In addition, although it is understood that researchers must acknowledge any uncertainty in their findings when publishing studies for the scientific community, popular science writing often emphasizes the uniqueness and importance of the research findings without such hesitancy.

Popular science writing requires active practice in a whole new form of communication that is free of **jargon** (highly specialized language) that would only confuse a lay (nonexpert) audience, but without compromising the accuracy or importance of the work. The more specialized a scientist is in his or her field, the harder it may be to remember what words are uncommon in everyday language. For example, "plasticity" is a word commonly used in science; in animal behavior, it is used to describe the degree to which a behavior can be modified by internal or external forces such as environmental conditions. To avoid terms such as this that may confuse nonscientists, the use of a more accessible synonym, such as "flexibility", or an analogy to describe the term is more appropriate. You could explain the concept of plasticity by having your reader imagine trying to squeeze a rock versus a ball of clay; the rock will be unyielding, whereas the ball of clay can easily be shaped into other forms. Although the field of animal behavior tends to inherently involve less abstract jargon than other fields, there are still many complex ideas and principles that need to be carefully explained.

When talking to nonscientists, animal behaviorists have an advantage in that many people are already very interested and engaged in this topic. In the ever-so-formal classification of science stories, animal behavior studies generally fall under the "cool stories" category (Zivkovic, 2011). Although it does, of course, vary by subject matter, in general, these stories tend to not alter the readers' daily habits, as would a study about health or fitness. They do, however, attract a wider audience, including individuals who wouldn't normally go out of their way to learn about science. There are television channels, Instagram accounts, and documentaries devoted to sharing information and videos about animals and nature. People love to learn about the similarities between human behavior and intelligence and animal behavior and intelligence. They also wonder about their emotional lives, for example, whether they show compassion for others and sorrow over loss. Comparative approaches, such as these, not only inform us how humans are and are not "unique" but also let us test hypotheses about why humans possess certain traits (Smulders, 2019). For example, comparisons with other long-lived species, such as killer whales, can help shed light on why women live well after menopause, even though they are no longer fertile (Smith, 2013). In both species, when done reproducing themselves, older females shift roles and become assistant caregivers to future offspring in their line. This is a relatable issue that people find compelling.

There are also special challenges in the field, including dealing with audiences with rigid beliefs or insufficient background knowledge. Many people who do not understand or do not accept the way evolution works may not see the value of animal behavior research, particularly when it has implications for human evolution. For example, they may question why researchers may be interested in the ways that great apes communicate during play (one of the author's dissertation topic) when in fact such research has important implications for the evolution of human language. Unfortunately, most scientists never receive proper training in how to communicate complex ideas to the general public, not just other experts, creating a gaping chasm between us. Being able to properly identify the prior knowledge, needs, and linguistic barriers of your audience is difficult, but it can make communication like this perfectly navigable with practice.

Purpose

In this exercise, you will practice communicating a scientific study to a general audience by taking a recent primary scientific journal article in animal behavior and translating it into a compelling science news story. First, you will need to develop a deep understanding of the study itself to gain insight into what is the "big picture". As you develop your story, you will find that writing for the public is very different from the careful, calculated manner in which you are taught to report your findings for other scientists. In the second part of the exercise, you will have the opportunity to receive and provide peer feedback to help you learn the balance between jargon-heavy overly detailed reports and oversimplified stories that diminish or misrepresent the importance of the study.

Methods

For this exercise, you will be asked to read and analyze the journal article chosen, translate the information into a compelling news story for a general audience, and exchange your news story with your peers for feedback. As this assignment is not an opinion editorial, the story should not include personal opinions regarding the study. The story should be approximately 500 words. Remember to use grammar, syntax, and spelling of publishable standards throughout.

Unless provided with an article by the teacher, each student should select a recently published primary journal article involving animal behavior in a species of their choice. Aside from a computer in which to locate your article and write your news story, this activity will only need the rubric handouts required for peer evaluation.

As the focus of the story is the selected study, if you encounter unfamiliar terms or concepts, you are encouraged to look them up to enhance your ability to understand and explain the elements of the study. Much of the study's information, related research in the field, and central concepts will be outlined in the introduction and discussion portions of the article and will be essential to help you put your news story in context. If you find you still need additional background for the article, you should do a little research, perhaps guided by the references listed in the publication for relevant studies.

Prewriting

Before you begin writing, you should first identify what you believe to be the significance or "big picture" of the study and its findings. Who is this important for and why should they care? Begin thinking about ways in which the reader might be able to relate to your animal behavior topic, whether immediately or in the long term. This could coincide with a bigger hot topic theme, such as the environment, climate change, or even public health. Is there something new or unusual about this particular study that sets it apart from other work being done in the field? These ideas can often be found in the discussion portion of the journal article. You must constantly remind yourself of who your audience is, how the information may be important to their own lives, and what background knowledge they already have versus what you need to supply for them.

Readers today want to receive information quickly and in a compact form; the stressors of modern life and a lack of free time can lead to impatience or a short attention span. Unlike writing primary literature, when writing for a general audience, you often must begin with the broader impact statement, giving away the big picture of the story rather than saving that for the end (Burke, 2015). This is a popular style of writing, given the overwhelming amount of digital content we have access to today, because readers may very well move on from your story as soon as they have gleaned sufficient information for their interest. Don't worry, you will still have plenty of curious minds that stick around for the finale. While well-established journalists who communicate through channels such as *National Geographic* can still successfully use a "narrative approach" in which the real juice of a much longer story comes at the end, here we will practice the shorter "inverted pyramid" method in which the most important information comes first.

Coming up with a title

In a news story, the very first thing your readers see will be your title. What makes a catchy title? You must draw your readers' attention, while avoiding any unfamiliar terms that may deter them from clicking on your story. It is important that you do not simply repeat the title of the original article, as it tends to be more detailed and technical than a general audience would find interesting. For example, consider the journal article title "Does the chimpanzee have a theory of mind?" (Premack & Woodruff, 1978). A general audience unfamiliar with cognition research may misinterpret this as scientists asking a bunch of chimpanzees whether they have their own "theory" about how minds work. While that may sound like a silly exaggeration, realistically when faced with unfamiliar concepts such as "theory of mind", a general audience will either use their imagination to make sense of such terms or simply decide not to read the article because they don't understand what it will be about. A more reader-friendly alternative for a news story could be "Chimps may be capable of comprehending the minds of others" (Caruso, 2016). This title, while being informative, also does not entirely reveal the whole story but encourages intrigued readers to click on your story and read on.

Ledes (leads)

Once you have a catchy title and have identified the importance of the study, you must find a way to capture the audience's attention to keep them reading. In journalistic terms, this is known as the "lede" of the article. Without this "hook", you may lose the reader right at the beginning. So what methods can we use to compose a good lede? There are several strategies that can be used, and the one most suited will likely depend on the topic you've chosen.

Make the reader smile

The tactful use of humor is an excellent way to introduce a topic while grabbing the reader's attention in a way we're not usually allowed to do in formal science writing. Think of a clever story or analogy that adds amusing context to the study and makes the reader wonder how a study could possibly be related. For example, depending on your topic this could be done by poking fun at the animalistic side of human nature that we often tend to ignore. Just be sure to quickly get to the point to give the reader a sense of what the story will be about (Meadows, 2015a). For example, Aron (2015) reported on the accuracy of science being portrayed in the movies:

> *Even black holes wear makeup in Hollywood. Last year's hit film* Interstellar *used real scientific equations to depict what happens when a team of space farers venture near a supermassive black hole. Now, a joint paper published in the journal* Classical and Quantum Gravity *from the movie's visual effects team and scientific consultant reveal that the real black hole (see above) was deemed too confusing for audiences, and some of the science had to be toned down.*

Leave the reader hanging

Another method is to use an attention-grabbing statement or short anecdote to include enough of the news to inform the reader overall what the topic is while piquing their curiosity and leaving them wanting more (Meadows, 2015a). You can explain the context the study takes place in without disclosing the actual discovery. For example, to begin a story regarding Australia's "firehawks", such as the black kite, that actively start fires to prey upon fleeing animals (Greshko, 2018), one could write

> *Bush fires have been shaping the Australian landscape for thousands of years due to human activity and natural phenomena such as lighting strikes. With recent increases in temperature and drier conditions, devastating bush fires have been ravaging throughout the continent, causing humans to flee their homes, taking lives and destroying animal habitats. However, the Aboriginal peoples of the region have correctly been assigning blame to a third culprit for years — a clever predator taking an active role in instigating the blaze.*

Find common ground with the reader

An excellent way to begin a story is by finding a relatable scene, dialogue, or scenario that can bring the reader a personal connection to the topic (Meadows, 2015a). This was the approach used to begin this chapter, by setting the relatable scene of being asked what you study and being unable to explain it to someone outside your field. If your species of choice for this exercise is a domesticated animal, perhaps consider appealing to pet owners for your lede. Ideally find a way to show rather than tell the readers about an experience.

Bullet lede

Another option would be to utilize the bullet lede preceding a summary of the study. This would entail listing three bullet points in succession that leave the reader curious about how each point or behavior is related and finishing by answering that question (Pechenik, 2007). Pechenik (2007) provides the following adept example regarding a study on octopi learning through observation of a conspecific:

> *We all know people who have trained their dogs to fetch the daily newspaper without tearing it. Similarly, we all know that horses can be trained to respond to the slightest movement of their riders. And we all know that goldfish can be trained to come to the front of the fish bowl at the sound of a bell. Now it turns out that even octopi (*Octopus vulgarus*) can be trained to perform certain simple tasks, and that they actually learn those tasks more quickly from each other than from a human trainer. (p. 253).*

Narrative lede

A lede that can stand alone as its own story can be particularly effective at engaging the reader, for example, with an intriguing personal experience (Meadows, 2015a; Pechenik, 2007). While this is an easier approach for researchers writing about their own work, you may still come up with a relevant story of your own that relates to the researcher's findings in some way as an uninvolved party. When considering the story about firehawks, perhaps one could narrate a story about walking onto the devastating scene of a forest decimated by wildfire. Describe firsthand the visible impact on wildlife and vegetation, finishing with a transition into what agent initiated the fire.

Surprise or paradox lede

A paradoxical statement can be used as your leading sentence by explaining a readily held belief about your selected species or a general animal behavior, followed by a sentence explaining that a new discovery contradicts prior convictions and then by a summary of the study (Pechenik, 2007). Grimm (2019) uses a similar tactic to describe a study showing new genetic evidence suggesting that early humans domesticated themselves:

> *Humans get a lot of credit for domesticating cats, dogs, and other animals. But the most important creature we may have tamed is ourselves. New genetic evidence suggests that our temperaments—and even our faces—could be the result of kicking bullies out of our early societies, making us a friendlier and more cooperative bunch.*

Body of the work

There are some basic principles for writing popular science that differ significantly from writing for a professional scientific audience. First, you generally should use much shorter paragraphs when writing for a general audience than for an academic audience. Additionally, as many readers will scan through your story rather than read it to completion, it is essential to use careful transitions to guide the reader from one paragraph to another (Burke, 2015). As you finish one section, you should lead the reader into the topic of the following paragraph, with each new paragraph essentially having its own topic sentence. Use an active voice whenever possible. For example, rather than stating "the fire was started by the hawks" write "the hawks started the fire".

When writing for other scientists in journal publications, detailed descriptions are incredibly important to demonstrate to other experts in the field that your work was thorough and accurate, that the study was well-developed, that it used appropriate methods, and that it needed to be done. Essentially, you put a lot of work into this, often months or even years on a single study, and you want some credit! However, the two worst enemies of good popular science writing are too much detail and the dreaded jargon monster. The heavy use of complex terminology or too much detail can quickly turn readers away because they are prone to lose interest or get confused. Thus you must simplify whenever possible. It is important not to approach this task as "dumbing it down", as the intelligence of your readers is not in question. Rather they may lack the background knowledge in a particular area. Ask yourself, is this term or complex procedure essential to understand the study? If it does not impact the main message of the story, then it is likely better to leave it out. This idea is comparable to the process of rounding mathematical numbers to just one or two decimal places for simplicity or trimming away portions of a phylogeny irrelevant to your study. While it may be slightly less accurate because it doesn't include every detail, it provides sufficient information at a level that won't confuse the reader.

There are of course instances in which specialized terms simply must be included so as to not misrepresent the science. In this case, be sure to thoroughly explain each term that is not part of everyday language for the reader. The most effective way to accomplish this is through the use of analogies, metaphors, imagery, and vivid language. If you can identify an image or analogy that can represent the unfamiliar concept or term, ideally something very familiar and relatable to the reader, this will be far more digestible than a wordy explanation (Ehrenstein, 2010). For example, to tackle the process of limb regeneration on a detailed scale in axolotls, a type of salamander, Dance (2020) begins with the concept of a symphony, which she carries throughout the article:

> *In a working limb, cells and tissues are like the instruments in an orchestra: Each contributes to actions like musical notes, to create a symphony. Amputation results in cacophony, but salamanders can rap the conductor's baton and reset the remaining tissue back to order – and all the way back to the symphony's first movement, when they first grew a limb in the embryo.*

Remember, people regularly elect to watch documentaries and TV series that center around the topic of animal behavior. It is harder to entertain them via a written format, especially when so much other content is just a click away. So compete with other compelling visuals by painting them a picture! Visual imagery can not only allow a story to come alive while requiring less background knowledge for the reader but also simplify the story (Ehrenstein, 2010). This is not always easy without a deep understanding of the scientific concept, so make sure you research thoroughly whenever an idea is unclear or unfamiliar. Although metaphors and mental images can enhance the reader's experience, it is essential that they are not used too fancifully at the expense of scientific accuracy.

Kickers

Although an engaging lede of a science news story is of paramount importance in roping in your reader, your audience won't feel entirely fulfilled by the end of your story without a good "kicker" (ending). Again, there are a few common methods that could help you find a good ending to your story. You may simply provide a summary of the results or a quote from an expert on the topic. Alternatively, you could try one of the following tactics, which can be more difficult, but extremely rewarding.

Circle back to the beginning

Depending on the type of lede chosen to begin your story, you can return back to the opening scene of your article and recount the main ideas in your story or leave them with future directions to think about, given the new insight your story provided (Meadows, 2015b). If you used a mininarrative to begin your article, then consider circling back to that story to give the audience some closure. If you decided to begin the story like a fairy tale, you could continue the theme throughout and tie everything together with the ending. You may be even able to identify the ending as you compose the lede to allow for a cohesive story that is connected in a crucial but unexpected way. If your topic has relevance to some environmental concern, perhaps try adding a call to action to leave the readers engaged after they have finished reading.

Make it personal

If your reader makes it to the end of your story, this means that you have successfully engaged the reader's interest. To bring the readers all the way home try to finish by engaging their emotions with a relatable statement or story. If you did not use a narrative lead consider ending your story with a stand-alone narrative depicting a personal experience. Even if the reader hasn't experienced something similar, describing it through your eyes and getting a sense of what you felt in the moment can still rope them in as they put themselves in your shoes. Take, for example, this excellent use of imagery and sensory detail that Ring (2013) used to conclude his report on salmon run research based on his personal experience swimming with salmon in the river:

> *As they undulated to keep themselves in formation, some brushed against me with their bodies, and some swiped their tails against exposed portions of my face. A few even wriggled under me, one by one forcing their way between my chest and the sandy bottom.*
>
> *Shifting slightly upstream, I turned to look at them head-on. Dozens faced me, just a few inches away, their jagged teeth exposed through gaping, elongated jaws—another physical change that occurs as they spawn, as if they're reverting to a completely primitive form that matches the landscape. Their gills flexed water in and out, extracting oxygen, and their eyes, eerie golden circles, gazed at me implacably, as if nothing else mattered except their instinct to spawn. I reached out, touched one, and then another, and another.*

Results/discussion

Peer review 1

After your news story is written, the next phase of the assignment is to exchange stories with a classmate. Read through their paper, marking where things are unclear and where your classmate has excelled. Fix any grammatical or spelling mistakes that you identify. Follow the rubric provided to score your classmate on the key elements of a good science story. Discuss your feedback with your partner. As you both have similar background knowledge in animal behavior and college-level science, but may be unfamiliar with the primary literature used by your partner, this is an excellent first step to locate areas where the story is overly detailed or lacking connection with the original study. Using your peer's feedback make a new draft of your story.

Particularly when starting out, the first draft on a task like this often ends up including far too much detail and more jargon than you realize. This is where the peer review can assist you. The second draft may err on the side of oversimplification, cutting out too much and compromising the important elements of the study. Take a moment to think about the challenges you faced in translating a detailed complex study into an easily digestible story.

Peer review 2 (optional)

Although your classmates can provide useful feedback regarding the overall quality and content of your science story, they do not represent a member of the general audience that you are writing your story for. To evaluate whether concepts are explained in a digestible way to a lay audience, whether you used engaging techniques to pique the reader's attention, and whether you found a balance between an overly detailed and oversimplified recount of the study, a review by a member of the general public is required. This could include any family member, friend, coworker, etc. who is not in a college-level biology program and thus lacks the background knowledge you and your classmates possess. Again, have them fill out the rubric, scoring you on how well they understood the science and noting any unfamiliar words or concepts. Complete any modifications required in a new draft and submit the final copy to your teacher.

Discussion questions during peer review

Although the style of writing for science stories is important, your main goal is still to evaluate whether you have appropriately captured the essence of the original study and its importance. Why was the study important? Why was this work done? What are the key new facts or ideas from the article? This should be very clearly stated somewhere in your article. Did you find yourself oversimplifying or "dumbing it down"? Did you need to do additional research on your topic beyond reading your primary article in order to complete the assignment?

Next comes digestibility. Does your story contain any unexplained jargon? Was it too difficult to identify and remove highly specialized terms? What is the basic knowledge in this area of science, and what background do your readers lack? Were your peers able to understand the concepts on the first try?

Why should the readers care, and is it relevant to any aspect of their lives? It is not always immediately clear how an animal behavior study could relate to your readers' lives. Many of us study behavior simply for the sake of learning something new about a species, and that is a sufficient reason to spend time doing so, but your readers will be more engaged if you can establish a personal connection.

Finally, discuss any difficulties you faced while trying to write about a complex topic in a compelling way. Was it difficult to come up with ledes and kickers for your story? Was your title engaging without being overly complex? Did your peers seem to find the story interesting? What did you think of the story written by your peers? Was it objectively interesting for a reader outside your discipline?

Acknowledgments

We would like to thank Sam Cooper for his helpful contributions in developing this chapter.

References

Aron, J. (February 13, 2015). Interstellar's true black hole too confusing. *New Scientist*. Retrieved from https://www.newscientist.com/article/dn26966-interstellars-true-black-hole-too- confusing/#.VVjT0_lVhBc.

Brownell, S. E., Price, J. V., & Steinman, L. (2013). Science communication to the general public: Why we need to teach undergraduate and graduate students this skill as part of their formal science training. *Journal of Undergraduate Neuroscience Education, 12*(1), E6–E10.

Burke, K. L. (July 31, 2015). Retrieved from https://www.americanscientist.org/blog/from-the- staff/12-tips-for-scientists-writing-for-the-general-public.

Caruso, C. (October 16, 2016). Chimps may be capable of comprehending the minds of others. *Scientific American*. Retrieved from https://www.scientificamerican.com/article/chimps-may-be-capableof-comprehending-the-minds-of-others/.

Dance, A. (January 31, 2020). What the axolotl's limb-regenerating capabilities have to teach us. *Discover Magazine*. https://doi.org/10.1146/knowable-012920-1.

Ehrenstein, D. (2010). Picture power. In C. Johnsen (Ed.), *Taking science to the people: A communication primer for scientists and engineers* (pp. 75–79). University of Nebraska Press. https://doi.org/10.2307/j.ctt1dfnrn5.13.

Greshko, M. (January 8, 2018). Why these birds carry flames in their beaks. *National Geographic*. Retrieved from https://www.nationalgeographic.com/news/2018/01/wildfires-birds-animals-australia/#close.

Grimm, D. (December 19, 2019). *Our favorite science news stories of 2019*. Retrieved from https://www.sciencemag.org/news/2019/12/our-favorite-science-news-stories-2019.

Meadows, R. (July 14, 2015a). *Good bbeginnings: How to write a lede your editor-and your readers-will love*. Retrieved January 8, 2020, from https://www.theopennotebook.com/2015/07/14/good-beginnings/.

Meadows, R. (November 24, 2015b). *Good endings: How to write a kicker your editor-and your readers-will love*. Retrieved January 8, 2020, from https://www.theopennotebook.com/2015/11/24/good-endings-how-to-write-a-kicker-your-editor-and-your-readers-will-love/.

Pechenik, J. A. (2007). Writing for a general audience: Science journalism. In *A short guide to writing about biology* (6th ed., pp. 251–258). New York: Pearson/Longman.

Premack, D., & Woodruff, G. (1978). Does the chimpanzee have a theory of mind? *Behavioral and Brain Sciences, 1*(4), 515–526. https://doi.org/10.1017/s0140525x00076512.

Rakedzon, T., Segev, E., Chapnik, N., Yosef, R., & Baram-Tsabari, A. (2017). Automatic jargon identifier for scientists engaging with the public and science communication educators. *PloS One, 12*(8). https://doi.org/10.1371/journal.pone.0181742.

Ring, R. (November 25, 2013). Hard lessons from the mighty salmon runs of Bristol Bay. *High Country News.*

Smith, R. (October 21, 2013). Hot flashes in cold waters? Killer whales undergo menopause. *National Geographic.* Retrieved from https://www.nationalgeographic.com/news/2013/10/131019-killer-whales-menopause-hot-flash-marine-mammals/#close.

Smulders, T. (December 23, 2019). *By studying animal behaviour we gain an insight into our own.* Retrieved January 8, 2020, from https://theconversation.com/by-studying-animal-behaviour-we-gain-an-insight-into-our-own-20001.

Zivkovic, B. (July 13, 2011). Retrieved from https://blogs.scientificamerican.com/a-blog-around- the-clock/httpblogsscientificamericancoma-blog-around-the-clock20110713telling-science-stories-wait-whats-a-story/.

Part II. Instructor notes

Classroom management/blocks of analysis

The time required for the activity as written is at minimum two 1-h class periods, with much of the work taking place outside the classroom. In the first session, the students learn about the project and identify their article. After the first draft is written, time should be allotted in the next class period for the peer review process and/or discussion with a partner. See later discussion for alternative modifications.

Teaching the activity

Should you prefer uniformity or less variation in article choice, you can provide one or multiple primary articles to choose from. However, the benefit of having them choose their own is they are more likely to be excited about the topic. Also, their peer will be unfamiliar with the original study, allowing them to objectively provide feedback where they do not understand concepts.

Modifications to the activity

To shorten the activity, you can modify the assignment by having only one or none of the peer-reviewing steps. However, to remove both is inadvisable, as the feedback is very important for helping the students find the balance between jargon and oversimplification. Alternatively, one of the activities could be done outside class (e.g., the article selection and writing or the peer review).

Another modification to shorten the activity would be to have the students select a story that's already in a science news/magazine article format rather than primary literature. Instead of translating an original study into appropriate writing for the general public, they can write up a summary or discuss with a partner, identifying the following similar ideas:

1. What makes this science news?
2. Why should readers care?
3. What was the most important finding of the study?

4. Does the story describe the methods used by the researchers?
5. What aspects of the article caught your attention when searching for stories?
6. If they choose to compare the news story to the abstract of the original article, then ask whether the news story accurately and appropriately covered the research.
7. What type of lede did the author use and was it effective?
8. How was the technical complexity handled—through simplification, analogies, or defining terms?

In-class preparation

As mentioned previously, the most difficult part of this task will likely be finding the balance between too much detail and/or jargon and oversimplification. If reducing jargon is particularly an obstacle for them, one excellent resource to use is the *De-Jargonizer* (http://scienceandpublic.com/), where students can directly input their written story into a free scientist-friendly online program that will assess their word choice instantly. This software has been thoroughly tested and found to effectively detect problematic vocabulary by color-coding the student's writing based on how commonly each word is used in our language (Rakedzon et al., 2017). The software will flag "jargon" (i.e., rarely used) in red and midfrequency words in orange for reevaluation, providing an overall score for how suitable the writing is for a general audience. Where these words are flagged, students can assess whether the terms are essential for understanding or can be removed or replaced with a more common equivalent or suitable analogy/description.

Recommendations for extensions or continuations for more advanced classes

If you have an advanced class, you can extend this assignment by adding the additional component of requiring the students to conduct an interview with a scientist involved in the study they chose (e.g., a principle investigator, graduate student, postdoc, etc.) and to incorporate their viewpoints and quotes into the news story. This of course would be most feasible when choosing a study conducted by individuals at your university to whom they would have ready access. However, doing so would limit the range of studies students are able to select. Alternatively, rather than the students writing out their news story, they can follow a similar format and deliver their story as an oral presentation in the form of an elevator pitch or a story pitch. Both written and oral communication appropriate for nonscientists are essential skills to acquire.

Samples of results

In the online supplementary materials, we have provided a citation to an example of an original journal article from a study on yellow-legged gull embryo communication, along with a citation for a published well-written news story about the research. For contrast, we have supplemented this with a bogus example that is very ill-suited for a general audience. The two pieces can be used for the students' reference or may be turned into a classroom activity where students work in pairs to identify areas of each story that were well done and/or poorly written, given the goal of communicating with nonscientists.

Part III. Supplementary material

Supplementary data related to this chapter can be found online at https://doi.org/10.1016/B978-0-12-821410-7.00004-2.

CHAPTER

Effective scientific writing 24

Megan Murphy

Department of Biology, Indiana University Bloomington, Bloomington, IN, United States

Chapter outline

Exploring Animal Behavior in Laboratory and Field. https://doi.org/10.1016/B978-0-12-821410-7.00012-1

Part I. Student instructions

Learning goals, objectives, and key concepts

- To examine standards of scientific writing in animal behavior.
- To develop scientific writing skills.

Background

Scientific writing is a crucial part of the scientific process, as it facilitates the sharing of information. No matter how novel or exciting your results are, they will have no impact unless you can communicate them effectively. Scientific writing is used in not only class assignments but also scientific publications and grant proposals.

A scientific writing style typically follows specific norms and conventions, including the format, organization, voice, and tense. While there are some aspects of a scientific writing style that are consistent, others vary by the type of writing. Some scientific articles are written for a relatively broad audience and others are much more specialized. Journals such as *Science* and *Nature* are written for a relatively broad audience, so their papers tend to be shorter, use less jargon, and often follow different layouts than specialist journals. Most journals have published Guidelines for Authors, which highlight the types of papers that they accept and any specific conventions that their articles follow. Similarly, grant proposals often follow very specific layouts and writing styles, as dictated by the funding agency.

Learning to write scientifically requires substantial practice and attention. One of the best ways to learn how to write in this style is to read publications by a variety of authors and in a variety of disciplines. Careful practice through writing exercises, revisions, and peer review are also needed to develop your skills. No writer is ever done learning to write—we can all improve our skills through continued work.

Purpose

This series of activities is designed to familiarize you with the norms of scientific writing and provide you practice with this style. These activities will require you to examine other people's writing, to write your own, and to discuss with classmates.

Methods

Basic principles of scientific writing

There are some aspects of scientific writing that vary between types of publications, but there are also some common rules. This first activity reviews some of those rules and provides you an opportunity to practice applying them in writing.

- Scientific writing must be rooted in the literature to provide credibility for your ideas. As much as possible, your references should be published, peer-reviewed works. When presenting information from another source, rather than quoting directly, you should always paraphrase the relevant ideas and use in-text citations (Peat et al., 2002). All sources that are cited in your text should also be present in your references list and vice versa.

In most cases, your statements should focus on the information that you're presenting, rather than on the authors (Peat et al., 2002). For example, the statement "Murphy (2020) highlights that it's important to paraphrase information from your sources" focuses on the fact that I said it, whereas the statement "It's important to paraphrase information from your sources (Murphy, 2020)" focuses on the information.

- Whenever possible use active voice. Active voice presents a clear actor and their action, whereas passive voice conceals the actor and often weakens the story structure (Schimel, 2012). In some cases, passive voice sounds more natural; in these cases, it is appropriate to use passive voice. The following examples highlight the differences between passive voice and active voice. The second example also demonstrates a situation where passive voice may sound more natural than active voice.
    - **Active**: Crickets call to attract mates. **Passive**: Calls are produced in order to attract mates.
    - **Active**: We scored individuals' behaviors for 5 min. **Passive**: Behaviors were scored for 5 min.
- Avoid unnecessary jargon. Your target audience determines what level of jargon is appropriate in your writing. If you are writing for a broad audience, you should avoid any technical terms or, if they are necessary to include, define them at your first usage. If you're writing primarily for specialists in your field, you can assume that they know more technical terms and use them accordingly. However, in all cases, you should use the simplest terms that accurately convey your ideas (Schimel, 2012).

Similarly, use acronyms judiciously. Acronyms are useful for abbreviating often-used, unwieldy terms but, when you use too many, they can make it challenging for your reader to follow your ideas. Always introduce the acronym by writing out the full term, then showing the acronym in parentheses; the one exception to this rule is if the acronym is more commonly used than the full term, such as DNA (Schimel, 2012).

Rewrite the following paragraph to better follow the abovementioned principles. Your goal is to convey the same ideas but to write more effectively.

Many females preferentially mate with males expressing energetically expensive signals (EES). While sexual selection selects for these EESs, natural selection is often thought to select against them because of their energetic costs. Legett et al. (2019) noted that "Females, however, are not the only receivers driving signal evolution. Mating signals also expose signalers to natural enemies, such as eavesdropping predators and parasites." The costs imposed by these predatory eavesdroppers may lead to antagonistic selection and the evolution of some otherwise non-adaptive traits. In Teleogryllus oceanicus *crickets, parasitism from an eavesdropping parasitoid is thought to have led to the loss of auditory sexual signals entirely (Zuk et al., 2006). However, in* Gryllus lineaticeps *crickets, parasitism from the same eavesdropping parasitoid led to them developing riskier signals that were more attractive both to female conspecifics and to the predator (Beckers and Wagner, 2018). These examples demonstrate that competing selective pressures can often have unpredictable results and drive the evolution of EESs.*

Pair up with a classmate and compare your paragraphs. First determine whether you identified the same issues with the abovementioned paragraph. Then compare how each of you rewrote it to make it more effective. Work together to produce a single rewrite using the parts that were successful in each of your paragraphs.

Concise writing

A scientific writing style is, in part, defined by clear and concise writing. Your goal is to state your ideas precisely, that is, to say what you mean, while also keeping your ideas clear and avoiding misinterpretations (Alley, 2018). To keep your ideas clear, you should avoid any unnecessary complexity in your word choice, your sentence structure, and the overall content of your paper (Alley, 2018). This exercise encourages you to focus on how to convey your ideas effectively and remove any unnecessary complexity.

1. Begin by reading the paper that your instructor assigns. Take notes on the authors' ideas and the major points that they are making.
2. For the focal paper write an abstract that is 200–250 words long. Your abstract should provide an overview of the paper and should highlight the research question, the significance or context of the work, an overview of the methods, and the findings and their significance. Note that in-text citations are not used in abstracts.
3. Exchange abstracts with your lab partner, either on paper or through Google Docs (as assigned). As you review their abstracts underline any ideas that are unclear or unnecessarily complex. Put a star next to any ideas that they included in their abstracts but you excluded from yours.
4. Work with your partner to rewrite a single abstract. To do so, you should discuss where you differed in the content that you included and determine what ideas are crucial to present the focal work. As appropriate, you should include pieces from each of your abstracts, while also working together to revise any sections necessary.

Reverse outlining

This activity is designed to help you identify the typical organization of ideas in the Introduction and Discussion sections of a scientific paper and to identify when/why they are particularly effective. First, you will independently "reverse-outline" a paper. Then you will work in groups to identify common themes and structures.

1. Begin by selecting a primary literature paper focused on animal behavior, or use one assigned by your instructor. You'll be reverse-outlining the Introduction and Discussion sections. On a separate piece of paper write a five to eight-word description of the major idea or topic of each paragraph, followed by a five to eight-word description of how that paragraph advances the ideas and arguments in the paper.
2. Work together in a group or as a class to answer the following questions and consider common themes in how scientific papers are organized.
   a. How do authors typically begin their Introduction? Is the research question immediately clear or explicitly stated?
   b. What background information do authors present first?
   c. At what point in the introduction do authors describe their study system or focal species?
   d. How do authors typically transition from their Introduction to their Methods section?
   e. How do authors typically begin their Discussion?
   f. Are there any parallels between how authors structure their Introduction and their Discussion?
   g. How do authors typically end their Discussion? What types of "take-home points" do they leave the readers with?

References

Alley, M. (2018). *The craft of scientific writing*. New York, NY: Springer.

Beckers, O. M., & Wagner, W. E. (2018). Males and females evolve riskier traits in populations with eavesdropping parasitoids. *Behavioral Ecology and Sociobiology, 72*, 174–184.

Legett, H. D., Page, R. A., & Bernal, X. (2019). Synchronized mating signals in a communication network: The challenge of avoiding predators while attracting mates. *Proceedings of the Royal Society B, 286*.

Peat, J., Elliott, E., Baur, L., & Keena, V. (2002). *Scientific writing: Easy when you know how*. London: BMJ Books.

Schimel, J. (2012). *Writing science: How to write papers that get cited and proposals that get funded*. New York, NY: Oxford University Press.

Zuk, M., Rotenberry, J. T., & Tinghitella, R. M. (2006). Silent night: Adaptive disappearance of a sexual signal in a parasitized population of field crickets. *Biology Letters, 2*, 521–524.

Part II. Instructor notes

These activities can be completed in a single class period, with the first steps of each activity completed outside class. However, it may be preferable to complete the activities throughout the semester and use them to supplement other lab activities. Most of these activities, particularly *Basic principles of scientific writing* and *Concise writing*, could easily be completed asynchronously online; *Reverse outlining* can be completed online with a synchronous discussion.

Preclass preparation

1. Before class, you should choose relevant papers for the students to analyze. As you're encouraging students to focus on the style and organization, rather than the content, these can be papers that they've read before in the class. Crucially, the content should be simple and/or familiar enough that students are able to follow the ideas and identify main themes in the paper.
2. *Concise writing*: For the activity, you should select a focal paper and provide students with a copy of the text WITHOUT the original abstract included.
3. *Reverse outlining*: This activity is most effective if students focus on at least four to five different papers, which will allow them to identify common themes across different types of papers.

In-class preparation

For many students, the most challenging part of scientific writing is the focus on clarity and conciseness. Beginners tend to mimic scientific writing by including jargon, even when they may not understand the terms, and relying on complex sentence structures.

You should watch for this in the *Concise writing* activity and encourage multiple rounds of revisions as necessary. For students who are really struggling, it can also be effective to have them verbally explain their ideas to their partners prior to putting them in writing. This practice highlights the direct approach that they use when talking about complex ideas and how that same direct approach can be useful in communicating complex ideas in writing.

Answer key

There are many approaches to rewrite the provided paragraph in *Basic principles of scientific writing*, but you should ensure that they've addressed the issues underlined and explained in the following. There are also many opportunities to improve conciseness and clarity.

> *Many females preferentially mate with males expressing energetically expensive signals (EES[1]). While sexual selection selects for these EESs, natural selection is often thought to[2] select against them because of their energetic costs. Legett et al. (2019) noted that "Females, however, are not the only receivers driving signal evolution. Mating signals also expose signalers to natural enemies, such as eaves-dropping predators and parasites."[3] The costs imposed by these predatory eaves-droppers may lead to antagonistic selection and the evolution of some otherwise non-adaptive traits. In* Teleogryllus oceanicus *crickets, parasitism from an eaves-dropping parasitoid is thought to have led[4] to the loss of auditory sexual signals entirely (Zuk et al., 2006). However, in* Gryllus lineaticeps *crickets, parasitism from the same eavesdropping parasitoid led to them developing riskier signals that were more attractive to both female conspecifics and to the predator (Beckers and Wagner, 2018). These examples demonstrate that competing selective pressures can often have unpredictable results and drive the evolution of EESs.*

For discussion questions in *reverse outlining*: These answers will differ based on the papers that students were reading, but the following answers can serve as a guide.

a. How do authors typically begin their Introduction? Is the research question immediately clear or explicitly stated?
Authors typically begin their Introduction by highlighting an open question in the field. While they generally don't include their specific research question at the beginning of the Introduction, they do highlight their broad research area.

b. What background information do authors present first?
Similarly, researchers generally begin by providing background information on the broader theme that they're studying.

c. At what point in the introduction do authors describe their study system or focal species?
In many cases, the authors don't introduce their study system until about halfway through their Introduction. After this point, they provide more specific information about their study system and why it's appropriate for addressing their broader question.

d. How do authors typically transition from their Introduction to their Methods section?

---

[1]Unnecessary abbreviation.
[2]Awkward use of passive voice.
[3]Inappropriate use of direct quote.
[4]Awkward use of passive voice.

The Introduction section typically ends with a statement of the researchers' question, their hypotheses, and/or their predictions. The authors may also include a broad overview of their methods.

e. How do authors typically begin their Discussion?
Authors typically begin their discussions by restating their research questions and highlighting their most important results.

f. Are there any parallels between how authors structure their Introduction and their Discussion?
While the Introduction often "funnels" ideas from a broad research question to a specific research system and method to test the question, discussions often reverse this funnel. Here, you frequently see authors discuss the significance and possible explanations for specific findings, then applying those ideas to the broader research area, and explaining how the ideas could be applied in other systems.

g. How do authors typically end their Discussion? What types of "take-home points" do they leave readers with?
Authors often end their discussions with a statement of their overall findings and how they apply to a broader research area; they may also include possible future research directions.

CHAPTER

25 Writing and reviewing grant proposals

Andrea M.-K. Bierema

Center for Integrative Studies in General Science and Department of Integrative Biology, Michigan State University, East Lansing, MI, United States

Chapter outline

Exploring Animal Behavior in Laboratory and Field. https://doi.org/10.1016/B978-0-12-821410-7.00002-9

Part I. Student instructions

Learning goals, objectives, and key concepts

- Based on observed traits develop hypotheses for how behavior influenced the selection of that trait resulting in adaptive evolution.
- Predict the niche of a species based on habitat and characteristics.
- Develop a literature review.
- Create a protocol and predictions to explore an animal behavior research question.
- Develop an animal behavior grant proposal.
- Critique animal behavior grant proposals written by one's peers.

Background

Writing research grant proposals is often a required task for researchers, and it is never too early to learn how to properly write one. Working on grant proposals allows students to practice creating research questions based on initial observations, researching the literature for related studies and possible methodology, writing methods, and making predictions.

The first task of writing a grant proposal is developing a research question that is testable, that is, it is specific enough to investigate (Ardehali, 2015; Proctor et al., 2012). For instance, the question "why do cowbirds lay eggs in some nests but not others?" is interesting, but what is being measured and which variable(s) is of interest? Rather, the question "Does egg size influence female cowbird nest selection?" uses egg size as the variable of interest.

Although the research question is placed at the end of the proposal's introduction, it is written first because the entire proposal relates to the question (Proctor et al., 2012). The introduction starts with a broad overview of the ultimate goals and the topic and narrows down to the specific research question (Gholipour et al., 2014). Additionally, it refers to the literature to explain what scientists know—or, more importantly, do not know (Seeman, 2015)—about the topic, how this research will fill the literature gap, and why this research is important (Wisdom et al., 2015). The importance of the research is more than just that there is a literature gap; how will stakeholders benefit from the completion of this research (Proctor et al., 2012)? Its significance is often either written in the introduction or included toward the end of the grant proposal.

The methods follow the introduction and research question. They are as detailed and concise as possible and consider potential complications and proposed changes that will be done to overcome those complications (Proctor et al., 2012). Each aspect of the methods should be justified (Seeman, 2015) and should align with the research question (Wisdom et al., 2015). The methods should also fall back to the literature (Drabble et al., 2014): why choose these methods? Has previous research used these methods? Why investigate this species? Related to the methods, possible conclusions and the meaning and significance of those conclusions are often included. Finally, create a budget that considers all needed materials.

To make the entire writing process easier, become familiar with the literature before starting the proposal (Ardehali, 2014). This helps in understanding why the research is important and which methods are common for that type of research question. Know the topic well enough so that it can be explained without much jargon. During the writing process check the grant application instructions to know what needs to be included. Adding additional information or a lot of extra words to increase the length often makes the proposal harder to follow rather than more informative (Ardehali, 2014).

Purpose

In this activity, you will produce a research grant proposal, using a real application form. There are several activities and milestones that will be accomplished to aid in your preparation, including (1) developing observation-inspired research questions, (2) writing a comprehensive literature review, (3) considering how class concepts apply to the research questions, (4) receiving feedback on an abstract, (5) critiquing peers' proposals, and (6) journaling about the process along the way.

Methods

Species selection

To start this project, you will make preliminary observations. Your instructor will explain how these observations are to be made. This may be during or outside class time and can consist of observations outside the classroom, at a park, zoo, aquarium, trails, or via videos. You should observe a variety of species and behaviors before deciding on which to focus your project.

Materials needed

You need access to peer-reviewed scientific articles to develop the literature review. There are no other materials required, but you may choose to use binoculars for outdoor observations or a dissecting microscope for observing micro- or macroinvertebrates.

Overview of activities and assignments

Choose a species and behavior

The first task is to choose a species and behavior. These observations can be made either outdoors or via videos and should include several species. During this process determine which ones appear particularly interesting. After choosing the species and behavior take a photo, if possible, and write a reflective essay that addresses the following:

- Describe and/or sketch the organism. It might also be helpful to take a picture of the organism for the project.
- Describe the environment where the organism was observed.
- Describe the trait of interest.
- Speculate about the function of the trait.

Journal assignments

A journal is a written reflection, and journaling is the process of reflecting. We will have journal assignments due several times as part of this project. The purpose of these journal assignments is twofold. First, each journal assignment is an application of the concepts learned during the course. Second, each one helps you reflect and apply new ideas to your research question, methods, and predictions.

These journal assignments are graded based on completeness (i.e., all questions are addressed), effort, and evidence of critical thinking. Because we are often making predictions, responses may or may not be correct, and for many of these assignments, the "right answer" may not be available because they have not been studied. Therefore do not worry about having the "right answer" for the journal assignments. Instead use the knowledge and skills that you obtained so far to make testable predictions that are biologically plausible.

Each journal assignment is about two paragraphs and will be completed at the end of the class. Instructions will be given in class. Make sure to include your name, your species, and the behavior on each assignment.

Milestone assignments

Milestone assignments are homework assignments that help you prepare for the final proposal. They are graded based on completeness (i.e., all questions are answered), accuracy of content, and grammar and spelling. The instructions and the grading rubric will be provided by your instructor.

In-class peer review assignments

You will review each other's proposals in a peer review process. Each proposal should be reviewed at least twice. Proposals are expected to be a final product. Rough drafts are not permitted because they make the review process more difficult for peers and require more work to be done shortly before the proposal is due. A list of questions will be provided to aid in the review process.

Student research grant guidelines (final product guidelines)

These guidelines are modified from the Animal Behavior Society (animalbehaviorsociety.org). Please note that these have been modified slightly for the course project, including some rewording and removal of some of the requirements. Animal Behavior Society grant proposals also require a formal budget, a curriculum vitae, and completion of the animal care questionnaire, which we are not doing for this project.

Preparing the grant

The research application must consist of the following components and all these pages must conform to the following formatting: single-spaced; Arial 11 pt. font; 1″ (2.5 cm) margins on the top, bottom, and two sides; and pages numbered successively from 1–5 at the bottom center. Each application must include the sections listed in the following. Each of these sections should have the section heading on a separate line, followed by the text for the section. Proposals that do not adhere to these specifications will not be reviewed. Please see additional details concerning each section below.

- Title
- Summary of Proposed Work
- Introduction
- Hypothesis(es) to be Tested
- Proposed Research
- Predicted Outcomes
- Importance of Proposed Research
- Bibliography
- Budget and Justification (one page max)

Content guide

Note that the following sections are to be on the first three pages: Title, Summary of Proposed Work, Introduction, Hypothesis(es) to be Tested, Proposed Research, Predicted Outcomes, and Importance of Proposed Research.

Title. Include a descriptive title with a maximum of eight words.

Summary of Proposed Work. This section should consist of a single paragraph providing an overview of the proposed work, a statement about the question to be addressed, a statement about the conceptual significance of the proposed work, and a statement about the intellectual merit and originality of your proposed study (up to 300 words).

Introduction. In this section provide the relevant background information pertaining to your proposed project, including both classic papers in the area of study and more recent work. Also provide background information about the organismal system to be used (note that we did this for the first milestone assignment) and why the system is appropriate for the question(s) to be addressed. Detail how and why the proposed work is important to the particular field of study, as well as the broader significance of the work. *Use at least four peer-reviewed resources* (this section should be about ¾ to 1 page).

Question(s) to be Addressed. In this section list the specific research question(s).

Proposed Research. In this section describe the methodology to be used to address the proposed hypothesis(es) and how the data will be statistically analyzed. You should provide sufficient detail for reviewers to understand the methods and analyses. Also provide any relevant information about your experience for the proposed research. Finally provide a timetable for the proposed work (about ½ to 1 page).

Predicted Outcomes. In this section detail in one to two paragraphs how alternative results will be interpreted with respect to your primary hypothesis(es) (use if/then statements).

Importance of Proposed Research. Describe in one to two paragraphs the conclusions that can be drawn from the study and their relevance to the field of animal behavior.

References Cited. On page 4 provide full reference information for all citations that appear in the body of the proposal. The citations you choose should cover the previous work on the topic and place your study in the broader context of the field. References are required in the introduction. If the method was previously published, then it should also be cited. The importance of the proposed research will also likely incorporate references (potentially the same ones used in the introduction).

Budget and Justification. On page 5 summarize the necessary resources needed to complete the project. This should include (1) an itemized accounting of all needed resources and (2) a justification for needing these resources. For this project, you do not need actual monetary values of each item (just note that grant proposals almost always require that information).

Here are the additional budget notes from the Animal Behavior Society that you may want to consider: research grants may NOT be used to provide salary supplements, to pay indirect costs (overhead), or to defray tuition fees. In addition, any request for a single item of equipment costing more than $500 will require special justification (i.e., a description of why the item is essential to the success of the proposed research and a statement that it is not available locally). Requests for travel money should be based on ordinary out-of-pocket expenses directly related to the research. Requests for travel money must be based on actual travel expenses rather than governmental mileage and/or per diem rates.

Results/discussion

In the peer review process provide at least nine suggestions and a summary of the feedback. Suggestions do not have to be provided for every section and you may provide more suggestions for one section than for another section.

The purpose of this review assignment is to provide feedback so that you can improve your proposals. Therefore reviewers are graded based on the quality of the reviews. The instructor will not use these reviews to grade the final grant proposal. Written evaluations will be seen by your peers, so please make sure to point out proposal merits as well as weaknesses. Try to provide encouraging advice for your colleagues! Use "I" statements (just like therapy!). For example, "I'm confused when you do ... Maybe you can clarify ..." Feedback needs to be detailed and critical. If there are aspects that you think are well done, then explain why it is well done and how it helps in your understanding of the proposal. Make sure to also explain ways that the paper can be improved.

As a guide, especially for first-time reviewers, some questions are listed under each section. You may address these questions in your written evaluation, but there is no need, of course, to address every question in your evaluation, just focus on the ones that seem most pertinent. Also feel free to provide additional feedback that may not be captured in these questions. Note that there is not a section for grammar; feedback should focus on content, not grammar.

Summary

- Was an overview of the proposed work provided?
- Was a statement about the question(s) to be addressed given?
- Was a statement about the conceptual significance of the proposed work provided?

Introduction

- Was the relevant background literature (both classic papers and recent work) covered?
- Did the applicant provide sufficient detail about the organismal system?
- Did the applicant describe the importance of the study to a subdiscipline of behavior and the broader significance of the work?
- Did the applicant use at least four scholarly references?

Question(s) to be addressed

- Did the applicant clearly state the question(s) to be addressed?

Proposed research

- Did the applicant discuss why the organismal system is particularly appropriate for the question(s) addressed?
- Did the applicant discuss the intellectual merit and originality of the proposed study?
- Did the applicant provide input on the strengths of the study, as well as constructive suggestions for improvement?
- Is the proposed work feasible?

Predicted outcomes

- Did the applicant provide valid statements for how the results will be interpreted (using if/then statements)?

Importance of proposed research

- Did the applicant discuss the relevance of the research to the field of animal behavior?

Bibliography

- Is coverage of the relevant literature for the field of study complete?
- Did the applicant use consistent formatting (e.g., APA, MLA, or other style)?

Budget and justification

- Is the budget request reasonable and properly justified?

After you are done reviewing a proposal meet with the "applicant" to discuss your feedback. Also meet with the person who reviewed your proposal (this may or may not be the same person whose proposal you reviewed). Then repeat this process with another "applicant."

The second-to-last journal assignment is based on peer reviews. After a class discussion write a reflection about the peer review feedback. What are the major points that the reviewers provided? Write a summary of the feedback. In your summary explain how you feel about the feedback received and what actions you plan to take to improve your final proposal.

The last journal assignment is completed after the final proposal is submitted. Reflect on the process of writing a research proposal. Which aspects were easy and which were challenging? Which aspects occurred as expected and which were surprising? How did the assignments help or hinder your final proposal? Now that class is almost done, what do you plan on doing with your proposal or the skills that you learned along the way? Write a reflection addressing these questions.

References

Ardehali, H. (2014). How to write a successful grant application and research paper. *Circulation Research, 114*(8), 1231–1234. https://doi.org/10.1161/CIRCRESAHA.114.303695

Drabble, S. J., O'Cathain, A., Thomas, K. J., Rudolph, A., & Hewison, J. (2014). Describing qualitative research undertaken with randomized controlled trials in grant proposals: A documentary analysis. *BMC Medical Research Methodology, 14*, 24. https://doi.org/10.1186/1471-2288-14-24

Gholipour, A., Lee, E. Y., & Warfield, S. K. (2014). The anatomy and art of writing a successful grant application: A practical step-by-step approach. *Pediatric Radiology, 44*(12), 1512–1517. https://doi.org/10.1007/s00247-014-3051-8

Proctor, E. K., Powell, B. J., Baumann, A. A., Hamilton, A. M., & Santens, R. L. (2012). Writing implementation research grant proposals: Ten key ingredients. *Implementation Science, 7*, 96. https://doi.org/10.1186/1748-5908-7-96

Seeman, E. (2015). The ABC of writing a grant proposal. *Osteoporosis International, 26*, 1665–1666. https://doi.org/10.1007/s00198-015-3085-2

Wisdom, J. P., Riley, H., & Myers, N. (2015). Recommendations for writing successful grant proposals: An information synthesis. *Academic Medicine, 90*(12), 1720–1725. https://doi.org/10.1097/ACM.0000000000000811

Part II. Instructor notes

Classroom management/blocks of analysis

This project is designed to take place across an entire semester. The following are possible in-class activities.

1. *Introduction and initial observations.* The project can be explained in about 10 min. Students can do the initial observations during class time (such as half of a lab period) or outside class.
2. *Journal assignments.* These require about 20 min per class (the supplement was created for nine classes total, but this will vary with the course because they are dependent on the course's topics). These could also be done as homework assignments.
3. *Peer review.* These take at least 1 h (possibly an entire lab period). These could also be done online via a discussion board.

Teaching the activity

1. The instructor prepares a timeline for the project.
   a. Note that this project was developed as a semester-long project (see Table 25.1), but it can be changed to a unit capstone instead. If it is a unit capstone, then provide students with a list of topics and methods that are relevant to the unit rather than having students investigate the behavior and methodology on their own.
   b. Determine how many journal assignments to incorporate into the course and when they will be completed. Their topics will vary according to the topics of the course.
   c. Assign when the milestone assignments will be due. It is advised to use all three milestone assignments to help students prepare for the final proposal.
   d. To reduce class time used for this activity assign the following as homework: the initial observations, milestone assignments, and peer review activity and assignment.
2. The instructor also determines if tasks are to be completed on paper or computer. To reduce paper, students can use the school's online learning management system for writing and submitting journal assignments (such as using a journaling, blogging, or discussion board feature). The milestone assignments and final proposal can be submitted online rather than as hard copies. The peer review activities can take place in an online discussion board rather than on

Table 25.1 Objectives and instructions for each milestone assignment as part of a semester-long project.

Task	Objectives and instructions
#1	**Objective**: Predict the niche of a species based on habitat and characteristics. **Instructions**: Research your chosen species and write a brief essay (three to four paragraphs long) that • Names the species (scientific and common) and selected trait; • Describes the abiotic factors of the species' habitat; • Predicts, based on the habitat and published literature, examples (at least three) of how the species interacts with other species; • Explains how the habitat description compares to the habitat that you originally found the species in; and • Explains why you selected the species and behavior. Use at least three references, including at least one peer-reviewed reference. Try to focus the essay on abiotic and biotic factors that you predict are important for your selected trait. Doing so will help you better prepare for the next milestone assignment and final proposal.
#2	**Objective**: Develop a literature review. **Instructions**: For the first milestone assignment, we researched the habitat and potential species interactions of the selected species. The present assignment focuses on the selected behavior of the species. Review peer-reviewed literature to find previous studies that have examined a similar trait in various species. It might be helpful to first search using the genus name of the species and then the order, and so on. Write a brief essay (three paragraphs long) that synthesizes four papers into a comprehensive review. This means that each paper should not be summarized in a separate paragraph. Instead, what are some similarities of the methods and findings of these papers? What can be learned from one paper that was not addressed in another paper? It will be helpful to reflect on how the authors of these articles wrote their literature reviews in the introduction sections of their articles. The purpose of this assignment is for students to review and reflect on various ways that the trait can be studied. Therefore synthesize both methods and findings of the studies.
#3	**Objective**: Create a protocol and predictions to explore an animal behavior research question. **Instructions**: Write the Summary of Proposed Work. This is the first section of the proposal and is a 300-word overview of the proposed work. See the proposal instructions for more information. Note that summaries typically do not contain references, but the full proposal will have references.

printed paper (students post their proposals by creating a new thread and then students reply to each other with their feedback; change the settings so that students cannot see initial posts until they post their own).

3. For the first day of introducing the project, if students are making initial observations during class time, then have binoculars and possibly dissecting microscopes available for student use. They can also bring their phones to class, with the *iNaturalist* app downloaded to help in identifying the species of interest (https://www.inaturalist.org/).

4. This project has students develop a research grant proposal, but they are not actually performing the investigation, so the Institutional Animal Care and Use Committee (IACUC) approval is not needed. This activity could be modified to have students develop the proposal and then perform the investigation. If this is the case, then IACUC approval or a taxon restriction is necessary.

The following are lesson plans for the in-class portions of this project.

First day lesson plan

This course project is introduced to the class toward the beginning of the semester. Provide instructions for the course project either online or as a hard copy handout to pass out during class. Summarize the project and the activities to be done along the way to develop the proposal.

Next, students will make observations to choose a species and behavior. This can be done either during class or outside class. Encourage students to spend some time outdoors for inspiration. Alternatively provide links to videos or webcams for students to observe. Afterward have students complete the first journal assignment, which is describing the species, behavior, and environment.

Weekly lesson plan

End all or some class periods with a journaling activity. Journal assignments apply the class concepts to the course project (except for the ones for the first day of class, the peer review assignment, and the last day of class). Share the journal question with students during the last 15–20 min of class, giving students about 15 min to write. Encourage students to write as much as possible and not to pay too much attention to grammar—this is a brainstorming activity.

Collect the assignments at the end of class by (1) having students turn in a paper copy and giving it back by the next class period, (2) using two-part carbonless paper so students can keep a copy, (3) taking a photo of the handwritten essay and uploading it to the school's online learning management system, or (4) having students type and submit them on the school's online learning management system.

Peer review practice lesson plan

Students practice providing reviews to each other and learn about each other's proposals before completing the full grant proposal. Each student brings in a copy of his or her third milestone assignment.

Begin the activity by explaining that the purpose of the activity is twofold: students receive feedback from their peers to help them with their papers and students practice providing feedback before doing the peer-review-graded assignment.

Ask students what kind of feedback would be helpful to receive from their peers. Write the ideas on the board. Discuss the feedback and emphasize that it is important that evaluations should not only include what needs to be improved but also how to improve it. Once a plan is determined have students trade their paper with another student. They write out their feedback and then once everyone is done, they discuss the feedback with each other. If the process goes rather quickly, then have students repeat this process. Wrap up the activity by asking students which types of feedback were most helpful and how they will use this feedback for their proposal.

Peer review assignment lesson plan

Each student needs two copies of their grant proposal. Begin the assignment by asking students to list what kind of suggestions would be useful, reminding them of the class discussion that took place after the peer review practice activity. List these on the board. Describe how the assignment will be graded. The reviewers are graded; the reviews are not used to grade the proposals themselves. Students are required to make nine suggestions and a feedback summary. Each suggestion must be detailed and critical. If a suggestion is a critique, it must also describe how to improve it.

Explain to students that they should not stop reading a proposal once they have nine suggestions. The purpose of this assignment is to offer feedback to their peers which can then be used to improve the proposal. An incomplete peer review may miss some key aspects that needed improvement.

If the instructor is going to collect the reviews at the end of the activity and there is not much time between the next class meeting and the due date, then the instructor needs to make copies of the reviews for grading. If this is the case, then tell students that the reviewers must write their names on every sheet and do not have sheets stapled (to make copying easier); otherwise, this step is unnecessary. Other collection options include writing on two-part carbonless paper, having students take photos of their written reviews, or typing the reviews and uploading them to the school's online learning management system.

Provide the review questions to students and have students trade their proposals. After students complete the reviews have them discuss the feedback with each other. These reviews usually take about 30 min each. Repeat this process again if there is time. Use the last few minutes of the class period to have students report on feedback that was particularly useful.

Areas of potential confusion or difficulty for students

1. **How to address and/or modify**

   Depending on the setup of the content of the course, students may need additional assistance on developing appropriate methods. The literature review milestone assignment focuses on methods, but if the methods used in the studies are never discussed in the class, then students may still struggle. Setting up meeting times with students to discuss their methods is a possible way to overcome this struggle.

2. **Recommendations for extensions or continuations for more advanced classes**
   This course project has students create a research grant proposal. This activity can be extended by having the proposal due sooner in the semester and requiring students to perform the investigation. Then students can develop a summary report that explains the results and conclusions of the study. Although this assignment was implemented in a junior-level undergraduate class, it is also appropriate for a graduate-level course. In that case, the literature review milestone assignment should require more sources and detail. This project can also be implemented in an ecology course rather than in an animal behavior course—just broaden the topics to either any physical traits or any biotic and abiotic factors.
3. **Samples of results**
   The following are examples of research topics proposed by students:
   - Alarm call mimicry in gray catbirds and impact on antipredatory responses by mimicked species
   - Abiotic effects on whirligig beetle activity
   - Defensive behaviors of American toads
   - Massasauga rattlesnake antipredator behavior in secluded and public areas
   - Cowbird influence on redwing blackbird and song sparrow survival

Part III. Supplementary material

Supplementary data related to this chapter can be found online at https://doi.org/10.1016/B978-0-12-821410-7.00002-9.

Appendices

Tools for observational data collection

Susan W. Margulis[1], Domenic Romanello[2]

[1]*Department of Animal Behavior, Ecology, and Conservation, Department of Biology, Canisius College, Buffalo, NY, United States;* [2]*Department of Anthropology, University of Texas at Austin, Austin, TX, United States*

Introduction

Think about the study of behavior and what do you envision? More likely than not, "animal behaviorist" conjures up the image of a disheveled, khaki-clad individual with binoculars and a clipboard, sitting in the midst of a jungle and jotting down notes about the fascinating behaviors he or she sees amidst a large and complex group of mammals. Field notes represent an invaluable resource for any behavior study, but this image misrepresents the fact that animal behavior research is dependent on specific and rigorous methods. The idea that behavioral observation is somehow a nonscientific, casual endeavor is far from true. With the expansion of ethology in the 1930s, the idea that animals could be observed in natural settings steadily grew in scientific importance. As the field of ethology and behavioral ecology expanded, there came an explosion of research methods, conventions, and practices. In 1974, the seminal paper published by Jeanne Altmann (Altmann, 1974) provided a critical conceptual framework and operational guide for behavioral data collection and quantification. Virtually all observational data ascribe to one of the methods outlined in this article. These methods were designed not only to provide some degree of standardization to the discipline but also to reduce bias by structuring observations such that an observer's choice of which subject to watch and what behaviors to record was based on a priori decisions and statistically valid procedures.

As technology has changed, so too have data collection methods. The image of the field researcher with a clipboard has been replaced by the researcher with a tablet or other handheld device; live observation may be replaced by digital video recording followed by playback, and analytical methods have grown in complexity as computers have become routine. A dizzying array of tools are now available, including many that are free or at nominal cost, to facilitate data collection. Although it is often advisable to have students collect data the old-fashioned way—paper and pen, then entered into a spreadsheet—there is also value in introducing students to the tools that may be

Table A.1 Data collection software tools for behavioral data collection.

Software	Price	Runs on	What's it good for?	Special features	Link
Animal Behaviour Pro	$0.99	iOS	The software supports continuous focal animal sampling, instantaneous scan sampling, and other behavioral data collection methods.	Simplicity: basic behavioral data collection software, inexpensive, and user friendly.	https://apps.apple.com/us/app/animal-behaviour-pro/id579588319
Animal Observer	Free	iOS	The software supports continuous focal animal sampling, instantaneous scan sampling, and other behavioral data collection methods. Includes audio, photo, and video recording.	Spatial data: GPS and compass features within this software record the coordinates of research subjects.	http://fosseyfund.github.io/AOToolBox/
Behavioral Observation Research Interactive Software (BORIS)	Free	Linux, Mac, and Windows	The software supports continuous focal animal sampling through coding of audio and video files.	Data visualization: plots and flow diagrams can be generated with this software.	http://www.boris.unito.it
CowLog	Free	Linux, Mac, and Windows	The software supports continuous focal animal sampling through coding of video files.	Simplicity: user-friendly software for coding digital video.	http://cowlog.org/
CyberTracker	Free	Windows	The software is set up primarily for wildlife monitoring and conservation, but it can be used to collect behavioral data.	Inclusivity: the software is able to run with pictographs so that data can be collected by individuals who cannot read or write a specific language.	https://www.cybertracker.org
EthoLog	Free	Windows	The software supports continuous focal animal sampling through coding of audio and video files.	Simplicity: basic behavioral data collection software, free, and user friendly.	http://www.ip.usp.br/ethog/ethohome.html
EVENT software	$50/h For custom programming. Average cost is $200–$500.	Android and iOS	The software supports behavioral data collection generally.	Customizability: software developers design the app for specific research projects upon request.	https://animalbehaviorassociateswa.com/behavior-collection-software/

The Janelia Automatic Animal Behavior Annotator (JAABA)	Free	Linux, Mac, and Windows	The software supports continuous focal animal sampling through coding of video files.	Machine learning: manually entered behavioral labels are converted into behavior detectors that are used to automatically classify behaviors.	http://jaaba.sourceforge.net/
JWatcher	Free	Mac and Windows	The software supports continuous focal animal sampling and behavioral data analysis.	Data reliability: the software includes algorithms for testing inter- and intraobserver reliability.	http://www.jwatcher.ucla.edu
The Observer	Variable, based on specifics, but generally is costly—must contact Noldus	Multiplatform	The software provides many options for types of behavioral data collection. Modules available for direct video scoring and use of automated devices such as eye trackers and facial recognition.	The most comprehensive (albeit most expensive) automated system available. Has many built-in features for data analysis, storage, and backup. Steep learning curve.	https://www.noldus.com/observer-xt
Prim8 Software	Free	Android and iOS	The software supports continuous focal animal sampling, instantaneous scan sampling, and other behavioral data collection methods.	Simplicity: basic behavioral data collection software, free, and user friendly.	http://www.prim8software.com
Solomon Coder	Free	Windows	The software supports continuous focal animal sampling and instantaneous scan sampling using video files to record behaviors.	Vocalization data: the software includes a spectrogram.	https://solomon.andraspeter.com/
ZooMonitor	$50.00/year Free for institutions accredited by the Association of Zoos and Aquariums.	Android, iOS, and Windows	The software supports continuous focal animal sampling, instantaneous scan sampling, and other behavioral data collection methods.	Data visualization: heat maps, activity budgets, plots and flow diagrams can be generated with this software.	https://zoomonitor.org/home

available to them for future studies. Here, we provide a brief guide to readily available software for behavioral observation. It is important to bear in mind that these tools are changing constantly, and a regular review of what is available, what is continuing to be supported, and what changes have been made to existing tools is critical.

Reference

Altmann, J. (1974). Observational study of behavior: sampling methods. *Behaviour, 49*, 227–266.

Basic statistics for behavior

James C. Ha, Renee L. Ha
University of Washington, Seattle, WA, United States

In this appendix, we provide definitions of statistical terms, familiarize you with some statistical notation, and provide a brief background in nonparametric statistical hypothesis testing.

Basic terminology

Animal behavior scientists are interested in questions such as the relationship between the number of zoo visitors and animal quality of life or the factors that influence the production of offspring in a species. In a perfect world, we would measure the number of visitors all year round at every zoo and the production of offspring by every member of a species, but obviously that is rarely possible. Thus while we would like to work with an entire population to know the complete set of individuals, objects, or scores of interest, it is much more common that we are forced to work with a subset of our population of interest. This subset is termed our sample. In fact, there is no reason to require us to determine the population of scores: statistics allow us to draw conclusions on the basis of much more feasible samples instead.

To study psychology, scientists must measure their subjects in some quantitative or qualitative fashion. Thus they measure properties, or variables, of behavior, physiology, socioeconomic status, and so on, which may take on different values at different times and change with various conditions.

In contrast to a variable, a constant is a value that does not change under different conditions.

Discrete variables do not have units in between them, such as hair color (blonde, brunette, red) or political party affiliation (democrat, republican, libertarian). Continuous variables have an infinite number of possible values between units, such as temperature (30 degrees, 31 degrees, 30.5 degrees, 30.54672 degrees) or concentrations of a hormone in the blood (2.343 picograms per milliliter). More difficult to fit into such clear categories are variables that are integer measures, such as the number of errors in navigating a maze (one error, two errors, but someone cannot make 2.3 errors). These are generally treated as if they were continuous variables.

As the measurements are collected, the resulting values are commonly called data. These values, or data, can then be summarized by, for instance, taking the average or examining the high and low values, and these resulting values that summarize something about the sample are referred to as statistics. If you were taking the average of the scores of all the individuals in the population of youths in sports, rather than just of the sample, you would be calculating a parameter or a descriptive summary of data from an entire population.

A classic experimental design is one in which the researcher randomly assigns individuals to the experimental group, which receives a treatment, and randomly assigns other individuals in the sample to the control group, which experiences all aspects of the experiment except the treatment. For example, in an effort to develop a treatment for nicotine addiction in a sample of smokers, you might assign some participants to an experimental group that views a video on the negative psychological and physiological effects of smoking and assign other participants to a control group that views a video on eye surgery (a topic presumably having nothing to do with smoking but a video viewing, to keep conditions as similar as possible).

In this experimental design, the property that the researcher either manipulates experimentally or records as a naturally changing condition is described as the independent variable. A clear distinction is that the independent variable is the predictor variable. The result of the experiment that the researcher is interested in measuring is defined as the dependent variable. The dependent variable is often referred to as the outcome variable.

Independent variable: A predictor property that the researcher either manipulates experimentally or records as a naturally changing condition (thus a quasi-independent variable).

Dependent variable: An outcome property that the researcher measures.

In theory, it is not only important to randomly assign participants to each condition but also necessary to randomly select participants from the population of smoking individuals. A failure to randomly select from the population as a whole could result in findings that cannot be generalized back to the entire population because the individuals are not typical of the population. It is rare for scientists to study a sample for its own sake: the goal of observing a sample while being able to generalize findings back to the population is a common one. Having said this, it is logistically nearly always impossible to randomly select from the (usually very broad) population that psychologists would like to incorporate into their conclusions. This is an important issue in some studies and one that needs to be considered in interpreting the results of any research study. Does the sample of participants considered in the study accurately reflect the important characteristics of the population to which the conclusions are being applied?

Descriptive versus inferential statistics

This appendix is organized so that you learn about descriptive statistics prior to learning about inferential statistics. Descriptive statistics are numbers that summarize your data; in other words, they are numbers that help describe your data set.

These include the average (or mean) and the range (largest score minus smallest score), which you have most likely encountered in earlier mathematics. It is important to understand that all these descriptive statistics attempt to describe a larger group of scores (your data) with a single number or category, such as an average, or visually by graphing the data as a frequency distribution.

For example, the following numbers are difficult to summarize verbally but can easily be summarized with descriptive statistics:

65	6	23	11	15	241	9	55	1	2	26	2	48
8	41	36	81	9								

The average, or mean, represents the central tendency of the data and is equal to the sum of the scores divided by the number of scores. In this case, the sum is 679 and the number of scores is 18, so 679/18 = 37.72222. The smallest score is 1, and the largest score is 241. If you subtract the smallest score from the largest score, 241−1 = 240, you get the range of scores. The mean is an example of a statistic that summarizes information about the "central tendency" of the data, while the range summarizes an aspect of the "variability" of the data around the center. The count ("*N*") is another descriptive statistic. Other statistics can summarize information about the "shape" of the scores or the pattern of the scores around the center of the distribution of scores.

Mean	37.72222
Range	240
Minimum	1
Maximum	241
Sum	679
Count	18

Measurement scales

Scientists collect different kinds of data. In this case, we don't necessarily mean different variables: obviously, they do that. Rather, there are different scales or types of data, and the scale on which data are recorded can significantly affect the hypotheses that can be tested and how to test those hypotheses.

As an example of similar dependent variables but different scales, you might be collecting data on the number of days in a month that a depressed individual felt overwhelmed or hopeless about his or her life. A similar measurement of depression, but on a different scale, would be to ask individuals in your sample to rate themselves on a scale from 1 to 5 on how depressed they felt on average, with 1 being not depressed and 5 being very depressed.

These sound like very similar measurements, and they are both attempts to measure underlying depression, but they are actually using two different measurement scales to collect the data for later statistical analyses.

It turns out that there are not two but four measurement scales that are used to collect data (dependent variable) and to design experiments (independent variable), and which ones you use determine the mathematical manipulations and the statistical tests that you can use on the data later. This distinction is not only critical to understand what inferential statistical tests you can perform later but also a concept that appears deceptively simple initially.

These measurement scales vary in their possession of three properties: magnitude, equal intervals, and absolute zero. Magnitude refers to having a quantified measure of the amount of something: a measure with the property of magnitude actually reflects a physical amount of the property in question. The term equal intervals refers to a scale in which the amount of the property being measured is the same at all places in the scale. For example, the difference in the amount of the hormone estrogen from 2.3 to 2.4 mg (0.1 mg) is the same amount of estrogen as the difference between 10.8 and 10.9 mg.

This cannot be said of, say, a rating system for football teams: first place, second place, and third place. The difference in ability between the first- and second-place teams is not necessarily the same as the difference in ability between the second- and third-place teams—that is what "a distant third" refers to.

Absolute zero refers to the property in which "zero" on a scale actually refers to the physical absence of something: zero degrees Fahrenheit does not reflect the absence of heat; it is an arbitrary zero point. Zero errors through a maze does reflect the absolute absence of errors, and therefore maze errors would be described as a scale having an absolute zero.

Examples of each measurement scale

Nominal data are collected whenever you ask how many students in a class belong to each political party (Democrat, Green, Independent, Republican, or Other), and you simply count the number (or frequency) of students that fall into each group. In this example, students are placed into categories of political parties. These categories cannot be meaningfully ranked, and they do not include information on magnitude (greater or less than) or a true zero point. The term nominal means "name," and nominal data are often referred to as "naming scales." Other examples of data collected on a nominal scale include the number (or frequency) of individuals in a sample who are wearing various types of running shoes (Adidas, Nike, Reebok, etc.), the number of individuals in a sample or population who belong to various ethnic groups (African, Asian, Caucasian, Middle Eastern, etc.), or the number of individuals in a sample or population who have different blood types (A, B, AB, O). Again, these are qualitative categories that cannot be ranked in order and do not include measures of magnitude.

Ordinal data do include information on rank or order. Some examples of this scale include the order that runners finish in a race (first, second, third, fourth, fifth, etc.). Although this scale obviously tells you who performed the best in this case, it does not contain any information on the magnitude of the differences between the ranks. You do not know whether first and second place was decided by 10 min or by 2/100th of a second. In the second scenario of our earlier depression example, the researchers asked participants to rank their average level of depression on a scale from 1 to 5. Data collected in this manner is on an ordinal scale and is common in the behavioral sciences, whether you are studying people or animals. An animal behaviorist might record the levels of aggression in a baboon troop by using a similar scale.

Interval data contains information on order or rank like the ordinal scale, but the data are more quantitative because they contain information on magnitude. Thus the scientist knows not only that one score is higher or lower than another but also exactly how much one score is higher or lower than another score. Marine biologists recording daily water temperatures (in Celsius) on the Great Barrier Reef are using an interval scale. Both Celsius and Fahrenheit temperature scales include information on magnitude: 23°C is warmer than 20°C, and it is warmer by 3°C. Another property of interval scales is that they have equal intervals between measurement points. In other words, the magnitude of difference between 1 and 2°C is the same amount of heat (or energy) as the difference between 30 and 31°C. However, while you can measure temperatures of 0°C or 0°F, these measures are relative zero points rather than absolute zero points. What does this mean? Absolute zero is the absence of temperature or, more accurately, the absence of energy. It turns out that although 0°C is the temperature at which water freezes, it is not the absence of all heat or energy. Likewise, 0°F does not measure the absence of energy. Thus both these temperature scales measure on an interval scale but not on a ratio scale.

Ratio scale contains a meaningful absolute zero point. You can measure temperature on a ratio scale, the kelvin scale. In this case, 0°K is the absence of heat or energy. Thus it contains a meaningful zero point as well as the properties of magnitude and equal interval. Distance can also be measured on a ratio scale (inches, centimeters, miles, kilometers, etc.). Distance is ratio rather than interval because you can have, at least theoretically, zero distance between objects. In the first scenario of the depression example mentioned earlier, where participants were asked to tell the researcher how many days a month they felt depressed, we were collecting data on a ratio scale. In this case, reporting more days per month means that a participant feels more depressed, and these data have equal intervals (1 day) and an absolute zero point (participants who are never depressed report 0 day feeling depressed). In practical terms, this is a better measure of depression than the earlier ranked (1–5) scale because it contains more quantitative information that allows a statistician more choices in later statistical analyses.

Overview of measurement scales

We have summarized the properties of each measurement scale in Table 1. Some of our students have found this method of presenting the information more useful than the definitions, and many students find this table and the examples to be invaluable in understanding the practical differences between the measurement scales. Fortunately, in practice, statisticians do not discriminate between interval and ratio scales because they both contain the necessary properties for all types of inferential statistics.

Table 1 Summary of the properties of the four measurement scales.

Scale	Order	Magnitude	Equal intervals	Absolute zero	Discrete or continuous
Nominal	No	No	No	No	Discrete
Ordinal	Yes	Some	No	No	Discrete
Interval	Yes	Yes	Yes	No	Continuous
Ratio	Yes	Yes	Yes	Yes	Continuous

Frequency distributions and graphs

We are presented with data in graphical form in newspapers and popular news magazines, as well as on television news broadcasts. Many of us know that these graphical representations can "lie" with the use of truncated axes that make small differences appear very large. This is one area of graphing that it is good to know about, but you can also use graphs (properly done) to aid in validly interpreting complex statistical data.

We will begin the section with some typical ways to organize or group data into categories in tabular format, and then we will turn to graphing. The graphs in this chapter are used to provide a summary of data.

Frequency distributions in table format

Frequency distributions are a useful way to summarize raw data. Typically, it is difficult to see patterns in pages and pages of raw data or scores, but those data can be easily summarized with a frequency distribution. The most basic frequency distribution is simply a list of each score or category and its frequency.

Example of a simple frequency distribution table using nominal data.

Eye color	Frequency
Blue	10
Brown	33
Green	2
Hazel	11
Other	1
Total	57

Example of a simple frequency distribution table using interval or ratio data.

Exam score	Frequency
50	1
49	2
47	5
45	3
44	10
43	19
40	8
38	3
37	1
26	1
Total	53

The problem with simple frequency distributions

Although these simple frequency distributions work well when there are not too many categories of scores, they quickly become less helpful when there are numerous categories. For example, assume that on a 50-point exam, every score from 0 to 50 was earned by at least one individual in the course. This would mean that 50 categories would be needed to present the data using a simple frequency distribution. This would not truly summarize the data because there would be 50 categories of scores and 53 individuals. Most categories would have a frequency of one, and only a few would have a frequency of two or more. The solution to this problem is to reduce the number of categories in the frequency distribution without losing too much information about the distribution of scores.

A solution to the problem: guidelines for grouping scores into categories

While grouping scores into categories or a range of values is a good solution to increase the degree of summarization in your data, it is also important to avoid oversummarizing to the point that it obscures the pattern in the data. Fortunately, there are some common guidelines for how to create categories.

General guidelines to create categories

a. Choose between 10 and 20 categories (or intervals).
b. It is best to have a minimum of five raw scores in each category (or interval) rather than a large number of categories (or intervals) with a low frequency of scores, so use this as a guideline for the absolute number of categories or intervals to use for your data.
c. The highest score should be at the top and the lowest score at the bottom of the table.

Specific instructions for creating a frequency distribution.

Step 1: Calculate the range of scores. Range = high score – low score.
Step 2: Calculate the size of your interval (or category).
Width of the interval (i) = range/number of desired intervals (be sure to round the answer you get for i).
Step 3: Make the lower limit of the lowest interval divisible by i.
Step 4: List the intervals with the lowest scores at the bottom.

Raw scores for example problem

17	44	43	8	17
16	13	16	36	23
30	7	37	24	28
19	22	42	39	43
33	41	38	12	34
18	14	48	50	38
25	47	35	11	9
27	15	21	49	10
32	40	26	46	29
31	5	44	32	9
45	20	50		

Example using raw scores

Step 1: Calculate the range of scores.

$$\text{Range} = \text{High score} - \text{low score}$$
$$\text{Range} = 50 - 5 = 45$$

Step 2: Calculate the size of your interval (or category).

$$\text{Width of the interval(i)} = \text{range/number of desired intervals}$$
$$\text{Width of the interval(i)} = 45/10 = 4.5$$

Note: Round the solution to a whole number, so 4.5 becomes 5.

Step 3: Make the lower limit of the lowest interval divisible by i.

The lowest score in the raw data is 5, which is divisible by i (which equals 5 when we round up in this example), so the lower limit of our lowest interval will be 5.

Step 4: List the class intervals, with the lowest scores at the bottom.

50–54
45–49
40–44
35–39
30–34
25–29
20–24
15–19
10–14
5–9

By adding in the frequency of the scores that fall into each interval, you get the following frequency distribution from the raw data.

Class intervals	Frequency
50–54	2
45–49	6
40–44	6
35–39	6
30–34	6
25–29	5
20–24	5
15–19	7
10–14	5
5–9	5
Total scores	53

Graphing

Many statisticians prefer to get a first feel for their data by looking at them visually. A frequency distribution graph can quickly demonstrate the range of the scores as well as the frequency of occurrence of each score in the sample or population and can even provide a preliminary indication of the central score. Before we go over sample graphs used in statistical analyses, we'll review some basic concepts in graphing data.

Graphing basics

Two-dimensional graphs have an X-axis (abscissa) and a Y-axis (ordinate). The X-axis is in the horizontal position, while the Y-axis is in the vertical position. It is traditional to show the score or independent variable on the X-axis and some characteristic of the score or dependent variable on the Y-axis. When the graph is a frequency distribution, the characteristic of the score shown on the Y-axis will be the frequency of that score in the distribution. If the graph is of interval or ratio data, you will often see the average score shown on the Y-axis.

General rules of thumb include (1) starting both axes at 0, (2) filling the area of the graph so that it is approximately $^3/_4$ filled, and (3) having the height at approximately $^3/_4$ of the width of the graph. If it is not feasible to have both axes begin at 0, then it is traditional to clearly show a break on the appropriate axis with "//" to indicate that part of the scale was dropped to avoid misinterpretation of the scale. Finally, it is important to label the *X*-axis and the *Y*-axis for the graph so that it is clear what is being represented.

Types of frequency graphs

Two major categories of graphs are based on whether the data on the *X*-axis are plotted as a discrete or continuous variable. For example, eye color can be categorized as a discrete variable and be plotted on the *X*-axis, with the frequency of each eye color plotted on the *Y*-axis. Alternatively, scores on an exam could be considered a continuous variable (0–100) and be plotted on the *X*-axis, with frequency on the *Y*-axis. We'll address how to plot discrete variables on the *X*-axis first (bar graphs) and then how to plot continuous variables on the *X*-axis (histograms, frequency curves, scatter plots).

Graphing a discrete variable on the *X*-axis

It is common to have discrete categories of data that you plot on the *X*-axis (e.g., blue, brown, green, hazel) and then plot the frequency on the *Y*-axis. This is known as a bar graph, and you'll see that the data on the *X*-axis are grouped in a way that is discrete rather than continuous. It can also occur with fewer discrete categories (e.g., week 1, week 3, and week 5 or smoking, nonsmoking, and unknown smoking status).

Graphing a continuous variable on the *X*-axis

There are more graphing options for a continuous variable on the *X*-axis. These include histograms, frequency curves, and scatter plots. Histograms are used to plot frequencies for continuous data and use a separate bar for each score or an interval of scores.

Alternatively, you could plot the same data using a frequency curve, where a smoothed line is plotted to represent the data shown in the histogram.

Scatter (or *X*–*Y*) plots are a useful way to compare the relationship between two variables.

Measures of central tendency

Central tendency is the property of a distribution of scores that describes the center of the data. We often want to describe our data with a single number, and the usual choice is to describe the location of the center of the scores. This makes sense, but it turns out to be more difficult than it might otherwise initially appear. The best number to describe the "center" of the data may depend on the measurement scale of the data as well as the shape of the data. Data that are very "pretty," that is, symmetrical, have a center that is relatively easy to describe. Data with a dramatic lack of symmetry (a distinct skew or shift to the left or right) or that have multiple peaks are not so easy to summarize in a single measure of central tendency.

The median: a useful measure of central tendency

The median is the value in the middle of your scores when they are rank-ordered; literally, it is the value where 50% of the scores fall below it (and of course, 50% of scores fall above it). The term rank order simply means placing the scores in order from the least to greatest magnitude. The median is important because it is not sensitive to extreme scores but only to the number of scores above and below it. We can calculate the median of a set of scores as in the following.

Example

Step 1: Rank the scores from smallest to largest.

4	6	8	10	11

Step 2: Find the center score.

If you have an odd number of scores, then your middle score is the median. In the above case, it is 8. If you have an even number of scores, then you must take the average of the two center scores. For example,

2	4	6	8	10	12

In this case, both 6 and 8 are the center scores. If you take the average of 6 and 8, you get 7: median = (6+8)/2 = 7.

Note that if we change the "12" in the data set, say, to a "14," it would have no effect on the median: the median would still be 7. Likewise, we could change the "12" to "35" and the median would remain the same. Thus the median is said to be relatively insensitive to exact scores.

Hypothesis testing

Researchers usually conduct an experiment because they have a hypothesis that an independent variable affects a dependent variable or that there is some relationship between the variables. This would be called the research, or alternative, hypothesis.

They wish to contrast this possibility with the possibility that the two variables are unrelated or that any patterns that are seen are due to chance factors rather than their experiment. This is the no-effect or null hypothesis.

These (hypothesized) effects can be directional (predicted to increase or to decrease) or nondirectional (the direction of the effect is not stated). Nondirectional hypotheses are more common, and more appropriate, because the researcher must have a prior scientific basis for predicting a specific direction to the effect. In fact, predicting a directional effect when the actual effect actually turns out to be in the opposite direction means that the experiment must be repeated. So while most researchers really have a directional hypothesis in mind, they usually actually test a nondirectional hypothesis to be conservative.

The statistical process that we have been describing asks us to make an inference about real effects. Real effects are the actual effects (or perhaps lack thereof) that the independent variable has on the dependent variable, out there in the real world. Researchers collect data to estimate the real effect of their independent or predictor variable, but we can never truly know the real effect of a variable. Instead, we can only make inferences about this effect. This inference is measured as the relationship between two variables or the effect of one variable on another, but because we do not know what the real effect is, it would be circular logic to try to say that what we see is the real effect.

Instead, we take a different approach: can we say that what we have recorded is not likely due to simple chance effects? An explanation based on simple chance effects is our null hypothesis, the flip or opposite possibility from our research hypothesis. This outcome, an outcome of no effect, can be tested! So we can only use statistics to test whether the null hypothesis is false, NOT whether the research hypothesis (our real interest) is true. This is a difficult but critical point for students.

We know how to deal with simply chance effects through the principles of mathematical probability. So we can assign a probability to an explanation that our results are simply due to chance. Based on that probability, there are two possible outcomes.

We can fail-to-reject the chance (null) explanation if the probability is high that it is true. If we fail to reject that explanation, then we must suggest that there is no real effect of our independent variable on our dependent variable, no relationship between the two variables.

Or we can reject the chance (null) explanation if the probability of it being true is very low. If we reject that null hypothesis explanation, then we must turn to the only other explanation (in a clean experiment where all other factors have been controlled) and suggest that there is a real effect of our independent variable on our dependent variable, or a real relationship between the two variables.

Rejecting the null hypothesis because there is overwhelming information suggesting that it is false still does not prove that the alternative hypothesis is true. All we can do is to fail to reject the null hypothesis, implying that the alternative hypothesis is true. Thus we can never state beyond a doubt that we have proven anything! This may make the scientific process seem pointless, but recall that part of the scientific method is in repeating experiments and testing alternative hypotheses. With this process, scientists are able to get closer and closer to the underlying truth. Nevertheless, because we rely on the laws of probability, and there is always a small probability that our results are due to chance factors alone, we cannot prove effects beyond all doubt. However, when specific effects are found repeatedly, and the likelihood that the effects are due to chance are small each time the experiment is conducted, the plausibility of a real effect of our independent variable becomes more and more accepted over time.

Critical level of probability

So now we need to determine the probability that our results are simply due to chance, i.e., that the null hypothesis is true. But how low a probability do we need before we reject the null hypothesis in favor of our research or alternative hypothesis? This is termed the critical level of probability.

To test hypotheses, researchers calculate a statistic from their data, which, in some way, describes the magnitude of the effect of the independent variable, and then ask how likely it would be to obtain that statistic, that effect, if chance factors alone were at work. If obtaining that statistic is very unlikely under chance conditions (≤ 0.05 or 5%), then the researcher can reject the null hypothesis (which states that there is no relationship between their two variables). Thus the researcher is suggesting that if their outcome would only occur in 5% or fewer cases under normal circumstances, then the likelihood that it occurred due to chance is low, and it is more likely that the variables in their experiment affected the outcome. Researchers must choose the critical level of probability for their experiment.

In this case, we have chosen 0.05 or 5% because that is the most commonly accepted critical level of probability in science. The critical level of probability is called the alpha (α) level. While 0.05 is the most commonly used alpha (α) level, 0.01 (1%) and 0.10 (10%) are used fairly often as well. For our purposes, we will use an alpha of 0.05.

Statistics, or measurements of effect

In this section, we describe just a couple common ways to measure the effect that has been recorded in an experiment, whether formal/controlled, or recorded using natural or uncontrolled independent variables, such as time of day, temperature, or group size. Some students may have learned about a category of effect-size tests called parametric tests, like the t-test or analysis-of-variance, if they have taken a statistics class. Unfortunately, these tests are not always the best choice for behavioral data. Parametric statistics have associated with them several mathematical assumptions about the characteristics of the data. In many cases, behavioral data violate these assumptions. Although there are other ways to deal with this, here we are focusing on statistics that do not make these rigorous mathematical assumptions, called "nonparametric statistics."

In the first case, we describe correlation, a measurement of the relationship between two numeric variables. In the second case, we describe a statistic to measure the differences (or lack thereof) across nominal or categorical data (the chi-squared statistic). In the third case, we describe a statistic to measure the difference (or lack thereof) between two groups of numerical (ordinal, interval, ratio) data, the Mann–Whitney U statistic. It should be noted that there are many statistical software packages that will calculate these tests for you, but here we go through the details of calculating them by hand for several reasons: (1) not all statistical software packages include nonparametric tests, (2) an advantage of nonparametric tests is that they are almost easier to calculate by hand if you are not comfortable with a software package already, and (3) to help you understand what these tests are actually doing, and in doing so, hopefully gain some insight into the process of science.

Correlation

Now we discuss how to measure linear relationships between two variables. For example, we can ask about the relationship between the number of hours that a student studies and his or her overall grade point average (GPA). In this case, both the independent and the dependent variables are measured on a noncategorical scale—ordinal, interval, or ratio—and the variables are typically referred to as X and Y. This is frequently the result of research that involves what some researchers call "quasi-independent" variables: independent variables that are not manipulated or set by the researcher, as in classical experimental designs, but rather are measured at the same time as the result or outcome, the dependent variable. To clarify this distinction, these designs frequently refer to what we have called "independent variables" as "predictor variables" and refer to "dependent variables" as "criterion variables." An example will help illustrate the concept behind this shift in research approach.

Suppose that we examine the relationship between the number of hours a week that 10 students spend on studying and their cumulative GPA. Both variables are continuous. We can graph those results as a scatter plot, using Excel, for instance.

Raw data

Student	Hours (X)	Grade (Y)
1	10	3.4
2	12	3.5
3	5	2.7
4	2	1.7
5	20	3.9
6	14	3.6
7	11	3.3
8	22	3.8
9	3	1.9
10	18	4.0

The paired data points from all of our 10 students are shown clearly in the figure. For each student, we went over X units and up by Y units to plot their data. If you do this for all your data points, you can get a visual estimate of the overall relationship between X and Y. In this case, it appears that as X increases, Y increases. In fact, there appears to be a linear relationship between X and Y.

In this example, the relationship between the X and Y variables is positive (direct). The alternative would be a negative or inverse relationship, such as the one you would expect between the amount of partying and the GPA (higher level of partying is associated with a lower GPA).

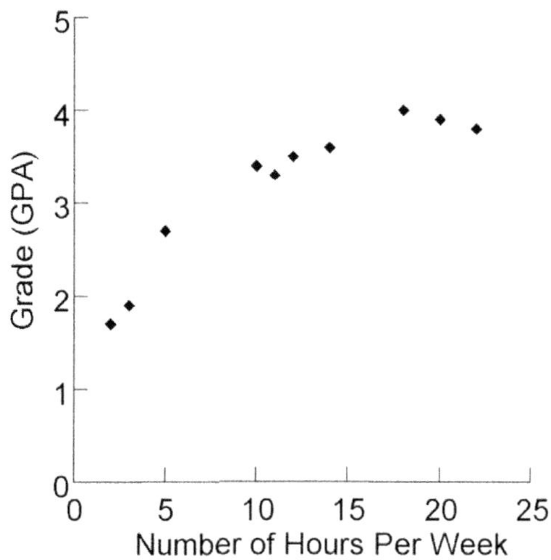

Graph of the raw data.

Correlation coefficient

A correlation coefficient is a statistic that expresses the degree of the fit of the data to a line and the type of relationship (direct or inverse). Correlation coefficients make no assumptions about the cause-and-effect direction of an X–Y relationship but rather simply measure the degree to which two sets of paired scores vary together in a consistent (linear) manner. So a "significant" correlation between two variables could be produced by (1) a cause-and-effect relationship between X and Y, (2) a cause-and-effect relationship between Y and X, or (3) the fact that a third, unmeasured variable is affecting both X and Y, producing the appearance of a relationship between X and Y! So be careful when interpreting a correlational relationship. This also implies that you can calculate a correlation coefficient for regression-type "cause-and-effect" data, but you cannot necessarily calculate a "cause-and-effect" regression line for correlational data.

Correlation coefficients have the following characteristics:

1. Values always range between -1 and $+1$.
2. A positive coefficient indicates a direct relationship (positive slope), whereas a negative coefficient (negative slope) indicates an inverse relationship.

3. A coefficient that is equal to zero indicates that there is no relationship between the two variables (supports your null hypothesis).
4. A coefficient that is equal to −1.00 indicates that you have a perfect inverse relationship between your variables, while a coefficient that is equal to +1.00 indicates that there is a perfect direct (positive) relationship between your variables.

The Spearman rank order correlation (called rho or r_s)

The Spearman rank correlation test is a simple measure of correlation.

Calculation steps include the following.

1. Independently rank each variable X and Y (strongest response = 1).
2. Give tied ranks an average ranking score, but then don't reuse the ranks that you averaged.
3. Calculate the "difference in rank" scores [$D_i = R(X_i) - R(Y_i)$] and the squared differences in rank.
4. Calculate rho (r_s) using the following formula:

$$r_s = 1 - \frac{6\sum d_i^2}{n(n^2 - 1)}.$$

5. Evaluate r_s using Table 2.

Table 2 Critical values of the Spearman correlation.

df	Critical *P* value
5	1
6	.886
7	.786
8	.738
9	.683
10	.648
12	.591
14	.544
16	.506
18	.475
20	.45

Complete example

The correlation between body weight (grams) and a high level of aggressiveness of an individual hamster (measured on an ordinal scale from 1–4, with 1 being the most aggressive) can be analyzed with the Spearman rank order correlation, or rho. The assumption was that high body weight would be correlated with high aggression.

Data

X (aggressiveness)	*Y* (body weight)
4	92
3	128
4	134
2	98
1	87
1	101
3	100
2	108

Step 1: Rank the two variables separately and handle tied ranks appropriately.

X_i (aggressiveness score)	Rank (X_i)	Y_i (body weight)	Rank (Y_i)
4	7.5	92	7
3	5.5	128	2
4	7.5	134	1
2	3.5	98	6
1	1.5	87	8
1	1.5	101	4
3	5.5	100	5
2	3.5	108	3

Note that lower scores on the aggressiveness scale mean that the individual is more aggressive, so the person gets top ranks (one and two, etc.). Similarly, higher body weight scores imply greater weight (g), so a high body weight score receives the top rank (rank = 1 for a body weight of 134). If you were to mix these, then you'd be trying to correlate low weight with high aggressiveness or high weight with low aggressiveness.

Step 2: Calculate "difference in rank" scores [$D_i = R(X_i) - R(Y_i)$] and square the difference scores as an intermediate step in preparation for using the rho formula.

X_i	$R(X_i)$	Y_i	$R(Y_i)$	D_i	D_i^2
4	7.5	92	7	0.5	0.25
3	5.5	128	2	3.5	12.25
4	7.5	134	1	6.5	42.25
2	3.5	98	6	−2.5	6.25
1	1.5	87	8	−6.5	42.25
1	1.5	101	4	−2.5	6.25
3	5.5	100	5	0.5	0.25
2	3.5	108	3	0.5	0.25
				Sum (D_i) = 0	Sum (D_i^2) = 110

Step 3: Calculate rho (r_s) using the following formula:

$$r_s = 1 - \frac{6(110)}{8(8^2 - 1)} = -0.31$$

Step 4: Interpret the correlation coefficient.

The Spearman rank order coefficient (r_s) for our body weight—aggressiveness example data is −0.31. A correlation coefficient this size suggests a fairly weak negative relationship between our two variables. But what we really need to do is to determine the probability that we could get a statistic of −0.31 or more extreme by chance. We also need to keep in mind that the negative correlation suggests that top-ranking weight scores are associated with bottom-ranking aggressiveness scores.

We evaluate the Spearman rho value using a table for critical values of rho based on the (adjusted) sample size, called degrees of freedom ("df", in this statistic it is equal to $N - 2$) and the desired level of alpha. We will evaluate the Spearman rank order coefficient in a similar way (see Table 2). In our case, the df are equal to 6 ($8 - 2 = 6$) and alpha = 0.05. The two-tailed critical value is 0.886. The absolute value of the r_s obtained (0.31) calculated in this example is less than the two-tailed critical value (0.886), and the critical value is positive, and your obtained value is in the opposite tail of the distribution.

Another way to express the relationship between the independent and the dependent variables is to square your r-obtained value to make it R^2.

$$R^2 = \text{proportion of the variability in } Y \text{ that is explained by } X.$$

The use of the capitalized R is intentional, as R^2 is the formal symbol for this statistic. If we apply this concept to our example, then R^2 is the proportion of the variance in body weight that is explained by how aggressive a personality the individual exhibits. Obviously, other possible (and potentially more likely!) factors are related to body weight. If $r_s = -0.3095$, then $R^2 = (-0.3095)^2 = 0.09579$.

This means that approximately 9.579% of the variability in body weight is explained by aggressiveness in the individual's personality. If this were the case, then only 9.579% of the variability in your dependent variable would be explained by your independent variable, with 90.421% of the variability in your dependent variable (Y) left unexplained. This would suggest that the factor you are studying does not have a strong relationship with your dependent variable.

How would these results be reported in a scientific journal article?

If the journal required American Psychological Association (APA) format, the results would be reported in a format something like this:

> There was no significant relationship between high body weight and a high level of aggressiveness in an individual, $r_s(6) = -.31$, $p > 05$; $R^2 = 0.10$.

This formal sentence includes the two variables (body weight score and aggressiveness score), as well as a statement about statistical significance, the symbol of the test (r_s), the degrees of freedom (6), the statistical value (−.31), and the estimated probability of obtaining this result simply due to chance (>0.05).

Chi-square

The chi-square test is appropriate when you have nominal independent variable(s) but only group membership (frequencies) as your raw data. For example, you might want to determine whether students in a class are randomly distributed in their political party affiliations. Conceptually, this is an appropriate analysis for a chi-square test because the purpose of a chi-square analysis is to compare the observed frequencies to the expected frequencies. The expected frequencies are determined by assuming that group membership is random. If membership in a group differs from random, then we can say that there are statistically significant differences between the actual data and the results that one would expect due to chance (or random factors).

Number of students who belong to each of the four political parties (observed frequencies):

Democrat	Green	Independent	Republican
35	20	7	50

Are students randomly distributed across political parties?

Calculation of expected values

If the expected frequencies are based purely on chance, then they are easy to calculate based on how many categories you have in your analysis. In the political party example, we know that we have four categories of political parties, and the probability of being in each category simply due to chance would be ¼ or 25%. This is the random model of calculating expected frequencies.

Calculation of the chi-Square (χ^2) test

f_e = expected frequency with random sampling from the null hypothesis population,
f_o = observed frequency in the sample,
k = number of cells (or categories).

$$\chi^2 = \frac{\Sigma(f_o - f_e)^2}{f_e}$$

The summation symbol (Σ) requires that you perform the calculations that follow it for each cell (or category) and then sum the numbers you calculated for each cell (or category) to produce the chi-square (χ^2) obtained value. Once that is done, you must compare the obtained value to a critical value to determine whether your results are due to chance.

What are you really doing in this calculation? It's very simple:

1. You are calculating a difference score that measures the deviation of the observed counts from the corresponding expected counts.
2. You are then squaring the difference score.
3. You are then dividing the squared deviation by the expected frequency to standardize the result for the size of the sample.
4. You are calculating this standardized squared deviation score for each category and summing them together to obtain χ^2.

If χ^2 (obtained) $\geq \chi^2$ (critical), then you reject the null hypothesis. If the obtained value is not greater or equal to the critical value, then you fail to reject the null hypothesis. A table of critical values for χ^2 can be found in Table 3. Critical values for χ^2 are based on the degrees of freedom ($k - 1$) and the alpha level (usually 0.05).

Complete example

An animal behavior lab student is observing a viewable beehive colony. The student observes that there seem to be four categories of activities: foraging (FORAGE), tending the queen (QUEEN), tending the larval young (YOUNG), and cleaning and defending the hive (HIVE). Of the 112 members of the hive that the student

Table 3 Critical values of the χ^2 statistic.

df	Critical *P* value
1	3.8415
2	5.9915
3	7.8147
4	9.4877
5	11.0705
6	12.5916
7	14.0671
8	15.5073
9	16.9190
10	18.3070
11	19.6751
12	21.0261
13	22.3620
14	23.6848
15	24.9958
16	26.2962
17	27.5871
18	28.8693
19	30.1435
20	31.4104

observed, the student noted 40 in the FORAGE category, 20 in the QUEEN category, 12 tending YOUNG, and 40 in the HIVE category. Is membership of an individual bee in a behavioral role distributed randomly across the four categories of roles?

Null hypothesis (H_0): number of individuals in each role are distributed randomly.

Alternative hypothesis (H_A): number of individuals in each role is not distributed randomly; some roles require more or less members.

Step 1: Determine the expected frequency (f_e).

Based on the story problem, we are supposed to compare the actual role affiliations of this observation sample with a random model. The f_e for this example would be ¼ or 25% (1 divided by the number of cells) because there are four possible cells, and if the results were random, then one would expect that each cell (or category) is equally likely.

Cell 1	Cell 2	Cell 3	Cell 4
25%	25%	25%	25%

To convert these percentages to a frequency, we need to multiply the percentage by the total *N*.

$$f_e = 112 \times 0.25 = 28.$$

Thus we have the same expected frequency for each cell.

Step 2: Calculate the χ^2 obtained.

	FORAGE	QUEEN	YOUNG	HIVE
f_o	40	20	12	40
f_e	28	28	28	28

$$\text{Category } 1 = (40 - 28)^2/28 = 5.142857$$

$$\text{Category } 2 = (20 - 28)^2/28 = 2.285714$$

$$\text{Category } 3 = (12 - 28)^2/28 = 9.142857$$

$$\text{Category } 4 = (40 - 28)^2/28 = 5.142857$$

$$\chi^2\text{obtained} = 5.142857 + 2.285714 + 9.142857 + 5.142857 = 21.71429.$$

Step 3: Evaluate the probability of obtaining this χ^2 value due to chance.

To evaluate your χ^2-obtained value, you must use the χ^2 distribution (see Table 3), and to do so, you need an alpha level (0.05) and degrees of freedom ($k - 1$). Thus the df for this problem are $4-1 = 3$. Compare the χ^2-critical value with your χ^2-obtained value and reject the null hypothesis when your χ^2-obtained value is equal to or greater than your χ^2-critical value. When alpha $= 0.05$ and the degrees of freedom are equal to 3, you should use 7.82 as your χ^2-critical value; $21.71 > 7.82$, so we reject the null hypothesis.

How should we interpret these data considering the randomness of behavioral role affiliation for the bees in this study?

These results suggest that you would expect to obtain these frequencies in role affiliation less than 5% of the time, assuming role affiliation was random. Thus it is not very likely that these frequencies all come from the random null hypothesis population. Of course, as always, there is some finite probability that you could get frequencies that deviate from the expected frequencies this much simply due to chance, but this probability is less than 5%.

How would these results be reported in a scientific journal article?

If the journal required APA format, the results would be reported in a format something like this:

> The frequency of bees in each behavioral role differed significantly from random, $\chi^2(3, N = 112) = 21.71, p < .05$.

This formal sentence includes the independent variable (roles), as well as a statement about statistical significance, the symbol of the test (χ^2), the degrees of freedom (3), the statistical value (21.71), and the estimated probability of obtaining this result simply due to chance (<0.05).

Test of independence between two variables: contingency table analysis

The contingency table analysis is a variation of the chi-square test that allows you to test the independence between two variables. Thus you can ask whether two categorical variables are independent from one another or related to one another. To perform this analysis, the variables must be classified into mutually exclusive categories. As we showed in the earlier chi-square discussion, the cells contain the frequencies.

What is a mutually exclusive category? Mutually exclusive categories are designed so that being a member of one category prohibits your inclusion in the other category. For example, consider biological sex as a mutually exclusive category. Being male prohibits one from being in the other category (female). If one of the independent variables is called Factor A and the second one is Factor B, then we can state the null and alternative hypotheses in the following way:

> Null hypothesis (H_0): Factor A is not related to Factor B.
> Alternative hypothesis (H_A): Factor A and Factor B are related.

Performing a contingency table analysis

The equation for the contingency table analysis is the same one that we used for the chi-square test, where you calculate a value for each cell that is determined by the observed and the expected frequencies. However, the f_e and the df are calculated differently for this analysis. The expected frequencies are slightly more challenging to calculate because you must do it for each cell, and the value is influenced by the subtotals in the rows and columns. Specifically, the f_e for each cell is calculated in the following way:

$$\frac{(\text{Row marginal sum for that cell}) \times (\text{Column marginal sum for that cell})}{\text{Total } N}$$

Every value or cell in the two-way table is part of a row and a column. The total frequencies in each row are called the row marginals, and the total frequencies for the columns are called the column marginals. So each cell has a corresponding row marginal value and a column marginal value. These are multiplied together, and the result is divided by the total sample size to obtain the expected value for each cell. We'll turn to an example to demonstrate the calculations and the marginal values.

Example

A holiday card company hires a statistician to determine whether or not men and women differ in their preference for a "mushy" Valentine's Day card (thank you to our tween-aged son for this example!). They obtain the following results from their survey of 400 adults.

Love 'em	Hate 'em	Don't care	Marginals
Men	100	50	50
Women	150	25	25
Marginals			

Note that the marginals are calculated by adding up the rows and columns. For example, the first-row marginal for men is 200 (100 + 50 + 50), the first column marginal for "love 'em" is 250 (100 + 150), and the total or overall N is 400.

Null hypothesis (H_0): Preference for "mushy" Valentine's Day cards is randomly distributed between men and women.

Alternative hypothesis (H_A): Preference for "mushy" Valentine's Day cards is not randomly distributed between men and women.

Step 1: Determine the expected frequency (f_e).

Based on the story problem, we are supposed to compare the actual frequencies of preferences for mushy Valentine's Day cards with the expected frequencies based on a random model. The expected frequencies will be based on row and column marginals, and there may be different expected frequencies for each cell.

Determine each row and column marginal by adding up the frequencies in the respective row or column.

	Love 'em	Hate 'em	Don't care	Marginals
Men	Cell 1 = 100	Cell 2 = 50	Cell 3 = 50	200
Women	Cell 4 = 150	Cell 5 = 25	Cell 6 = 25	200
Marginals	250	75	75	400

Note that cells are numbered from left to right (marginals do not count). Calculate the expected frequency values.

$$\text{For each cell, } f_e = \frac{(\text{Its row marginal}) \times (\text{Its column marginal})}{\text{Total } N}$$

$$\text{Cell 1 } f_e = \frac{250 \times 200}{400} = 125$$

$$\text{Cell 2 } f_e = \frac{75 \times 200}{400} = 37.5$$

$$\text{Cell 3 } f_e = \frac{75 \times 200}{400} = 37.5$$

$$\text{Cell 4 } f_e = \frac{250 \times 200}{400} = 125$$

$$\text{Cell 5 } f_e = \frac{75 \times 200}{400} = 37.5$$

$$\text{Cell 6 } f_e = \frac{75 \times 200}{400} = 37.5$$

Step 2: Calculate the χ^2 value obtained.
f_e Values for each cell are indicated in the parentheses.

	Love 'em	Hate 'em	Don't care	Marginals
Men	100 (125)	50 (37.5)	50 (37.5)	200
Women	150 (125)	25 (37.5)	25 (37.5)	200
Marginals	250	75	75	400

$$\text{Cell } 1 = (100 - 125)^2/125 = 5.0000$$

$$\text{Cell } 2 = (50 - 37.5)^2/37.5 = 4.1666$$

$$\text{Cell } 3 = (50 - 37.5)^2/37.5 = 4.1666$$

$$\text{Cell } 4 = (100 - 125)^2/125 = 5.0000$$

$$\text{Cell } 5 = (50 - 37.5)^2/37.5 = 4.1666$$

$$\text{Cell } 6 = (50 - 37.5)^2/37.5 = 4.1666$$

$$\chi^2(\text{obtained}) = 5 + 4.1666 + 4.1666 + 5 + 4.1666 + 4.1666 = 26.6664.$$

Step 3: Evaluate the probability of obtaining this χ^2 value due to chance. To evaluate your χ^2-obtained value, you must use the χ^2 distribution (see Table 3), and to do so, you need an alpha level (0.05) and degrees of freedom [df = $(r - 1)(c - 1)$], where r is the number of rows and c is the number of columns. Thus the df for this problem are $(2 - 1)(3 - 1) = 2$. Compare the χ^2-critical value with your χ^2-obtained value and reject the null hypothesis when your χ^2-obtained value is equal to or greater than your χ^2-critical value. When alpha = 0.05 and the degrees of freedom are equal to 2, you should use 5.99 as your χ^2-critical value; 26.67 > 5.99, so we reject the null hypothesis.

How should we interpret these data?

These results suggest that you would expect to obtain these frequencies in Valentine's Day card preference less than 5% of the time. Thus it is unlikely that these frequencies come from the null hypothesis population where card preferences are distributed randomly by sex. It appears that sex and preference for mushy Valentine's Day cards are related. In fact, it appears that women tend to prefer mushy valentines (!).

How would these results be reported in a scientific journal article?

If the journal required APA format, the results would be reported in a format something like this:

> There was a significant relationship between sex and a preference for mushy Valentine's Day cards, χ^2 (2, $N = 400$) = 26.67, $p < .05$. It appears that women tend to prefer mushy valentines.

This formal sentence includes both the independent variables (card preference and sex), a statement about statistical significance, the symbol of the test (χ^2), the degrees of freedom (2), the statistical value (26.67), and the estimated probability of obtaining this result simply due to chance (<0.05).

Mann–Whitney U test

The Mann–Whitney U test (U) is appropriate when you have an independent (or between-groups) design with two groups (e.g., experimental vs. control groups). This test is designed to evaluate the separation between the groups. A larger separation suggests that the populations are unlikely to be the same, while a smaller separation suggests that both groups come from the same underlying null hypothesis population. This test is very similar to the parametric equivalent, the independent t test.

Calculation steps

Assume that there are two groups of scores, experimental and control, for instance. We'll call them Group 1 and Group 2 scores.

1. Assign a rank, in the order from least to greatest, to all the scores in both samples regardless of the experimental condition; that is, ignore which group they belong to and rank all the scores as 1, 2, 3, 4, etc.
2. If there are tied scores, then assign both (all?) of them the average of their ranks. But then, do not reuse those ranks.
   For example, say you have three identical scores: 20.3, 20.3, and 20.3. If you were to ignore the ties, they would have ranks, say, 4, 5, and 6.
   Average their ranks (that would be (4+5+6)/3 = 5) and assign each of them the rank 5.

BUT do not continue the next rank at 6. You used the rank 6 and the next rank in the order is 7. This is by far the most common error performed in calculating this test statistic.

3. Now, sum the ranks separately by experimental group (let's call these two sums: R_1 and R_2).
   Can you see that if the sets of scores overlap a lot (no effect, no difference between groups), your two summed ranks will be very similar and if there is a big difference in your two sets of scores, the summed ranks would be very different, one with all the low ranks (*1*'s, *2*'s, *4*'s, etc.) and one with the high ranks (*12*'s, *15*'s, etc.)?
4. Calculate two test statistics (called U-obtained values) by using the following equations:

$$U_1 = R_1 - \frac{n_1(n_1+1)}{2}$$

$$U_2 = R_2 - \frac{n_2(n_2+1)}{2}$$

   The lesser of the two results is the true U-obtained value.
5. Evaluate the U-obtained value using the Mann–Whitney table (Table 4) and reject the null hypothesis if $U_{obt} \leq U_{crit}$.

Table 4 Critical values of the Mann–Whitney U test.

n1\n2	2	3	4	5	6	7	8	9	10	11	12	13	14	15	16	17	18	19	20
2							0	0	0	0	1	1	1	1	1	2	2	2	2
3				0	1	1	2	2	3	3	4	4	5	5	6	6	7	7	8
4			0	1	2	3	4	4	5	6	7	8	9	10	11	11	12	13	14
5		0	1	2	3	5	6	7	8	9	11	12	13	14	15	17	18	19	20
6		1	2	3	5	6	8	10	11	13	14	16	17	19	21	22	24	25	27
7		1	3	5	6	8	10	12	14	16	18	20	22	24	26	28	30	32	34
8	0	2	4	6	8	10	13	15	17	19	22	24	26	29	31	34	36	38	41
9	0	2	4	7	10	12	15	17	20	23	26	28	31	34	37	39	42	45	48
10	0	3	5	8	11	14	17	20	23	26	29	33	36	39	42	45	48	52	55
11	0	3	6	9	13	16	19	23	26	30	33	37	40	44	47	51	55	58	62
12	1	4	7	11	14	18	22	26	29	33	37	41	45	49	53	57	61	65	69
13	1	4	8	12	16	20	24	28	33	37	41	45	50	54	59	63	67	72	76
14	1	5	9	13	17	22	26	31	36	40	45	50	55	59	64	69	74	78	83
15	1	5	10	14	19	24	29	34	39	44	49	54	59	64	70	75	80	85	90
16	1	6	11	15	21	26	31	37	42	47	53	59	64	70	75	81	86	92	98
17	2	6	11	17	22	28	34	39	45	51	57	63	69	75	81	87	93	99	105
18	2	7	12	18	24	30	36	42	48	55	61	67	74	80	86	93	99	106	112
19	2	7	13	19	25	32	38	45	52	58	65	72	78	85	92	99	106	113	119
20	2	8	14	20	27	34	41	48	55	62	69	76	83	90	98	105	112	119	127

Complete example 1

A zoo wanted to investigate whether providing behavioral enrichment (e.g., such as providing toys or forage) to zoo animals while they were on exhibit (rather than behind the scenes) also benefited zoo visitors in terms of increased engagement. They evaluated the duration spent by visitors at the exhibit (in seconds) when animals were provided enrichment on exhibit (experimental group) versus behind the scenes (control group).

Experimental group vs. control group (observation of enrichment)

Experimental (Exp)	Control (Cont)
610	340
600	410
585	365
625	315

$n_1 = 4.$
$n_2 = 4.$

Step 1: Place scores in rank order and rank scores from both samples regardless of the experimental condition.

Group	Cont	Cont	Cont	Cont	Exp	Exp	Exp	Exp
Raw scores	315	340	365	410	585	600	610	625
Rank	1	2	3	4	5	6	7	8

Step 2: Sum the ranks separately by the experimental group (R_1 and R_2).

$$R_1(\text{control group}) = 1 + 2 + 3 + 4 = 10.$$

$$R_2(\text{experimental group}) = 5 + 6 + 7 + 8 = 26.$$

Step 3: Calculate the two U-obtained values by using the following equations:

$$U_1 = 10 - \frac{4(4+1)}{2} = 0$$

$$U_2 = 26 - \frac{4(4+1)}{2} = 16$$

Remember that the smaller value becomes the true U-obtained value; in this example, U-obtained is 0.

Step 5: Evaluate the U-obtained value using the Mann–Whitney table.

To look up the U-critical value, you need to know the sample sizes for each group and your alpha level (usually 0.05).

Decision rule: If U-obtained $\leq U$-critical, then reject the null hypothesis.

How would these results be reported in a scientific journal article?

If the journal required APA format, the results would be reported in a format something like this:

> There was a significant difference in the time spent observing zoo animals depending on the presence or absence of behavioral enrichment on exhibit, $U(N = 8) = 0.00, p < .05$. Visitors who saw animals with behavioral enrichment spent more time at the exhibit than visitors who did not observe enrichment.

This formal sentence includes the dependent variable (duration at the exhibit), the independent variable (behavioral enrichment presence vs. absence), as well as a statement about statistical significance, the symbol of the test (U), the sample size (8), the statistical value (0), and the estimated probability of obtaining this result simply due to chance (0.05).

Complete example 2

A researcher rates individual macaque monkeys based on the number of individuals in their social network. He/she then tests whether there is a statistically significant sex difference in social connectivity.

Males (M)	Females (F)
5	10
6	10
4	8
7	5
5	8
	9
	9

$n_1 = 5$
$n_2 = 7$

Step 1: Place scores in the rank order and rank scores from both samples regardless of the experimental condition.

Ties on ranks—if two numbers are tied:

1. Get an average rank.
2. Don't reuse ranks.

Group	**M**	**M**	**M**	**F**	**M**	**M**	**F**	**F**	**F**	**F**	**F**	**F**
Scores	4	5	5	5	6	7	8	8	9	9	10	10
		(2)	(3)	(4)			(7)	(8)	(9)	(10)	(11)	(12)
Ranks	1	3	3	3	5	6	7.5	7.5	9.5	9.5	11.5	11.5

Average ranks used to calculate assigned rank: raw scores of 5 get a rank of 3 ((2+3+4)/3=3), *8*'s get a rank of 7.5 ((7+8)/2=7.5), *9*'s get a rank of 9.5, and *10*'s get a rank of 11.5 each.

Step 2: Sum the ranks separately by experimental group (R_1 = Group M and R_2 = Group F).

$$R_1(\text{males}) = 1 + 3 + 3 + 5 + 6 = 18.$$

$$R_2(\text{females}) = 3 + 7.5 + 7.5 + 9.5 + 9.5 + 11.5 + 11.5 = 60.$$

Step 3: Calculate the two U-obtained values by using the following equations (remember that the smaller value becomes the true U-obtained value):

$$U_1 = 18 - \frac{5(5+1)}{2} = 3$$

$$U_2 = 60 - \frac{7(7+1)}{2} = 32$$

So the U-obtained value = 3.

Step 4: Evaluate the true *U*-obtained value of 3 using the Mann–Whitney table (Table 4).

Under these conditions, your *U*-critical value is 5.

Decision rule: If *U*-obtained $\leq$ *U*-critical, then reject the null hypothesis.

If we use the *U*-obtained and *U*-critical values, then $3 < 5$, so we can reject the null hypothesis.

How would these results be reported in a scientific journal article?

If the journal required APA format, the results would be reported in a format something like this:

> Males were significantly less socially connected than females, $U\ (N = 12) = 3.00$, $p < .05$.

This formal sentence includes the dependent variable (ranking), the independent variable (sex), as well as a statement about statistical significance, the symbol of the test (U), the sample size (12), the statistical value (3.00), and the estimated probability of obtaining this result simply due to chance (<0.05).

Kruskal–Wallis Test

The Kruskal–Wallis test (*H*) is used when data are in three or more groups, and it replaces the one-way (independent) ANOVA (analysis of variance) when either or both of the assumptions are broken. This test is very similar to the Mann–Whitney test just described.

Calculation steps

1. Rank all scores from the smallest to the largest.
   - a. Disregard groups initially when ranking.
   - b. Handle tied ranks as you did with the previous tests.
2. Sum ranks for each group (R_i).
3. Calculate the H-obtained value using the following formula.
   First add up the following for each group:

$$R = \sum \frac{R_i^2}{n}$$

   That is, for each group, square the sum of the ranks and divide by *n* (sample size of the group) and then add to the total (*R*).
   Now calculate

$$H = \frac{12}{N(N+1)} R - 3(N+1)$$

where *N* is the total sample size of the data set.

4. Evaluate using chi-square table, df $= k - 1$, where k equals the number of groups in the experiment. Note that the decision rule returns to the typical pattern where you reject the null hypothesis if and only if $H_{obt} > H_{crit}$.
5. If the null hypothesis is rejected, pairwise combinations of Mann–Whitney U tests should be used to determine the post hoc pattern of differences among the groups.

Complete example

A college instructor asks whether the performance of his/her students varies with the time of day the class is held. The instructor compares the GPAs of the same course topic held during three different periods to determine if there is a difference in performance based on time of day alone.

GPA scores for one course held at three different periods

Early a.m.	Early p.m.	Late p.m.
2.5	2.8	3.8
3.1	3.2	3.2
2.2	3.0	4.0
2.6	2.7	3.7
2.2	2.3	3.7

Step 1: Rank scores regardless of the group/condition and handle ties appropriately.

Raw scores	2.2	2.2	2.3	2.5	2.6	2.7	2.8	3.0	3.1	3.2	3.2	3.7	3.7	3.8	4.0
	(1)	(2)								(10)	(11)	(12)	(13)		
Rank	1.5	1.5	3	4	5	6	7	8	9	10.5	10.5	12.5	12.5	14	15

Step 2: Sum the ranks separately by the experimental group.

Early a.m.	(R_1)	Early p.m.	(R_2)	Late p.m.	(R_3)
2.5	4	2.8	7	3.8	14
3.1	9	3.2	10.5	3.2	10.5
2.2	1.5	3.0	8	4.0	15
2.6	5	2.7	6	3.7	12.5
2.2	1.5	2.3	3	3.7	12.5

Sum ranks for each group (R_i):

Sum $R_1 = 4 + 9 + 1.5 + 5 + 1.5 = 21$.

Sum $R_2 = 7 + 10.5 + 8 + 6 + 3 = 34.5$.

Sum $R_3 = 14 + 10.5 + 15 + 12.5 + 12.5 = 64.5$.

Step 3: Square each of the summed ranks and divide by sample size. Add the results together.

$$R = (21^2 / 5) + (34.5^2 / 5) + (64.5^2 / 5) = 88.20 + 238.05 + 832.05 = 1158.3$$

Step 4: Calculate H-obtained value using the following formula:

$$H - \text{obtained} = \frac{12}{15(15+1)}(1158.3) - 3(15+1) = (0.05)(1158.3) - 48 = 9.915$$

Note that the capitalized N is the total N for all subjects participating in the experiment, regardless of the condition, while the small n_i refers to the sample size in each group/condition, respectively. Confusing these two values is a common error, so be very careful as you work through examples on your own that you keep track of which sample size is required in the formula.

Step 5: Evaluate *H*-obtained using the chi-square table (Table 3).

You need to know the number of groups in the experiment (k) and your alpha level to look up the H-critical value. In this example, you have three conditions, so the df $= k - 1$ or $3 - 1 = 2$, and alpha $= 0.05$, as usual. Under these conditions, your *H*-critical value is 5.99.

Decision rule: If *H*-obtained > *H*-critical, then reject the null hypothesis.

Since $9.915 > 5.99$, we can reject the null hypothesis. The three conditions vary significantly in GPA scores, suggesting that the time of day influences course performance.

APPENDIX 3

Citation formats for the sciences

Heather Zimbler-DeLorenzo[1], Andrea M.-K. Bierema[2]

[1]*Department of Life and Earth Sciences, Perimeter College at Georgia State University, Decatur, GA, United States;* [2]*Center for Integrative Studies in General Science and Department of Integrative Biology, Michigan State University, East Lansing, MI, United States*

Citing and documenting your sources is a critical component of using research in your writing. It gives credit to the experts upon whom you have built your argument. From an ethical standpoint, citing correctly, accurately, and thoroughly strengthens your credibility and the validity of your ideas.

Citing and referencing sources include using an in-text marker that directs the reader to the complete bibliographical reference at the end of the document. This process contains two main elements:

- In-text citations: When you (1) refer to a source; (2) quote, paraphrase, or summarize a source; or (3) use data or graphics from a source, you place an in-text citation referring to that source within the paragraph.
- Reference list: Include a list of all the sources you have cited in your paper, documented properly, at the end of your paper. A reader familiar with academic conventions can tell what kinds of sources you are referencing and can find the source based on the information included.

In-text citation guidelines

When citing your sources, your instructor specifies which citation style they prefer. For each source that you use, you need to keep track of certain types of information to include in your citations.

Several styles exist. The CSE (Council of Science Editors) style is used primarily in the physical sciences, life sciences, and mathematics. The APA (American Psychological Association) style is used primarily in the social sciences. The MLA (Modern Language Association) formatting style is commonly used in the humanities. Details on all these styles can be readily found online. We recommend https://owl.purdue.edu/ as an excellent source for the details of these formatting styles. Here, we focus on CSE style. They recommend using one of two systems (Council of Science Editors, 2014):

- a *citation-sequence system*, which lists sources in the reference list according to the order in which they appear in the document;

- a *name-year system*, which is similar to the author-date system used by APA.

Citation-sequence system

This system relies on superscripted numbers: a raised numeral located at the end of a clause or sentence identifying the content as outside-source material. There is no parenthetical information required. The superscripted number flags the reader's attention to a correspondingly numbered entry on a reference list located at the end.

Name-year system

For the name-year system, the citation can be placed at several places:

- At the end of the sentence, if the entire sentence is a quotation, paraphrase, or summary of the source's idea: *Our current knowledge of interesting phenomena such as dominance hierarchies, mating behavior, and habitat preferences has been based on careful, systematic observations* (*Martin & Bateson 1993; Siiter, 1999*).
- Directly after the name of the source: *Martin & Bateson (1993) claims....*

NOTE: Your citation should NOT go inside the quotation marks; it is not part of the quotation. However, punctuation must be placed AFTER the citation, as that citation is part of that sentence (or clause), not part of the next sentence.

If you are discussing the ideas in a source at length (e.g., in a summary), you do not need to cite every consecutive sentence. Cite the first time you mention the source. As long as the following sentences clearly indicate that the ideas come from the same source—for example, you are using signal phrases, such as "the author further clarifies the problem by..."—you do not need to keep citing. If you stop using signal phrases, then be sure to include a citation. If you introduce material from another source or add your own analysis between references to that source, you will have to re-cite the source when you refer to it again. Always make sure your reader knows which ideas come from a source and which come from you and when you shift from one to the other. If in doubt cite.

If the source you are citing has one or two authors use their names in your signal phrase:

- Siiter (1999) States that ... or ... (Siiter, 1999).
- Martin and Bateson (1993) demonstrates that ... or ... (Martin & Bateson, 1993).

If the source has three or more authors use the name of the lead author, followed by *et al.*, the Latin term meaning "and the others." Like all Latin words, *et al.* should be italicized:

- Carvaggi *et al.* (2020), in their study on camera traps, in their study on baboons, found that ... or ... (Carvaggi *et al.*, 2020).

It can be tricky to know exactly where to place the in-text citation in your sentence. Generally, the default position for a citation is at the end of the sentence, unless placing it there would create confusion. For example, where should the citation go in the following sentence?

> *Smith claims that "[insert a quotation here]," but other scientists argue that her conclusions are flawed.*

If you place the citation for Smith at the end of the sentence, you are saying that Smith acknowledges that many scientists think her conclusions are flawed. That wouldn't make sense. You would have to cite like this:

> *Smith (2020) claims that "[insert a quotation here]", but other scientists argue that her conclusions are flawed (Jones & Adams, 2020).*

In this example, Jones and Adams have argued that Smith's claims are flawed, so their citations are at the end.

Citing sources at the end of the paper

Whether using the citation-sequence system or the name-year system, the paper ends with a list of all the sources you have cited in your paper, alphabetized by the first word of each source, which is most often the last name of the first author. Each reference must provide thorough and complete documentation so that readers can identify the kind of source and retrieve it if they want to read it. It is important to use the correct formatting for each type of source, as readers familiar with academic conventions will expect this and they will be able to tell what kinds of sources you are referencing based on what information is included and how it is formatted. If you use conventions incorrectly (such as failing to italicize, abbreviate journal names, or bold volume numbers), you can confuse and mislead your readers.

To cite, you may need to collect the following pieces of information:

1. author(s),
2. date when the item was published,
3. title,
4. title of journal/book,
5. volume and issue number (if a journal article),
6. page numbers where the article or chapter appears,
7. DOI (digital object identifier, a unique alphanumeric code assigned to most journal articles) or URL (website address) of the article,
8. publisher and location (if a book).

- Scholarly journal article
  Example: Adams CJ, Smith RW, Hones MR. 2020. Interobserver reliability. Animal Behaviour 123(1):1–10. *[add DOI if known or published in online only journal].*

- Book
  Example: Adams CJ, Smith RW, Hones MR. 2020. Interobserver reliability. 1st ed. New York (NY): Oxford University Press.
- Edited book
  Example: Zimbler-DeLorenzo HS and Margulis S, editors. 2021. Exploring Animal Behavior in Laboratory and Field. San Diego (CA): Elsevier.
- Website
  Example: The Biology Project [Internet]. c2001-2004. University of Arizona. [cited 2020 Dec. 20]. Available from http://www.biology.arizona.edu/DEFA ULT.HTML.
  NOTE: In your reference section at the end of your paper, it is a courtesy to list the names of ALL the authors who contributed to the source (rather than using *et al.*, which is used for the in-text citations). However, if there are 10 or more authors (for CSE format), it is acceptable to use *et al.* in your reference list.

Fig. 1 shows the beginning of a journal article. It helps you determine the various elements of an academic article that must be included in your reference. Note the difference between the database (such as Elsevier, EbscoHost, JSTOR, etc.) and the name of the journal (in this case, *Energy*).

An academic article will have information about the database, title, volume, issue, date, pages, authors, DOI, and summary, but the database is not listed in the citation.

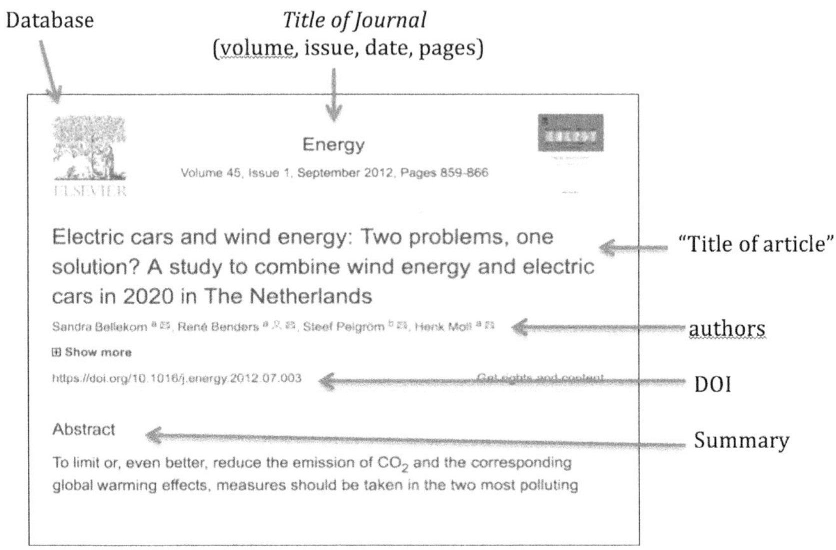

FIGURE 1

Elements of an academic article. *DOI*, digital object identifier.

Bellekom, S., Benders, R., Pelgröm, S., & Moll, H. (2012). Electric cars and wind energy: Two problems, one solution? A study to combine wind energy and electric cars in 2020 in The Netherlands. Energy, 45*(1), 659–666.*

Plagiarism

Missing either the in-text citations or the reference list may result in plagiarism. Plagiarism is the use of someone else's ideas, results, equipment design, visuals, wording, or even sentence structure as if they were your own.

Plagiarism can be intentional.

- You use someone's ideas or results without citing the source.
- You copy something word for word without using quotation marks, even though you cite the source.
- You use all or part of a visual without crediting the source.

Or it can be accidental.

- You do not realize what is considered plagiarism.
- You can't think of a better way to say it and so copy sentences, phrases, or even sentence structure from the original without using quotation marks.
- When you took notes, you did not put the exact wording in quotation marks and now you plagiarize without realizing it.

It is important to avoid plagiarism. In the United States and many other countries, plagiarism is considered academic misconduct, and you are expected to avoid plagiarism, either intentional or accidental. Also, plagiarized work can result in a failing course grade, expulsion, rejection of a paper submitted for publication, denial of an advanced degree, loss of job or fines. It is increasingly serious now that the Internet has made plagiarism easier than ever before.

To avoid plagiarism practice paraphrasing (i.e., putting someone else's ideas into your own words) because it's often difficult to do. Avoid the temptation of paraphrasing too many details. Focus on the main idea or evidence that you need to cite. Once you have determined what you need to paraphrase reread the source and then physically cover it up. Write the main idea from memory and then check to verify that you haven't used the exact wording or sentence structure. Simply changing the verb tense or substituting one adverb for another, but leaving the sentence structure essentially the same, is still considered to be plagiarism.

If you simply cannot figure out a different way of saying it, just use quotation marks to indicate that you are quoting exactly. (Because few writers in science or engineering use quotations, generally preferring paraphrases, paraphrasing is a skill you must learn in those fields. In contrast, writers in the humanities often use quotations to illustrate key points, but they also paraphrase when exact wording is not essential.)

Always cite your source, whether for text, visuals, or ideas. If you cannot remember the source, you cannot use the information. Put citations in as you write your first draft so that you don't have to go back later when identifying the source may be difficult.

References

Council of Science Editors. (2014). *Scientific Style and Format: The CSE Manual for Authors, Editors, and Publishers* (8th ed.).

Carvaggi, A., Cole, A. C., Clark, D. A., Fisher, J. T., Grass, A., Green, S., Hobaiter, C., Hofmeester, T. R., Kalan, A. K., Rabaiotti, D., et al. (2020). A review of factors to consider when using camera traps to study animal behavior to inform wildlife ecology and conservation. *Conservation Science and Practice, 2*(8), e239. https://doi.org/19.1111/csp2.239.

Siiter, R. J. (1999). *Introduction to animal behavior.* Pacific Grove (CA): Brooks/Cole Publishing.

Martin, P., & Bateson, P. P. G. (1993). *Animal behaviour: An introductory guide* (2nd ed.). Cambridge (UK): Cambridge University Press.

Attributions

This appendix is a modified derivative of the following articles:

"Citing and Documenting Sources in IEEE" by Suzan Last, *Technical Writing Essentials*, CC BY 4.0. https://pressbooks.bccampus.ca/technicalwriting/part/citing/.

"Recognizing and Avoiding Plagiarism" by OpenStax College, *Plagiarism and Scientific Writing*, CC BY 2.0. Download for free at http://cnx.org/contents/93e3eb4d-5745-4daf-b7c4-8080a8f59b12@1.1.

Index

'*Note:* Page numbers followed by "f" indicate figures and "t" indicate tables.'

F

G

H

I

K

L

M

N

O

9780128214107